Third Edition
MATHEMATICS
A GOOD BEGINNING
Strategies for Teaching Children

Third Edition
MATHEMATICS
A GOOD BEGINNING
Strategies for Teaching Children

Andria P. Troutman
University of South Florida

Betty K. Lichtenberg
University of South Florida

Brooks/Cole Publishing Company
Pacific Grove, California

Brooks/Cole Publishing Company
A Division of Wadsworth, Inc.

Printed in the United States of America
10   9   8   7   6   5   4   3

Library of Congress Cataloging-in-Publication Data
Troutman, Andria, [date]
    Mathematics, a good beginning.

    Bibliography: p.
    Includes index.
    1. Mathematics—Study and teaching (Elementary)
I. Lichtenberg, Betty K., [date].   II. Title.
QA135.5.T76   1987   372.7   86-14802
ISBN 0-534-06984-3

Consulting Editor: *Robert J. Wisner*
Sponsoring Editor: *Jeremy Hayhurst*
Editorial Assistant: *Amy Mayfield*
Production Editor: *S. M. Bailey*
Manuscript Editor: *David Hoyt*
Permissions Editor: *Carline Haga*
Cover Design: *Katherine Minerva*
Cover Photo: *Richard Bellamy*
Art Coordinator: *Judith MacDonald*
Interior Illustration: *Ryan Cooper*
Typesetting: *Omegatype Typography, Inc., Champaign, Illinois*

To Don and Bill

Preface

In the past 50 years, three principal concerns of the educational world have had profound influence on the mathematics taught in our elementary schools. Each concern, in its own time, has dominated the curriculum and produced shifts in emphases, which have resulted in inadequate elementary mathematics programs and caused confusion for all concerned—children, parents, teachers, and administrators. Consequently, many are questioning what has been and is being taught in our schools.

In the 1920s, '30s, and '40s, arithmetic instruction was based on a stimulus-response psychology that focused on analyzing the processes to be learned in the minutest detail. An algorithm for computing was broken into a series of discrete steps, each involving a separate skill to be learned independently. Learning was based on memorization of processes, and instruction often neglected to provide relationships between these processes and their associated applications.

In the 1940s and '50s, a social-utility "theory" of arithmetic instruction was superimposed on the existing instructional pattern. As a result, some progressive schools instituted an integrative approach to the curriculum and provided arithmetic instruction only as it was apparently "needed" in science, social studies, and so on. Other schools decided what arithmetic to teach solely on the basis of whether it was commonly used in society.

In the late 1950s and '60s, the influence of the Sputnik launching dominated. The mathematics curriculum was rewritten in numerous projects, most of which were financed by the National Science Foundation. The emphasis of these curriculum-writing projects was on mathematics as the mathematician sees it. The resulting instructional programs for elementary schools focused on the development of rational numbers as a mathematical system and the use of structure properties to justify computational procedures. Precise language and extensive symbolism were characteristic of these materials. Little consideration was given to the cognitive and affective development of children.

In the 1970s, emphasis shifted to accountability and assessment and a cry for "back to the basics." Tremendous constraints on teachers' time were often the result, as well as more narrowly defined curriculum objectives. At this point, the National Council of Supervisors of Mathematics issued a widely endorsed position paper that called for a broad definition of basic skills, including these ten vital areas:

1. Problem solving
2. Applying mathematics to everyday situations
3. Alertness to the reasonableness of results
4. Estimation and approximation
5. Appropriate computational skills
6. Geometry
7. Measurement
8. Reading, interpreting, and constructing tables, charts, and graphs
9. Using mathematics to predict
10. Computer literacy

The 1980s began with an emphasis on problem solving and an acceptance of this broad interpretation of basic skills. Developments in computer technology are having a significant influence on the mathematics curriculum of this decade. The position paper of the National Council of Teachers of Mathematics makes the following eight recommendations for school mathematics of the 1980s in *Agenda for Action:*

1. Problem solving be the focus of school mathematics in the 1980s
2. Basic skills in mathematics be defined to encompass more than computational facility
3. Mathematics programs take full advantage of the power of calculators and computers at all grade levels
4. Stringent standards of both effectiveness and efficiency be applied to the teaching of mathematics
5. The success of mathematics programs and student learning be evaluated by a wider range of measures than conventional testing
6. More mathematics study be required for all students, and a flexible curriculum with a greater range of options be designed to accommodate the diverse needs of the student population
7. Mathematics teachers demand of themselves and their colleagues a high level of professionalism
8. Public support for mathematics instruction be raised to a level commensurate with the importance of mathematical understanding to individuals and society.

In the middle of the decade, we are seeing definitions for being mathematically powerful, and we are focusing on standards of excellence.

Regardless of the prevailing influence on curriculum and the associated shifts in emphases, we believe that a sound theory of arithmetic instruction must be based on the needs of mathematics as a subject, the needs of children as learners, and the needs of the society in which these children will live. We have written this book for preservice and in-service teachers to help them develop the abilities necessary for implementing such an instructional program.

The content of this book was selected on the basis of its direct applicability to the mathematical ideas usually introduced in grades K–6. The presentation of this content is made throughout in conjunction with appropriate teaching strategies. The selection of specific topics and teaching strategies was determined by research findings and relevant contemporary trends. For each concept presented, a variety of its interpretations and possible applications are given, followed by related activities for children to develop specific concepts and skills. Formal language and symbolism are kept to a minimum, and concrete and pictorial representations for mathematical ideas are emphasized. A chapter is included with diagnostic techniques for the concepts and skills that are developed. Both error patterns in computation and conceptual misinterpretations are discussed, and strategies for remediation are developed.

In addition to content that is usually found in an elementary-school mathematics program, topics are included in this book that are just beginning to appear in the curriculum, and appropriate teaching strategies related to these topics are provided. The reading of mathematical exposition is given a high priority. Many mathematical difficulties that children experience are related to their comprehension of mathematical text. These difficulties are examined, and teachers are provided with strategies for their prevention or remediation. Each chapter contains "fun" or realistic applications. Estimation techniques are emphasized. Behaviors of children that relate to problem solving are described, and teachers are given opportunities to classify activities that elicit these behaviors as well as opportunities to design classroom activities that require the demonstration of these behaviors. The chapters on measurement use the metric system exclusively. Decimal notation is emphasized. The approach to geometry is informal, and ideas from transformational geometry are developed intuitively. Ideas from statistics and probability are introduced as tools for the analysis of data. The use of computers is considered with respect to all topics, and a complete chapter is devoted to computers and mathematics instruction.

The presentation of activities at three levels provides a way for teachers to deal with individual differences. Any of the Level Three activities can be used directly with talented or gifted children. Level One activities, with some modification, can be used with special children who are not mathematically talented.

We would like to thank Craig Barth for his constant support and professional insight, and Bob Wisner for his perceptive review of the manuscript.

We also wish to thank the student programmers at the Florida Center for Instructional Programming, especially Doug Woolley and Ben Cash; they spent countless hours working the bugs out of the computer programs for this third edition.

Finally, our sincere appreciation is extended to the reviewers of the first and second editions as well as to the following who reviewed this third edition: James Babb (Minot State College, North Dakota), Joann Mueller (Indiana University, Pennsylvania), Beverly J. Kochmann (University of Wyoming), Dan Farris (State University of New York, College at Cortland), Dorothy Spethmann (Dakota State College, South Dakota), Willis Johnston (Murray State University, Kentucky), Ralph Gremillion (Richols State University, Louisiana), Rosemary Roehl (St. Cloud State University, Minnesota), Miriam Leiva (University of North Carolina—Charlotte), and Jane McLaughlin (Trenton State College, New Jersey).

Andria P. Troutman
Betty K. Lichtenberg

Contents

2

. . . 100s, 10s, 1s . . . The Best Yet
Our Base-Ten Numeration System   35

3
Better Ways to Use the Best
Addition and Subtraction of Whole Numbers   64

4
Better Ways to Use the Best
Multiplication and Division   103

5

Some Theory about Numbers
Factors, Multiples, Primes, and Composites   146

6

Not All Numbers Are Whole Numbers
Representing, Adding, and Subtracting Rational Numbers 168

7

Security Is Knowing Why
Multiplying and Dividing Rational Numbers 209

8

Diagnosis, Before and After
Common Mathematical Difficulties 240

9
Putting It All Together
Problem Solving   280

10
The Shape of Things
Geometric Figures and Relationships   309

11
Seeing Is Believing
Children's Development of Geometric Ideas   331

12
Before You Teach Measurement
Attributes of Measurement . . . The Metric System . . . Expectations   361

13
Superstitious? Not Us . . .   382

14
Sizing It Up
The Measurement of Attributes   383

17
The End . . . Your Beginning
Toward Efficient Instruction **473**

Third Edition
MATHEMATICS
A GOOD BEGINNING
Strategies for Teaching Children

Zero, Our Starting Point

■ *Have you ever heard of a Chapter 0? We hadn't until we wrote this book! Then we realized that before we write about teaching children, we need to write about how we're going to write about teaching children. That is, we must acquaint you with the book's organization, some of its special features, and the ways in which you can use it. Chapter 0 is for you. It will get you ready to begin.* ■

If you're interested in methods for teaching mathematics to elementary school children, then this book is for you. If you are preparing to become a teacher, this book will acquaint you with mathematical concepts children should learn and specific strategies and materials for teaching these concepts. If you are already a teacher, the book will help you extend and update your present knowledge. Even if you are a seasoned teacher, the book will provide a valuable resource for classroom activities and teaching aids that have been successfully used with children.

To get started, you should consider four important questions:

1. What is mathematics?
2. How do children learn mathematics?
3. What mathematics should children learn?
4. How will computers in the classroom affect your teaching of mathematics?

What is mathematics? Your answers to the first question are important because they influence how and what you will teach. Suppose you think of mathematics solely as adding, subtracting, multiplying, and dividing numbers; no doubt, then, you'll stress computation and perhaps neglect important applications. Or, if you think that mathematics is a set of relationships that enables people to solve everyday problems, you'll probably stress applications and avoid problem-solving activities that are not specifically related to everyday problems. Or suppose you feel that mathematics is primarily a set of logically organized deductive systems. Then you'll be tempted to teach topics as they are logically developed, and you may ignore the fact that, psychologically, children cannot always handle the logical development of mathematical ideas. For example, many children learn and use the numbers 1 through 10 before considering the number 0, although 0 logically comes before any of the others. Many children also learn to use the symbol "10" to represent ten objects before they can appreciate the associated place-value ideas. In fact, children more easily learn place-value ideas associated with a symbol such as "12," because 2 ones are more concrete to deal with than 0 ones.

Obviously, how a teacher views mathematics is important. So one of the purposes of this book is to help you develop a view that will enable you to provide broad and flexible mathematical experiences for children. These will include a wide range of mathematical ideas from topics such as geometry, measurement, and statistics and probability, as well as from arithmetic. They should help children develop concepts, acquire skills, explore relationships, apply generalizations, solve problems, and extend their knowledge.

How do children learn mathematics? Answers to the second question are equally important. Research indicates that children must develop special thinking abilities in order to interpret certain mathematical situations. Because the development of some of these abilities is so natural, teachers often fail to consider that children in certain stages of development may not have acquired them. For example, it's unrealistic to require some 6-year-olds to interpret number sentences such as "$5 < 7$," because they don't know "left" from "right" and aren't able to distinguish consistently between the signs $<$ and $>$. This doesn't mean that these children aren't ready to learn ideas related to such sentences. They can learn to compare sets of objects to find which has more or to compare pictures of objects to tell which shows more. (But even these tasks involve the development of abilities that may be overlooked. We'll have more to say about this later.)

To be an effective teacher, you must know *when* children can be introduced to a given concept and *at what level of abstraction* they can deal with the concept. Basically there are three levels at which concepts can be represented: concrete, pictorial, and symbolic.

- Concrete level. A concept can be represented by the appropriate manipulation of objects.

Example: Placing three beans into each of four margarine tubs and finding the total number of beans illustrates that $4 \times 3 = 12$.

- Pictorial level. A concept can be represented by appropriate pictures.

Example:

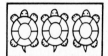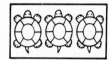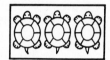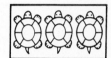

This picture illustrates that $4 \times 3 = 12$.

- Symbolic level. A concept can be represented by symbols.

Example: $4 \times 3 = 12$

The essential difference between the first two levels (concrete and pictorial) is that meaning can be discovered on the first two levels intuitively—that is, by examining the concrete and pictorial representations. The meaning of symbolic representations, however, must be learned.

So, another purpose of this book is to describe how children develop mathematical ideas and to introduce you to examples of appropriate classroom activities that are written at different levels of abstraction. You will also find that it is entirely appropriate to use materials that represent concepts at more than one level. For example, a 5-year-old may benefit from this combination of all three levels.

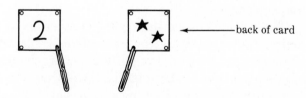

What mathematics should children learn? Okay, what about this question? It's easy to come up with general answers, but getting down to specifics is difficult. It's easy to say that the concepts a child should learn must be useful, entertaining, or both. But how do we know what concepts will be useful to children in the future? How can we predict what ideas will be entertaining? The very best we can do is to carefully examine the global kinds of problems children encounter while growing up and the kinds of mathematical skills they'll need as adults.

Surely, children will need efficient computing and estimating skills, although calculators and computers will revolutionize the way we think about teaching these skills. Children will need measuring skills, and the adoption of the metric system will influence how we present them. The growing use of both metric concepts and calculators will increase the need to learn decimal notation earlier. (After all, you can't plug "2 lb, 5 oz" into a calculator!) Children will also need to develop geometric ideas that enable them to interpret and organize their environment. Will they need ideas from probability and statistics? Of course! How else will they be able to evaluate claims based on biased data? Will they need problem-solving skills? No doubt about it.

How will computers in the classroom affect your teaching of mathematics? This one is a toughy to answer. In fact, there's no way to predict what changes will occur. We do know, now, that these machines have a tremendous potential. They can make your life easier, and judiciously used, they can make your teaching of mathematics significantly better.

Careful selection of software by you and your mathematics supervisor or curriculum specialist is crucial. You must help in the continual evaluation of commercially available computer materials. Your professional judgment will be extremely important to ensure that choices will be made on a sound mathematical and pedagogical basis.

We could go on and on, but Chapter 0 isn't supposed to contain the entire book! We'll wrap up this section by saying that the purpose of this book is to help you (1) find workable answers that are compatible with all four questions and (2) develop the flexibility to revise your answers when events or circumstances require it.

→ Some special features of this book

To accomplish our purposes, we provide certain features that you should anticipate. First, we describe mathematical ideas we think children should learn, then ways in which these ideas can be presented, and, finally, difficulties children may encounter while learning them. Our choices of ideas are based on current educational goals and research as well as on current social trends.

The concepts discussed in each chapter are translated into instructional objectives for children. These objectives are listed at the end of the chapter. In Appendix A (p. 483) you'll find a chart that classifies each instructional objective by one of three levels. Level One consists of content for children *normally* in grades K–2, ages 5–7; Level Two encompasses grades 2–4, ages 7–9; and Level Three covers grades 4–6, ages 9–11.

MORE LEVELS?

	Level One	*Level Two*	*Level Three*
Ages	5–7	7–9	9–11
Grades	K–2	2–4	4–6

The chart enables you to identify both the typical age at which children attain a given concept and the different concepts that should be developed within the same age period. The overlap is realistic and must be considered when planning instruction for a class of children. Notice that this is a different use of the word "level," one similar to the developmental levels described by the Swiss psychologist Jean Piaget and his followers.

To help children accomplish a set of objectives, the teacher must achieve some instructional goals related to those objectives. So, each chapter also specifies teacher goals that are appropriate to the objectives for children.

The objectives listed for teachers and children will help you decide where you want to go. To give you a good start on getting there, we describe about 300 field-tested activities. These are provided as examples of the types of high-interest activities you can prepare. The level of each activity is identified by a numeral in the upper-right corner of the activity. The one below, for example, is designed for developing multiplication concepts for children at Level Two.

Concept Development — Level Two

C-2

Example 4-2
Additive-type problems

Materials: Three sets of cards—cards with story problems, cards showing objects in sets having the same number, and cards with multiplication sentences.

Sample cards:

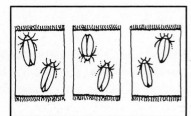 

3 rugs.
2 bugs on each.
How many bugs?

$3 \times 2 = 6$

Teacher directions: Match the cards so that the right story problem, the right picture, and the right number sentence are together in a stack.

Some of the activities are *concept*-developing activities—that is, they are designed to help children learn new concepts or to reinforce concepts previously learned.

Example:

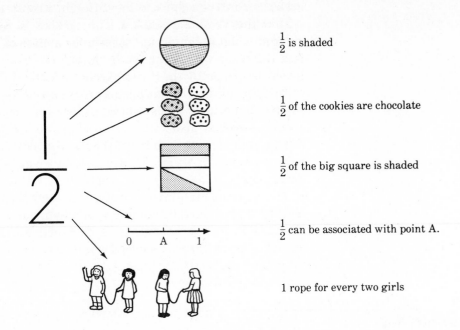

$\frac{1}{2}$ is shaded

$\frac{1}{2}$ of the cookies are chocolate

$\frac{1}{2}$ of the big square is shaded

$\frac{1}{2}$ can be associated with point A.

1 rope for every two girls

Other activities in which children perform skills are *skill*-developing activities. In these activities, children may be developing a new skill or practicing a skill they have already begun to acquire.

Example: Pick out all the names for $\frac{1}{2}$.

$$\frac{2}{4}, \frac{3}{8}, .5, \frac{2}{3}, \frac{3}{10}, \frac{12}{24}, \frac{200}{100}$$

Still other activities allow children to use concepts and skills they have learned previously to *solve problems*. We believe that this application of mathematical concepts and skills is an extremely important aspect of instruction.

Examples: Find the thickness of a sheet of paper.
Find the distance between your ears.
Find the average height for children in our class.

Although many of the activities involve a combination of these three types, each has a primary focus. Identifying the primary focus of each activity will allow you to build a conceptual framework for integrating purpose, strategy, and content. To aid you in this identification, we have labeled each activity in the upper right with a C or an S. The C indicates that the activity's primary focus involves either learning new *concepts* or using concepts to find relationships or solve problems. An S indicates that an activity's primary focus is developing or reinforcing a *skill*.

Materials for most of the activities can be prepared inexpensively and repeatedly used. This will help you avoid the dull routine of textbooks and "dittos," a rut that's easy to fall into but hard to climb out of. Some activities are games, and others are puzzles. Some are for large groups, some for small groups, and some are for individual children. Together, they illustrate the many ways to prepare nontextbook experiences for children, which often are a neglected part of schooling. Although each activity is associated with specific objectives, minor changes will make each suitable for helping children achieve other learning objectives.

In addition to the activities, we provide a set of *material sheets* (see Appendix C). These can be reproduced to provide inexpensive but valuable teaching aids to help you get a good start.

At the beginning of each chapter is a short preview of the chapter, telling you what the major emphases are and providing a brief rationale or justification for their use. At the end of each chapter is a short summary, called "In Closing," which once again identifies the main focus of the chapter. These two important sections will give you a sense of the organization of the chapter's content as well as reasons for the content selection.

In some chapters, there are short sections titled "When to Teach What." These give you specific information on when to introduce and how to order concepts in a broad category, such as division. Some sample calculator activities are provided throughout the book. Every chapter provides sample problem-solving activities; in addition, there is a special chapter devoted to this important topic.

Preventing and correcting mathematical difficulties that children encounter is also extremely important, so Chapter 8 is devoted to these topics. Here again, we have discussed related ideas throughout the book. The list of references at the end of a chapter provides a source for further study. About one-third of these reflect relevant research and two-thirds present practical and sound ideas that can be used directly in the classroom.

At the end of most chapters we have included a section titled "Ideas for Enrichment." We believe that enrichment is appropriate for everyone and that horizontal enrichment for talented students is usually preferable to vertical acceleration. So, we have given some topics that you can use to broaden the experiences of everyone and challenge those students who can achieve excellence.

Many of the activities in this book call for children to use "higher-order thinking skills" to solve problems (the term comes from Bloom's Taxonomy of Educational Objectives). The six levels, in order from low to high, are *knowledge, comprehension, application, analysis, synthesis,* and *evaluation.* In each chapter exercise set, you will be asked to identify those activities that involve higher-order thinking skills. So, by the time you're finished, you'll have a fine accumulation of "good" (worthwhile and useful) problems.

Chapter 16 gives you an overview of the current status of instructional computing and also a discussion of the use of microcomputers in mathematics instruction. In short, the chapter concerns how computers can help you be a better teacher. At the end, there is even a glossary of computer terms so that it will be easier for you to remember and understand all of this new material. Most of the chapters contain a computer-corner activity and discussion that you can use directly to teach or reinforce the concepts in the chapter.

One more thing: Near the end of each chapter is a set of exercises titled "It's Think-Tank Time." Before you read a specific chapter, it might be helpful to read through the "Think-Tank" section. Then try to complete the exercises as you read the chapter. This not only will reinforce concepts you acquire but also will help tie specific concepts together and identify those that aren't clearly understood.

Enough said. It's time you got on to the really important things!

Getting Ready for a Good Beginning
Pre-Number Concepts

■ *This chapter is about children between ages 4 and 7 and their development of number ideas. During this period, children should develop the necessary pre-number concepts for establishing appropriate number-numeral association, for acquiring appropriate order concepts, and for learning to count to ten. We describe these pre-number concepts and trace their development through the development of appropriate counting concepts. We also illustrate how counting concepts can be used to build readiness for addition and subtraction. The early development of number ideas is more* *subtle than most adults think, so we describe special teaching strategies and suggest activities that aid children in this development. These activities emphasize the use of motivating materials that children can manipulate. We hope that after reading this chapter, you will be left not only with many specific teaching ideas related to number concepts but also with an important overall impression: Children must experience a lot of seeing, doing, and saying before reading and writing numerals will have meaning for them!* ■

Objectives for the child

1. Selects objects when a specific attribute is given.
2. Sorts objects on the basis of a general attribute.
3. Identifies some of the specific attributes that an object has.
4. Duplicates a pattern that has been made with objects.
5. Extends a pattern.
6. Determines when a one-to-one matching between two sets of objects can be achieved.
7. Determines by one-to-one matching whether one set of objects has as many as, more than, or less than another set.
8. Conserves an as-many-as, more-than, or less-than relation.
9. Produces a set of objects with as many objects as another set.
10. Uses the properties of the as-many-as, more-than, and less-than relations to compare sets.
11. Classifies a group of sets on the basis of the as-many-as relation.
12. Uses sets to show the meaning of more, less, one more, and one less.
13. Orders a group of sets on the basis of the more-than or less-than relation.
14. Orders a set of two or more whole numbers less than 11.
15. Identifies the ordinal position of an object in a set with respect to a given reference (first, second, third, . . .).
16. Counts up to ten objects.
17. Selects a set of objects appropriate to a given numeral (less than 11).
18. Selects the correct numeral when a number name is spoken (less than 11).
19. Writes numerals for the numbers from 0 to 10.
20. Gives the number for each of two sets and tells how many in all (less than 11).
21. Tells how many objects are left when some are taken away.
22. Tells how many more are needed to make a set with a certain number (less than 11).
23. Tells how many more are in one set than another.

Teacher goals

1. Be able to identify and explain pre-number and number concepts that young children should develop.
2. Be able to describe methods for teaching children to count.
3. Be able to identify and explain readiness concepts for addition and subtraction that young children should learn.
4. Be able to identify and carry out activities that will aid the development of pre-number, number, and addition- and subtraction-readiness concepts.

Helping the child develop pre-number concepts

Just because children can parrot the words "one," "two," "three," "four," and so on doesn't mean that they can compare two sets of objects and determine whether one set has more, less, or as many as the other. Neither does it mean that they can associate these sounds with finding "how many." Repeating these sounds may be equivalent to repeating the letters of the alphabet or reciting the Pledge of Allegiance to the Flag. A child may even choose the appropriate symbol for replacing □ in an arithmetic sentence such as 3 + 4 = □ and yet insist that set A below has more stars than set B.

Actually, there are many pre-number concepts that children must acquire before they can use whole-number concepts correctly. These pre-number concepts enable children to recognize some significant features of whole numbers:

> With whole numbers, we can determine how many objects are in a set.
> With whole numbers, we can compare two sets and find whether one set has as many as, more than, or less than another set.

What are these pre-number concepts? How do we help children develop these concepts? To appreciate the answers, it is helpful to take a look at quantitative thought in the early years.

■ The child's view

Children feel the need to use quantitative ideas early in life. "My brother is bigger than I am," "The door bell is higher than I am," and "My friend has more pennies than I do" are all typical statements that children may use to describe situations important to them.

Early development involves measurement concepts as well as whole-number concepts where important differences among these concepts are not distinguished. Children may use general terms such as "bigger" or "smaller" to compare attributes of objects, such as length, area, or volume. One child may compare the lengths of two boxes to decide which is bigger, while another may compare their heights. Obviously, their conclusions may differ. Children often assume that one object is heavier than another because it has greater volume or that a container holds more because it is taller.

Children also confuse measurement concepts with whole-number concepts. It would not be unusual for a child to decide that there are more elephants than butterflies pictured below simply because the elephants are bigger.

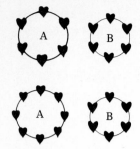

Children might also decide that circle A at the left has more hearts than circle B because of the difference in distance between the hearts. Some children may correctly decide that circle A has more hearts than circle B. It is possible, however, that the correct response would be given for the wrong reason: because circle A is larger than circle B.

The fact is, in their early years, children are not sure what strategy to use when deciding "which set has more," "which set has less," "which stick is shorter," "which glass holds more," and so forth. Perception is their guide. The role of the teacher is to help children sort and refine their initial quantitative ideas. By providing the right kinds of experiences, the teacher can help children develop consistent strategies for answering many of these questions.

In this section we discuss six important abilities that children should acquire before being introduced to numbers. These abilities are: (1) classifying, (2) finding the relation between two sets, (3) conserving, (4) finding and using properties of relations, (5) classifying on the basis of the as-many-as relation, and (6) ordering on the basis of the more-than or the less-than relation.

■ Classifying

It would be almost impossible for any of us to get through the day without classifying. We classify when we decide that an object is a fork and not a spoon, that an animal is a cow and not a dog. We classify when we purchase "essentials" and "nonessentials." We classify when we marry! Not only do we constantly classify, but we are able to maintain our sanity because others have classified. Can you imagine looking for a telephone number in an unalphabetized directory?

Children must learn to perform classification tasks early, not only because the ability to do so leads to the development of sound number ideas but because these abilities are necessary for organizing the environment. In what follows, we describe some ways in which children can learn to classify.

We classify objects by designating a *general attribute,* such as shape, size, color, or type of material. Once an attribute has been designated, the objects can be put into *specific attribute classes.* Suppose a set of objects is to be classified and the general attribute is color. The objects would then be placed into specific color classes: red, blue, purple, and so on. The following are classification tasks that children should learn to perform. They are listed from simple to more difficult.

- Selects objects having a specific attribute. The child is given a set of colored objects and is asked to pick those that have a specific color (the color need not be named, merely indicated).
- Sorts objects on the basis of a general attribute. The child is given a set of differently shaped objects and is asked to separate the objects into groups so that all the objects in a group have the same shape.
- Duplicates a pattern. The child is given a set of objects grouped in a particular way and is asked to use other objects to make a similar arrangement.

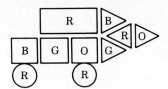

Example: Use your geo-pieces to make an engine like this one (see Material Sheet 1). [Note: The letters in the drawing indicate the color of the construction paper used.]

- **Extends a pattern.** The child is given a row of objects and asked to find the one that comes next.

Example: Find the geo-piece that comes next in this row.

Materials for classification activities are easy to collect. Usually discarded objects can be used: small plastic bottles, bottle tops, margarine tubs, toothpaste caps, and so on. Inexpensive objects can be purchased: bags of toy animals, planes, cars, trucks, or beads. Other objects can be made with little effort and cost (see Material Sheets 1 and 2).

Children should be encouraged to find as many ways as they can to classify a set of objects. Attributes that might be considered are:

General attribute	Specific attribute
Texture	smooth, rough, fuzzy
Function	cuts, covers, writes
Material	wood, plastic, paper
Color	red, blue, green, orange
Shape	square, triangle, circle
Size	big, little

Games also can be devised in which children guess their classmates' classification schemes. After children have had ample experience carrying out these tasks, they should solve problems or riddles based on classifying. At first, the riddles should require classification on the basis of one attribute. Later, the riddles should require classification on the basis of more than one attribute.

Example 1-1   C-1
An activity requiring classification on the basis of one attribute

Materials: Toys.

Teacher directions: Look at these toys, then see if you can solve this riddle: They are fuzzy. Which ones?

☞

Example 1-2   C-1
An activity requiring classification on the basis of more than one attribute

Materials: Geo-pieces (see Material Sheet 1).

Teacher directions: Look at these geo-pieces, then solve this riddle: It's red, and it has corners. Which ones?

The preceding activities are appropriate for children in kindergarten through second grade. With modification, however, they can be used at any grade level.

■ Finding the relation between two sets

Learning whole-number concepts is based upon the ability to decide when the objects of two sets can be *matched one-to-one*; that is, the ability to decide whether a matching can be achieved so that for every object in the first set, there is exactly one object in the second set. The ability of children to make such decisions leads to the formation of three important concepts. These concepts allow children to compare two sets and determine the relation that exists between them.

When two sets of objects can be matched one-to-one, then one set has *as many as* the other set (*as-many-as relation*).

When two sets of objects cannot be matched one-to-one, then one set has *more than* the other set (*more-than relation*), and one set has *less than* the other set (*less-than relation*).

As we discussed earlier, perception can mislead a child who is trying to decide which relation exists between two sets. The size of the objects and their geometric arrangement can lead to inconsistent judgments. If children are to consistently find the correct relation between two sets, then they must have many experiences using the *one-to-one matching test.* First, they should work with two sets having the same number of objects. The objects should be somehow related. After these experiences, more difficult situations should be provided, where obvious characteristics are not shared by the objects. Questions should be posed that encourage children to use the one-to-one matching test as a basis for establishing the correct relation. The children should also be encouraged to "tell what they find out" by using the correct vocabulary; for example, "this set has more," or "this set has less."

Children use the phrases "more than" and "as many as" frequently, even before entering school. They are not as likely to use "less than" since there is no perceived need for it. For example, if Willie has three cookies and Maria has five, "Maria has more than Willie" covers the situation. But often children will describe the situation as, "Willie doesn't have as many as Maria." Thus, the transition from the child's language to more refined expression must be a gentle and careful one.

The child's transition

Comparing sets	Comparing numbers	Symbols
I have *more* cookies than you. ⟶ more than	5 is greater than 3 ⟶	5 > 3
I have *as many* cookies *as* you. ⟶ as many as	3 + 2 is the same as 5 ⟶ 3 + 2 is equal to 5 ⟶	3 + 2 = 5
I do *not* have *as many* cookies *as* you. ⟶ less than	3 is less than 5 ⟶	3 < 5

The symbols for greater than and less than probably should not be introduced until the second grade. When they are taught, they need to be accompanied by appealing little stories that are designed to help children memorize what the symbols look like. A story about "Greedy the Gator" who always gobbles up the one with the most in it can make the connection between the sign for greater than and Greedy's mouth and tail. The less than sign is pointing to the smaller number. Children will continue to need review all through elementary school.

The following activities illustrate initial experiences in finding the relationship between two sets.

Example 1-3
An activity in which the child matches objects of two sets

C-1

Materials: Funny faces and mouths (see Material Sheet 2).

Teacher directions: Match the funny faces and mouths.

Questions: Can you put a mouth on each face? Is there a face for each mouth? Tell whether there are more, less, or as many faces as mouths.

Example 1-4

C-1

An activity in which the child matches two sets of objects that have no apparent common features

Materials: Ten unrelated objects. Five of the objects are on a blue piece of paper, and five are on a red piece of paper.

Teacher directions: Match each object on the blue paper with an object on the red paper.

Questions: Are there as many objects on the blue paper as there are on the red paper? Are there as many objects on the red paper as there are on the blue paper?

on blue paper   on red paper

■ Conserving relations

Suppose a child can perform a one-to-one matching test and determine whether there are as many objects in one set as in another. Does this mean that the relation is viewed as an unchanging one? No! Once the sets are rearranged, the child's perception may take over. When a child remains secure that the relation does not change because the objects are rearranged, he or she is said to *conserve the relation*. The activities suggested earlier can be extended so that the child is required to demonstrate this ability. Here are some examples.

Example 1-5

C-1

An activity requiring the use of the as-many-as relation

Materials: Four baseballs and four bats (see Material Sheet 2).

Teacher directions: Match the balls and the bats.

Questions: Are there as many balls as bats? As many bats as balls?

New directions: Now put the balls in a separate pile.

Questions: Now are there as many balls as bats? As many bats as balls?

Example 1-6

C-1

An activity requiring the use of the more-than relation

Materials: Nine funny faces and eight mouths (see Material Sheet 2).

Teacher directions: Match the funny faces and mouths.

Questions: Are there more funny faces than mouths? How do you know?

New directions: Put the mouths in a separate pile.

Question: Are there still more funny faces than mouths?

Example 1-7

An activity requiring the use of the more-than relation

Materials: Pictures like the one shown below.

Teacher directions: Make up a story about the picture. Use the words *more* and *than*.

Example:

Picture

Story

"Elmer has more shoes than socks."

Example 1-8

An activity that reinforces the more-than relation

Materials: Cards showing pictures of sets.

Sample cards:

Teacher directions: Play the "more-than" game. Cards are shuffled and distributed among the children. Each child plays one from his or her stack. The child with the "more-than" set takes the trick. When a tie occurs, each child plays another card. The player with the "more-than" set takes all the cards. The game is over when all but one player is out of cards. The winner is the child holding the most cards.

■ Finding and using properties of relations

The as-many-as relation.

It's important that children know how relations behave. Such knowledge provides them with the tools for making decisions based on logical thinking. For example, will spreading a cup of candy out on a plate create more candy? Of course not, but as we've seen, children may think so. It's important that they learn this idea about the as-many-as relation. Mathematically, we say:

$A = a$ then $a = A$

1. If A is a set, then A has as many as A. (*reflexive property*)

Suppose A and B are sets and you know that A has as many as B. What else do you know? You know that B has as many as A. In mathematical language, we say:

2. If A and B are sets, and A has as many as B, then B has as many as A. (*symmetric property*)

$A = B$
$A = n$
$B = n$
$B = A$

Do you think this idea is obvious to children? Not always! When children complete a matching test, it's necessary to emphasize this idea. Note that, in the activities in which funny faces and mouths are matched one-to-one, the questions asked are: "Are there as many mouths as funny faces?" "Are there as many funny faces as mouths?"

Suppose you know that Julie has as many tapes as Larry and Larry has as many tapes as Shane. What can you say about the way Julie's and Shane's tapes are related? Of course! Julie has as many tapes as Shane. Mathematically, we say:

$$A = B, \quad B = C \quad \text{then} \quad A = C$$

3. If A, B, and C are sets, and A has as many as B and B has as many as C, then A has as many as C. (*transitive property*)

This idea about the as-many-as relation is a very important one for children to learn. First, they should practice verifying the idea. Then they should learn to use the idea for establishing the relation between two sets. The following is an example of an appropriate activity.

Example 1-9
C-1
An activity requiring the use of the transitive property

Materials: Three funny faces, three mouths, three toothbrushes (see Material Sheet 2).

Teacher directions: Match the funny faces and the mouths.

Questions: Are there as many mouths as brushes? As many brushes as mouths? Do you think there are as many funny faces as there are brushes? How can you tell?

Questions: Are there as many faces as mouths? As many mouths as faces?

New directions: Match the mouths and the brushes.

The more-than and less-than relations
Suppose you have more college credits than your brother Sam. Then how would you say his credits compare to yours? Would you say that he has less than you do? Sure you would! Mathematically, we state this idea this way:

If A has more than B, ←——— The more-than and less-than
then B has less than A.   relations are not symmetric.

$$\text{if } A > B \text{ then } B < A$$

Suppose Lucy hits more home runs than Manard, and Manard hits more than Henry. Then how do Lucy's home runs relate to Henry's? You're right, it's a more-than relation. Suppose A, B, and C are sets. Mathematically, we say:

$A > B, \quad B > C$
$\text{then } A > C$

If A has more than B,
    and B has more than C,
    then A has more than C.

or

> The more-than and less-than relations are **transitive**.

If A has less than B,
    and B has less than C,
    then A has less than C. $A < B, \; B < C$
$\text{then } A < C$

As you can see, the more-than and less-than relations are also characterized by certain properties, which children should learn. Understanding these properties of the more-than and less-than relations leads to learning how to **order numbers** between 1 and 10 inclusively, as well as learning to **count** to 10. Many activities such as the ones given for the as-many-as relation can be used that not only will enable children to verify the properties but also will require them to use the properties when establishing the relation between two sets. The following are examples of appropriate activities.

Example 1-10   **C-1**
An activity demonstrating that more-than (less-than) is not a symmetric relation

Materials: Cutouts of five bowls and three balls.

Teacher directions: Match the bowls and the balls. Are there more or less bowls than balls?

Are there more or less balls than bowls? Give the reason for your answer.

Example 1-11   C-1
An activity demonstrating that the less-than relation is transitive

Materials: Cutouts of three shoes, four socks, and five little feet.

Example:

Teacher directions: Match the shoes and socks.

Question: Are there as many shoes as socks?

New directions: Now match the socks and the feet.

Questions: Are there as many socks as feet? Are there enough shoes for those poor cold feet? How can you tell?

■ Classifying on the basis of the as-many-as relation

The next skill the child should acquire is that of classifying sets on the basis of the as-many-as relation. A collection of sets is sorted into piles (stacks, categories, and so on) so that each set in a pile has the same number of objects. The following is an example of such an activity.

Example 1-12   C-1
An activity requiring children to classify on the basis of as-many-as

Materials: For this activity you'll need 30 *set cards* and 10 *set envelopes.* Make three set cards and one envelope for each of the numbers 1 through 10. To make the cards, draw or paste figures on colored file cards or pieces of construction paper. To make the envelopes, paste or draw happy faces on envelopes. The envelopes should be big enough to hold three set cards.

Example:

Teacher directions: Mix these cards. Then put them in envelopes. When you put a card in an envelope, make sure it shows as many objects as the envelope shows.

When having children complete this activity in Example 1-12, you should first work with just a few set cards and envelopes. However, always use more than one set card for each envelope you choose. It's also possible that children may not be able to tell by inspection when a card shows as many objects as an envelope. A set of loose objects such as beans or buttons may be needed so that children can compare the sets shown on the card and on the envelope.

■ Ordering sets on the basis of the more-than or less-than relation

Classifying sets on the basis of the as-many-as, more-than, or less-than relation leads to ways for ordering sets. Children first find the relation between two given sets. A third set is then presented, and children are required to "put it where you think it belongs," between the original sets, to the right of both sets, or to the left of both sets. Eventually, such activities should include more than three sets and involve sets that increase by more than one when ordered according to number. Finally, children should have experience sorting a collection of sets into categories (stacks) according to the as-many-as relation and then arranging the categories according to the more-than or less-than relation. The following is an example of this kind of activity. Some children may need to perform simpler but similar tasks that lead to successful completion of this activity.

Example 1-13
An activity requiring children to classify and to order sets

C-1

Materials: Set cards, set envelopes (see Example 1-12).

Teacher directions: Mix the cards. Then put each card in the envelope that shows as many objects as the card shows.

New directions: Look at these two envelopes. Do the cards in this envelope show more or less objects than the cards in the other envelope? How do you know?

Question: Where should we put this envelope? (Teacher gestures.)

 Here? X   Here? X     Or here? X

Question: Now where do you think we should put this envelope? [Etc.]

Teacher comment: Now we have put our sets in order.

■ Summary

In this section we've discussed six important tasks that children should be able to perform before being introduced to numbers. These tasks are: (1) classifying, (2) finding the relation between two sets, (3) conserving, (4) finding and using properties of relations, (5) classifying on the basis of the as-many-as relation, and (6) ordering on the basis of the more-than and less-than relations. Teachers should involve children in activities to determine whether they can do these tasks and should aid them in developing the ability to do them. Often, appropriate activities are not commercially available, so teachers must depend on personal resources to construct them.

Learning the numbers 1 through 10

■ Naming a family of sets with a number

Children are ready to be introduced to the numbers 1 through 10 whenever they can perform the pre-number tasks discussed in the last section. First they learn to associate a number with a family of sets. For example, they must associate the number 1 with any set having as many as this set { ⊙ }. They must associate the number 2 with any set having as many as this set { ⊙ ⊙ }. Appropriate association must be made for the numbers 1 through 10. Since children have previously learned to order sets, learning these number associations allows them to grasp concepts involving order relations of numbers. The following activities require associating a number and a family of sets.

Example 1-14   C-1
An activity requiring children to classify and to order sets

Materials: Set envelopes marked with the numerals 1 through 10 (picturing the appropriate set) and set cards.

Teacher directions: Mix the cards. Then put each one in the right envelope. Put the envelopes in order.

Example 1-15   S-1
An activity requiring children to classify sets and to order numbers

Materials: Envelopes marked with the numerals 1 through 10. (These envelopes should not picture sets.) Set cards.

Teacher directions: Mix the cards. Then put them in the right envelope. Put the envelopes in order.

☞
Example 1-16 S-1
An activity requiring children to associate sets with numerals

Materials: A cardboard shirt front with ties at the neck and felt pockets marked with numerals (or use an apron). Cards picturing sets of objects.

Teacher directions: One student wears the shirt. Another student puts each card in the right pocket.

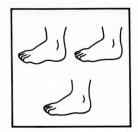

☞
Example 1-17 S-1
An activity requiring children to order numbers

Materials: Envelopes marked with the numerals 1 through 10. Set cards. The teacher puts appropriate cards in the envelopes and places the envelopes on a surface so that the numerals can't be seen. Hiding the numerals will keep the children from selecting only known numerals.

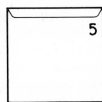

etc.

Teacher directions: Take some of the envelopes. Look at the numerals. Put the envelopes in order. Take the cards out of the envelopes and check yourself by looking at the cards.

Important skills to be considered along with number-set associations include writing the numerals 1 through 10 and learning their written names. Activities should be provided that not only allow children to practice writing these numerals and reading their word names but also reinforce associated concepts. Here is an example.

☞

Example 1-18
An activity requiring children to write numerals and number names

S-1

Materials: Teacher-made ditto such as the following.   *Teacher directions:* For each set, write the number. Write the number name. Use the marks to help you.

 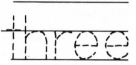

etc.

■ What about 10 and 0?

So far we have not mentioned that 0 and 10 are special numbers. The number 0 is special because it's associated with the family of all sets that have no members. At first this number idea is hard for children to understand, so it is wise to introduce the numbers 1 through 10 before introducing 0. Once children are accustomed to using these numbers to tell "how many," they can more readily accept that 0 tells "how many" when a set has no members. Let them tell you "How many elephants are in the room?" (Zero.)

The number 10 is special because of the way it's represented in our numeration system. The numeral for 10 is the first one children encounter that involves place-value ideas. However, children can learn to write the numeral "10," say "ten," and associate the symbol appropriately with sets before they can appreciate that the symbol indicates 1 set of ten and 0 ones. So, at first the number 10 should be developed similarly to each of the numbers from 1 through 9. Once children learn these numbers and can count from 1 to 10, the teacher can begin to develop place-value ideas.

→

It's time to count

Teaching children to make appropriate number associations before teaching them to count is more important than it may seem. If children are allowed simply to associate the numbers 1, 2, 3, . . . , 10 with a set of ten objects by pointing to the objects one at a time and saying the number names, they may confuse ordinal position with the idea of "how many." By *ordinal position* we mean the relative position of an object in a set with respect to some reference—first, second, and so on. It's important that children realize after counting ten objects that the number 10 is associated with all ten objects and not just the last one counted.

To avoid possible confusion, teachers should construct activities in which children are required to associate the numbers 1 through 10 with "small" sets they build from the set to be counted. For example, if children are to count the number of Happy Faces in a set having 10, they should be instructed in the following manner:

Teacher's Comments and Gestures   *Happy Faces*

Here are some Happy Faces. We're going to find how many by counting.

Teacher moves one Happy Face aside and says "one."

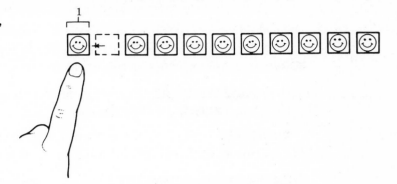

Teacher moves one more Happy Face aside and says "two."

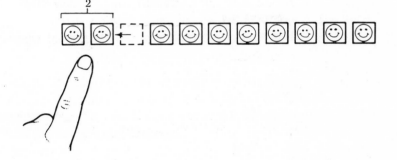

Teacher moves another Happy Face aside and says "three," and so on.

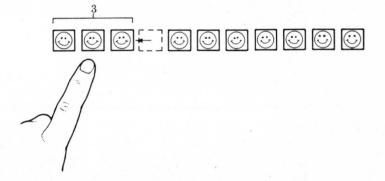

The importance of this strategy is that as each number is referred to, it is associated with a set of objects belonging to the family of sets named by the number. As children become confident that by counting they find the number of objects in a set, the method for counting can be relaxed; children may then simply match the numbers one at a time with the Happy Faces to be counted and can decide that the last numeral spoken refers to the entire set. Finally, children must memorize the names and the standard order for the numbers 1 through 10. Once children can count to 10, it's time to introduce ordinal terms such as "first," "second," "third," and so forth.

There are many activities that can be used to give children counting experiences. The teacher can use Happy Face pieces (see Material Sheet 3), beans, bottle tops, a counting board with beads on it, or even children.

→ Getting ready to add and subtract

Once children can count to ten, related readiness concepts for addition and subtraction should be acquired. Appropriate readiness concepts for addition grow out of experiences requiring children to answer these questions:

1. How many are in each of these two sets of objects?
2. How many objects are there in all?

Subtraction is more complex. Appropriate readiness concepts grow out of experiences that require children to answer a variety of questions:

1. How many objects are left in this set when this many objects are removed? (*take-away question*)
2. How many more objects must be put with this set to make a set with this many? (*missing-addend question*)
3. How many more objects are in this set than in the other? (*comparison question*)

Activities stressing these questions should be very informal. The addition symbol, the subtraction symbol, and number sentences should not be introduced. Children should describe observations orally or should represent observations by manipulating concrete objects or by indicating numbers. Activities for addition should not lead children to believe that sets of objects must be combined in order to find the total number. This is because many real-life applications do not involve combining sets of objects. For example,

Lonnie has four pennies in one pocket.
He has two pennies in another pocket.
How many pennies does Lonnie have?

Activities for subtraction should not emphasize the "take-away" question any more than the other questions. Such an emphasis leads children to believe that subtraction is synonymous with "take-away." This limited notion of subtraction may greatly inhibit the development of skills for solving subtraction word problems that involve separating a set into two smaller subsets.

In Chapter 3 we discuss the teaching of addition and subtraction more thoroughly. For subtraction, we discuss a fourth question that involves the partitioning of a set of objects. We omit this question for children at this level, since its concrete illustration would not appear to be different from concrete illustrations given for the other subtraction questions. In Chapter 8 we describe common difficulties children experience when computing and solving problems. After reading these chapters you will more fully appreciate the "do's" and "don'ts" given in this chapter. Until then, have faith! Read the following sample activity for addition, then see if you can design a sample activity for each of the subtraction questions given in this section (see Exercise 8 in "It's Think-Tank Time" later in this chapter).

Example 1-19   C-1
An activity requiring children to use addition concepts

Materials: Several sets of cards labeled with numerals from 0 to 10, with holes punched in all four corners of each card. Plastic chain links or large paper clips.

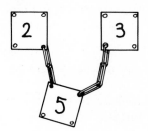

Teacher directions: Here are two cards. (Teacher arranges some of the cards in pairs so that the sum of indicated numbers is 10 or less.) Build chains for each card. Then count all the links (clips) you used. Find this number card and fasten it to your other cards. Have children give sentences such as "two and three more make five."

The computer corner

Computer programs for instruction can be powerful tools at every level of instruction especially when they involve concept development that appeals to several senses. Programs used in the early years should require little knowledge of the keyboard.

Here's a program that fills the bill. It provides kinesthetic, audio, and visual stimuli as children construct sets with ten objects.

Example 1-20   C-1
A concept development activity for learning to count to ten

Materials: The COUNT program on the MGB-Diskette Program on the MGB Diskette.

Teacher directions: This program will help you learn to count to ten. You can press almost any key on the keyboard. As you press keys the computer will draw pictures to show "how many." The computer will also tell the number of keys you have pressed. When you get to ten you must start over. The computer will let you count backwards too.

Sample computer screen:

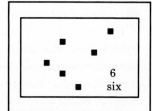

Here is another activity that illustrates a very important idea: To learn computer concepts you don't have to sit behind a computer keyboard constantly punching keys.

Example 1-21   C-1, 2
This activity encourages children to look for number patterns in connection with the computer terms "input," "output," and "memory"

Materials: One set of cards labeled with numerals 1 through 9. These are the "input" cards. Another set of cards labeled with numerals through 18. These are the "output" cards. A set of computer "memory" cards. These cards are labeled with rules such as the following:

The number that comes before
The number that is 6 greater
The number that is 5 less

Teacher directions: Students, we are going to play "computer." One student will play the computer, and will be in charge of the memory and output cards. The computer player will choose a memory card. Other students will take turns choosing input cards to give to the computer. The computer must use the rule on the memory card to make changes on the input card. To indicate changes the computer must show the correct output card. Other students must guess the rule.

You also need a brown paper bag with a computer drawn on the front and holes should be cut in the bottom sides for a student's head and arms.

■ Computer tips

In the early grades it is particularly important that furniture, heights of monitors, and heights of keyboards are adjusted to the sizes of children. Without such adjustments, neck and wrist strain can result. *Monochromatic* or *RGB* monitors are much easier on the eyes than *composite* color monitors especially when computer instruction involves detailed reading and writing. It is an exercise for you to research the characteristics of each of these types of monitors.

Throughout school experiences, students should have opportunities to *play computer*. These experiences allow children to concentrate on important concepts without the technical complications that accompany the actual operation of a computer.

Ideas for enrichment: Classification games

Most of the activities in this chapter can be used to extend the knowledge of more talented children. By changing the questions that are suggested in the activities, you can challenge the brightest ones. For example, in any lesson that deals with classification, it is appropriate to ask, "What else can you tell me?" or "How could you do that another way?" Lessons dealing with larger numbers need not assume that children know how many. In numerous situations, children can answer a question such as, "Are there more chairs than children?" A two for one situation leads naturally to discussing "two times as many." A kindergarten class of 23 with four at a table can be asked, "How many tables do we need?" The answer is clearly "six," but a few children may decide, "We have four (fives) and three more." The idea of

$$23 = (4 \times 5) + 3$$

is possible for them.

An activity that involves a classification game uses the premise that someone likes certain things and doesn't like others. We're supposed to guess which (and why). Try this!

This illustration of the game is too difficult for children at an early level. It can be used easily with geo-pieces or commercially available materials, such as attribute blocks.

Sally likes	*but doesn't like*
Bill	Bob
Annie	Jane
vanilla	chocolate
cookies	cake
Missouri	Iowa
Betty	Andria

Hint: Sally adores Mississippi and Tennessee (answer on page 33).

	Tony likes	*but doesn't like*
Game 1	(Hold up a) red	(Hold up) any other color
Game 2	(Hold up a) triangle	(Hold up) any other figure
Game 3	words that start with a "B"	words that start with anything else

Give everyone who wants to play a chance to find something that Tony likes. (Did you figure out what kind of things Sally likes yet? If you give up, look at the end of the Think-Tank section that follows.)

In closing . . .

In this chapter, concepts have been introduced that we think children should acquire before they begin work with formal arithmetic concepts and before they learn to name numbers greater than ten. These concepts fall into four main categories: pre-number concepts, naming and ordering number concepts, counting concepts, and addition- and subtraction-readiness concepts. The position that we take in this chapter is based on two considerations. First, acquiring the concepts that are stressed here makes learning arithmetic more than merely learning to manipulate symbols. Second, these concepts underlie many important mathematical ideas that children will eventually encounter but that are not related to arithmetic.

We describe several activities that aid in the development of the concepts discussed. Most depend on manipulation of paper cutouts. That's because we want to furnish you with sheets that can be used later. It's important, however, that children carry out other activities in which they must manipulate objects other than cutouts.

In later chapters you will encounter many other important concepts that children should acquire in the early years. These concepts are prerequisite for the development and application of ideas in arithmetic, geometry, and measurement.

It's think-tank time

1. Determine which of the activities in this chapter involve higher-order thinking skills and could be used to promote problem solving by elementary school children.
2. Six pre-number tasks are discussed in the text. See if you can describe them without rereading. Then go back through the chapter to check yourself.
3. In the section on classification, several tasks are described that children should learn to do. For each situation below, find the tasks that the child is doing or is asked to do. Be careful: sometimes more than one task is involved.
   a. The child finds all the red stars in a set of different colored stars.
   b. The child separates a set of toy barnyard animals into groups, such as chickens, cows, ducks, and pigs, according to directions.
   c. The child is given some geo-pieces and asked to duplicate a set of designs.

   geo-pieces (see Material Sheet 1), each duplicated in four colors: red, blue, orange, and green.

   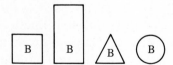

possible design

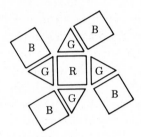

   d. The child is given the geo-pieces and is asked to choose all the pieces that are orange and square.
4. In the section on finding properties of relations, several properties for the as-many-as, less-than, and more-than relations are described. Read the situations below. Identify the property that the child is using, is asked to use, or should but fails to use.
   a. A group of children is given a set of socks and a set of shoes, and is told that there are more socks than shoes. They are asked whether there are more shoes than socks or as many shoes as socks. The children are unable to answer until they have matched shoes and socks.
   b. Children are given three sets of stick figures: moms, dads, and daughters. By one-to-one matching, they find that there are

as many daughters as moms and as many moms as dads. The children decide that there are as many daughters as dads.

c. A small boy says he would rather have his candy on a paper plate than in a cup because that way he'll have more. What's wrong?

d. A little girl finds by using a matching test that the blue cup has more pennies than the red and the yellow more pennies than the blue. She is then asked "Which has more pennies, the yellow cup or the red one?" She is unable to answer until she uses the matching test on the pennies in the red and yellow cups.

5. Assume that children understand the six pre-number concepts; explain how you could teach them to associate the appropriate number with a set. Explain also how you would teach children to order sets of numbers between 0 and 10 inclusively.

6. Outline methods for teaching children to count once they are able to make appropriate set-number associations.

7. How good are you at matching? Let's find out. Do you agree that if the members of one set can be matched one-to-one with the members of a second set, then the two sets have the same number of members? Check this out:

the whole numbers: 0 1 2 3 4 . . .
the even numbers: 0 2 4 6 8 . . .

If these lists were continued indefinitely, would there be a whole number for every even number? An even number for every whole number? Are there as many even numbers as there are whole numbers?

8. Identify four questions that can be used to aid children's development of readiness concepts for addition and subtraction. Tell what precautions you must take when developing addition concepts and subtraction concepts.

9. Study the objectives for the child listed at the beginning of this chapter. Make sure you understand each one. Make a card file of activities for the objectives, including at least one activity for each objective. Use activities given in this chapter, find ones in other sources, or make up your own.

10. Articles on computer products can be found in a number of computer magazines. Brochures and demonstrations can be obtained from computer dealers or popular computer stores. Find three sources of information on computer monitors that describe the features of RGB monitors and composite monitors. List the differences. Explain why it is often necessary to use a monochromatic monitor when the monitor is also composite.

11. Find and abstract at least three magazine or journal articles that discuss issues related to the use of microcomputers. Identify principles that will help guide educators who are preparing computer experiences.

12. Prepare an activity in which children *play* computer. Extend the input, output, and memory idea developed in Example 1-21.

13. Take a rest; you deserve it!

The answer to the puzzle on page 31 presented in Ideas for Enrichment is *Sally likes words with double letters.*

➡ ━━━━━━━━━━━━━━━━━━━━━━━━━━━━━━━━━━━

Suggested readings

Biggs, John B. "The Development of Number Concepts in Young Children." *Educational Research, 1*(1958–1959): 17–34.

Clements, Douglas Harvey. "A Comparison: The Effect of a Logical Foundation vs. a Number Skills Curriculum on Young Children's Learning of Number and Logical Operations." (Doctoral dissertation, State University of New York at Buffalo, 1983.) *Dissertation Abstracts International, 43A*(1983):94A

Clements, Douglas H., and Callahan, Leroy G. "Number or Prenumber Foundational Experiences for Young Children: Must We Choose?" *The Arithmetic Teacher*, November 1983, pp. 34–37.

Copeland, Richard W. *How Children Learn Mathematics* (3rd Ed.). New York: Macmillan, 1979.

Hollis, L. Y. "Mathematical Concepts for Very Young Children." *The Arithmetic Teacher*, October 1981, pp. 24–27.

Horak, Virginia M., and Horak, Willis J. "Let's Do It, Developing Mathematical Understanding with Bead Strings." *The Arithmetic Teacher*, December 1982, pp. 6–9.

Horak, Virginia M., and Horak, Willis J. "Let's Do It, 'Button Bag' Mathematics." *The Arithmetic Teacher*, March 1983, pp. 10–16.

Huey, J. Frances. "Learning Potential of Young Children." *Educational Leadership*, November 1965, pp. 117–120.

Jensen, Rosalie, and O'Neil, David R. "Let's Do It, We've Got You Pegged." *The Arithmetic Teacher*, October 1981, pp. 10–16.

LeBlanc, John F. "The Performance of First-Grade Children in Four Levels of Conservation of Numerousness of Three I.Q. Groups when Solving Arithmetic Subtraction Problems." In *Piagetian Cognitive-Development Research and Mathematical Education.* Proceedings of a conference conducted at Columbia University, October 1970. Washington, D.C.: National Council of Teachers of Mathematics, 1971.

Leidtke, W. W. "Rational Counting." *The Arithmetic Teacher*, October 1978, pp. 20–26.

Payne, Joseph N. "Number and Numeration." In *Mathematics Learning in Early Childhood*. Thirty-seventh Yearbook of the National Council of Teachers of Mathematics. Washington, D.C.: The Council, 1975.

Piaget, Jean. *The Child's Conception of Number*. London: Routledge and Kegan Paul, 1964.

Prasitsak, Prapa Poowaton. "A Study of the Relationship between Piagetian Conservation of Number Tasks and the Ability in Counting of Young Children." (Doctoral dissertation, University of Houston, 1983.) *Dissertation Abstracts International, 43A* (1983):1364A.

Rundell, Judith Lipscomb. "A Comparison of the Effectiveness of Two Instructional Models Used with Mathematically High-Risk First-Grade Students." (Doctoral dissertation, Northwestern State University of Louisiana, 1983.) *Dissertation Abstracts International, 44A* (1984):3312A.

Steffe, Leslie P. "The Performance of First-Grade Children in Four Levels of Conservation of Numerousness and Three I.Q. Groups when Solving Arithmetic Addition Problems." In *Piagetian Cognitive-Development Research and Mathematical Education*. Proceedings of a conference conducted at Columbia University, October 1970. Washington, D.C.: National Council of Teachers of Mathematics, 1971.

Van Engen, Henry. "Epistemology, Research and Instruction." In *Piagetian Cognitive-Development Research and Mathematical Education*. Proceedings of a conference conducted at Columbia University, October 1970. Washington, D.C.: National Council of Teachers of Mathematics, 1971.

Wadsworth, Barry J. "The Development and Learning of Mathematics and Science Concepts." In B. J. Wadsworth, *Piaget for the Classroom Teacher*. New York: Longman, 1979.

Yvon, Bernard R., and Dopheide, Jane Dallinger. "Iggies Come to Kindergarten." *The Arithmetic Teacher*, January 1984, pp. 36–38.

... 100s, 10s, 1s ...The Best Yet

Our Base-Ten Numeration System

■ *This chapter briefly traces the evolution of the base-ten system of numeration and emphasizes the advantages of that system. An analytical look at the way we represent numbers can provide much information concerning how we should present these ideas to children.*

It is apparent that children have an inadequate understanding of place value and that the most effective approach to developing an understanding of our base-ten numeration system is to emphasize this system. Place value is one of the most important topics in arithmetic but probably one of the least understood.

We supply examples of conceptually sound materials for children to manipulate as you reinforce place-value ideas. We provide for extensive work with tens and hundreds before attempting to represent larger numbers. Varied interpretations of numerals are presented that will foster a broader understanding of the numeration system. The logical extension to decimal fractions also is examined. ■

Objectives for the child

1. Groups objects to illustrate the number of tens and ones that are possible.
2. Physically represents the base-ten numeral for numbers less than 100 with the appropriate number of tens and ones.
3. Symbolically represents a physical illustration that uses tens and ones for numbers less than 100.
4. Identifies the place value of digits in a base-ten numeral for numbers less than 100.
5. Expresses numbers less than 100 as so many tens and so many ones.
6. Groups objects to illustrate the number of hundreds, tens, and ones that are possible.
7. Physically represents the base-ten numeral for numbers less than 1000 with the appropriate number of hundreds, tens, and ones.
8. Symbolically represents a physical illustration that uses hundreds, tens, and ones for numbers less than 1000.
9. Identifies the place value of digits in a base-ten numeral.
10. Expands a base-ten numeral to illustrate the multiplication and addition components of the numeration system.
11. Places a number between two multiples of 10 and decides which multiple of 10 it's closer to.
12. Interprets a numeral in various ways using nonstandard names.
13. Indicates applications for numbers greater than 1000.
14. Identifies place value of digits in a numeral with up to three decimal places.

Teacher goals

1. Be able to identify and explain place-value concepts that are important to the development of children's mathematical ideas.
2. Be able to diagnose the performance of children with respect to place-value concepts.
3. Be able to provide instructional procedures and activities that will aid in the development of concepts related to place value.

Understanding numeration systems

■ Introduction

The development of a place-value numeration system is one of the most significant and far-reaching events in human history. The results of this development affect every one of us every day. Do you believe this? Stop and think for a moment how you are affected. How old are you? What year is this? How far is it to the sun? Or to Chicago? How many freckles on your nose? How much do you weigh? How much does the question mark at the end of this question weigh? How did you use our place-value numeration system to answer these questions?

The numeration system that we use *is* capable of handling answers to these questions. In fact, it can be used to represent a wide variety of numbers with a comparatively simple and systematic procedure for interpreting the numerals. This system of numeration is called the *Hindu-Arabic numeration system,* because the Hindus are credited with its development and the Arabs with its transmission to Western Europe. The exact details of how this base-ten numeration system came to be formed are not known. It *is* clear that it didn't evolve overnight and that it wasn't invented by any one person. The development was slow, and there were many contributors. We're so used to our numeration system that we often overlook many of its subtleties. But by doing so, we miss out on many opportunities for teaching important ideas that will lay a foundation for elementary school mathematics.

A numeration system consists of a set of symbols and rules for combining these symbols. Its purpose is to make it possible to communicate number ideas. The symbols, or any appropriate combination of them, are called *numerals.* Thus, numerals are names for numbers and are used to talk or write about numbers.

Example: Bird Black 3

| A | B | C |
| All of these deserve the name *bird.* | All of these have a property that deserves the name *black.* | All of these sets have a property that deserves the name *three* or 3. |

Items A, B, and C demonstrate the connection that the child must make between the symbols and what the symbols represent. Which of these connections do you think is the most difficult? Admittedly, the association of a numeral with a number is difficult. That is one of the reasons why the content and activities in Chapter 1 are so important. In this chapter, we'll see how the base-ten system of numeration can facilitate that association for numbers greater than 10.

There are only ten symbols (called *digits*) in our numeration system: 0, 1, 2, 3, 4, 5, 6, 7, 8, and 9. With these ten digits and rules for combining them provided by our base-ten numeration system, we can write names for any numbers we need. Just as we put together letters of the alphabet to form words to communicate thoughts, we put together digits to communicate ideas about numbers. The digits, then, form the "alphabet" for numeration.

Each digit has to be memorized and associated with a number. At the early stages of the development of children's ideas about numbers, the number is usually associated with a set of things. Each *numeral* represents the *number* of things.

Examples:

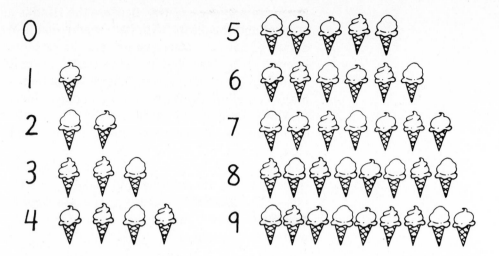

■ Early numeration systems

Thousands of years ago, people were using the same kind of procedure to communicate number ideas. A symbol was used to represent a certain number. At the most primitive level, a simple tally mark was the symbol, and the rule for using it was to mark | for each thing.

Thus, we have the following numeration system:

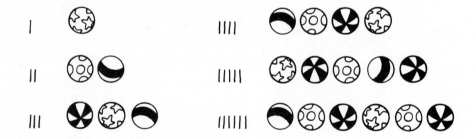

A slight modification of this system resulted in the following:

The rule change was, "When you get ||||, make the next mark sideways: ⫫." so the numeral ||||||| was represented like this: ⫫||.

The Egyptian and Roman systems used the same simple idea; that is, symbols stood for numbers, and the symbols were repeated as many times as necessary to represent a given number. Therefore, systems such as these are called *repetitive systems of numeration*. For example, in both the Egyptian and the Roman systems, to represent the number for

the symbol I would be repeated to result in III.

Roman		Egyptian
I	one	I
V	five	
X	ten	∩
L	fifty	
C	one hundred	9
D	five hundred	
M	one thousand	⌀

In the Roman system, to represent the number for

✳✳✳✳✳✳

we use a V and two I's.

If we have IIIIIIIIIIIIIIIII, we use an X, a V, and three I's to get XVIII. (The Egyptian representation would have been ∩ I I I I I I I I .) A slight modification did change the Roman system so that "IX" did not mean the same thing as "XI." The position of the "I" with respect to the "X" indicated one more or one less. The purpose of this kind of change was to make the numerals more compact. It takes less room to write "IX" than it does to write "VIIII" and less room for "XL" (ten less than fifty) than for "XXXX" (four tens). So, this modification did shorten some numerals, but there were many others that remained cumbersome.

■ A base-ten numeration system

Notice that in our system we have a different symbol to represent each number from 0 through 9. Fortunately, we stopped inventing new symbols and decided on rules for using the ones we had. In other words, after 9, it's a whole new ball game.

We introduced the idea of grouping ten objects and thinking of 1 ten. Since we already had a symbol for 1, that was all right, but something had to be done to distinguish 1 ten from 1 one. What we did was change the position (place) of the 1. With the invention of the symbol 0 to represent the number zero, we could write "10" to mean 1 ten and 0 ones. Thus, the symbol 0 was necessary for the complete development of a place-value numeration system.

Now our numeration system becomes a kind of coding system where 13 indicates 1 ten and 3 ones. Children can appreciate the idea of a code if there are physical objects involved and if the grouping procedure is an integral part of instruction.

In our numeration system, 12 is easier to think of and to communicate to children than 10 is because we can group 12 things into 1 ten and have 2 ones remaining. In the numeral 12, it's clear that the position of the 1 is significant and that the entire numeral means 1 ten and 2 ones. The child's physical (perceptual) experience is represented symbolically, and the symbols make sense. The physical representation for ten is one group of ten, and it requires special attention to convince young children that 10 means 1 ten and 0 ones. Thus, initial place-value activities should involve the numbers from 11 to 19. Then special attention can be given to 10, 20, 30, and so on.

Since our numeration system is based on ten, we group in tens. Thus, 10 tens will be grouped as 1 hundred, 10 hundreds will be grouped as 1 thousand, etc. (That's a pretty important "etc.")

The powers of 10 tell us the value of each place:

$$10^0 = 1$$
$$10^1 = 10$$
$$10^2 = 100$$
$$10^3 = 1000$$

.

.

.

The small 0, 1, 2, and 3 are called *exponents*. They indicate how many tens are to be multiplied to determine the value of a place (except in the case of 10^0, where it is 1 by definition). They also tell how many 0's are to the right of the 1. For example, $10^2 = 10 \times 10$ or 100, and 100 has two 0's to the right of the 1. Thus we know that in a base-ten numeral for a whole number, the place on the right is the ones place, the place to the left of that is the tens place, to the left of that is the hundreds place, to the left of that is the thousands place, and so forth. For any given place, the value of the place at its left is tens times as great.

To interpret a numeral in our base-ten numeration system, we use the two operations of multiplication and addition, combined with the previously discussed idea that each place has a specific value. In interpreting "539," for example, we know that "9" is in the ones place, "3" is in the tens place, and "5" is in the hundreds place. So "539" means "$(5 \times 100) + (3 \times 10) + (9 \times 1)$." Another way of writing this expanded form is "$(5 \times 10^2) + (3 \times 10^1) + (9 \times 10^0)$."

Let's look at some other whole numbers less than 1000 and pay attention to what the numeral means.

$$803 = (8 \times 100) + (0 \times 10) + (3 \times 1)$$
$$= (8 \times \underline{10 \times 10}) + (0 \times 10) + (3 \times 1)$$
$$= (8 \times 10^2) + (0 \times 10^1) + (3 \times 10^0)$$

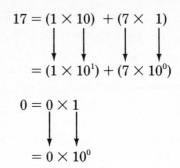

$$17 = (1 \times 10) + (7 \times 1)$$

$$= (1 \times 10^1) + (7 \times 10^0)$$

$$0 = 0 \times 1$$

$$= 0 \times 10^0$$

A "0" in the ones place tells how many ones we have. Right? Right! (We'll have more about that later when we need it.)

■ Reading and writing with larger numbers

Children should be taught that there are systematic procedures for reading numerals. Deciding how to read 1245602879 can be a motivational topic in fifth and sixth grades. The ideas will be presented again in junior high school, so they can be approached in a more relaxed manner here.

Current recommendations are to omit commas. If there are only four digits, they aren't separated at all. This implies that we all need a good understanding of place value and quite a bit of familiarity with numbers such as 4327. By the sixth grade, we should expect children to immediately recognize this as 4 thousands, 3 hundreds, 2 tens, and 7 ones, and be able to read it as "four thousand, three hundred twenty-seven." The number in the preceding paragraph, 1245602879, presents a different story.

With 1245602879, all of us need help in determining the value of the largest place. (From then on, we're in good shape.) Starting with the ones, we set off periods of three digits. Traditionally, this was done with commas, but current recommendations are to separate the three-digit periods with a blank space, which is a common practice throughout the world. Thus, we have:

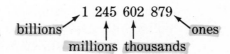

Then, we determine the names for these groups of three: in this case, billions, millions, thousands, and ones. (The left group may have one, two, or three digits.) At this stage, children can be shown that the most we have to worry about is reading three-digit numerals. All we need to know is the name of each group. Commas are helpful when reading the words: one billion, two hundred forty-five million, six hundred two thousand, eight hundred seventy-nine.

■ Extension to decimal fractions

The place-value patterns for decimal fractions are consistent with those for whole numbers. There is a specific value assigned to each place in the decimal fraction, and this value is a power of ten. The following chart indicates several important place-value ideas.

	Thousands	Hundreds	Tens	Ones	Tenths	Hundredths	Thousandths	
ETC.	1000	100	10	1	.1	.01	.001	ETC.
←	10^3	10^2	10^1	10^0	10^{-1}	10^{-2}	10^{-3}	→

Notice those two very important *et ceteras*. They mean that the same pattern continues in both directions. If we use the decimal point to locate the ones place, then it is possible to focus on a kind of symmetry in the chart with tens and tenths being the same distance from the ones; hundreds and hundredths are each two boxes over from the ones, and so on. Putting this together, we can interpret "5643.207," for example, as meaning:

$$(5 \times 1000) + (6 \times 100) + (4 \times 10) + (3 \times 1)$$
$$+ (2 \times .1) + (0 \times .01) + (7 \times .001)$$

or

$$(5 \times 10^3) + (6 \times 10^2) + (4 \times 10^1) + (3 \times 10^0)$$
$$+ (2 \times 10^{-1}) + (0 \times 10^{-2}) + (7 \times 10^{-3})$$

The expressions with exponents, in all probability, are not very useful for children in the elementary school. They *will* need decimal fractions, however, and they should be readily convinced that these ideas are just extensions of our base-ten numeration system.

■ Applications

Applications of this system can be related to our monetary system and also to the metric system of measurement. Children are easily convinced that it takes 10 one-dollar bills to exchange for 1 ten-dollar bill and that 10 ten-dollar bills are needed to get 1 one-hundred-dollar bill. They will undoubtedly have had some experience with these ideas—if not in real dollars, then at least in game situations. We can also expect that familiarity with the metric system will soon be a part of the real-life experiences of children. The use of metric units can reinforce concepts related to place value.

Relationships between money, metric units, and place-value concepts must be explicit and conceptually sound. It's tempting to resort to "moving decimal points," so be careful to relate your classroom activities explicitly to the base-ten, place-value numeration system and to operations on numbers. For example, 200 centimeters is the same as 2 meters. The "2" in "200" tells how many hundreds there are. Since we know that 100 centimeters is the same as 1 meter and that we have 2 hundreds, it should be easy to conclude that this is 2 meters. We could compute 200 ÷ 100 with paper and pencil to find how many hundreds are in 200, but a knowledge of place value means we don't have to resort to that. Similarly, 246 centimeters can be expanded to:

$$200 \text{ centimeters} + 40 \text{ centimeters} + 6 \text{ centimeters}$$

$$2 \text{ meters} \quad + \quad 4 \text{ decimeters} + 6 \text{ centimeters}$$

As we become more familiar with the metric units, it will be easy to go from 246 centimeters directly to 2 meters, 4 decimeters, and 6 centimeters, just as we easily make the transition from 3 ten-dollar bills to 30 one-dollar bills. We could compute 3×10, but here again, a knowledge of place value tells us that we can write 3 tens as 30 ones.

■ A few advantages

Our place-value numeration system has many advantages. Some are illustrated below.

Alpha Centauri, the nearest star other than the sun, is about 4×10^{13} kilometers from Earth.
This is a very large number, but there is no number so large that we can't represent it!

The weight of an electron is about 9×10^{-27} grams.
This is a very small number, but there is no number so small that we can't represent it!

If a book is 500 pages long and you've read the first 200 pages, you've read .4 of the book.
This isn't a whole number, but we can represent numbers that aren't whole numbers!

How many rhinoceroses are you holding in your hand?
Zero, but we can tell how many even when there aren't any!

So, we can easily represent very large or very small numbers, we can represent fractional numbers, and we have a symbol to indicate "not any." In short, our numeration system makes it easy for us to communicate number ideas.

In addition to these advantages, the place-value numeration system facilitates the computational processes, and the numerals are easy to interpret. By looking at "405" (and knowing the appropriate rules), we can tell that it represents an odd number that is not prime and that it is a multiple of 5 and 9. We can make 4 groups of 100 with 5 left over, or 40 groups of 10 with 5 left over. It's less than 406 and greater than 404. If we subtract 5, we'll get a multiple of 100; adding 5 will produce a multiple of 10. Etc., etc., etc.!

→ Helping children understand place value

Material Sheet 3 contains a model that has been used effectively with children to demonstrate some of the concepts related to place value. The 10×10 array of Happy Faces should be duplicated so that each child has at least 10 hundred-squares, 20 ten-strips, and 20 ones.

The smallest pieces, the single Happy Faces, represent ones. They can be used with some of the activities in Chapter 1 and in the early development of the whole-number operations discussed in Chapter 3. They also can be used in the introduction and extension of the place-value concepts discussed in this chapter.

The horizontal ten-strip contains 10 Happy Faces. This is easily verified by counting or by matching one-to-one. After being convinced that 1 ten equals 10 ones, children are willing to exchange 1 ten-strip for 10 ones or 10 ones for 1 ten-strip.

The 10 × 10 square contains 100 Happy Faces; this can be verified by the children. It also contains 10 ten-strips, and this, too, can be checked by the children. After this verification, children are much more willing to accept an exchange of 1 hundred-square for 10 tens, or 10 tens for 1 hundred-square. The big square, being a single piece of paper, is more clearly identifiable as 1 hundred than are other place-value models. A bundle of 100 sticks, for example, is clumsy even when the sticks are grouped in 10 bunches of 10—and children aren't convinced that there are *exactly* 100 sticks in the bundle.

With children at the first-grade level, it may be preferable to use the term "singles" to refer to the small pieces, or even "little smilies." It is confusing for young children to be asked how many ones are in 12. Place-value concepts are not developed yet, and they are very likely to answer "one," referring to the symbol "1" that is in "12." However, if they see that 12 is represented by a ten-strip and two single pieces, they can focus on the ten-strip as 1 ten. The two singles are 2 and, later, 2 ones. At this stage, the designation of 12 as 1 ten and 2 units is also acceptable. Many textbooks refer to the "tens place" and the "units place." Certainly children will be exposed to terminology such as the "tens digit" and the "units digit." For most children, the choice of words is no longer a problem by the second grade.

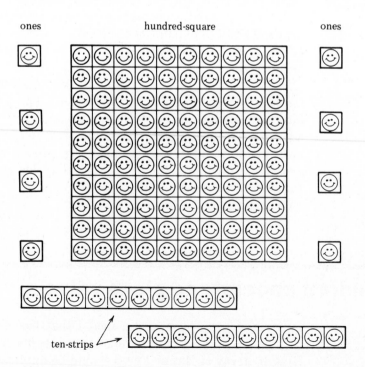

With these Happy Face models, children can be shown explicit place-value representations for numbers as large as 999. For example, to represent 723, they can select 7 of the hundred-squares, 2 of the ten-strips, and 3 of the ones pieces. Conversely, when shown 4 of the hundred-squares, 1 of the ten-strips, and 5 of the ones pieces, they can determine that this represents 415. They can also represent addition, subtraction, multiplication, and division and physically illustrate computational processes. The models provide a necessary emphasis on place value and lay the foundations for the understanding of operations, properties, and processes.

■ Activities for teaching place-value concepts

The following activities are samples to be used with the Happy Face pieces. Some of the activities are developmental; others are more practice-oriented. The pacing should depend on the purpose. Diagnosis of individual progress is essential in each activity. A pause or a quizzical look, as well as an incorrect response, can indicate a need for immediate attention. Mechanical, consistently correct responses may sometimes indicate that a change of direction is needed to more difficult concepts and questioning. Try these activities! Extend them! Make up your own!

The activity in Example 2-1 is a concept-development activity. It assumes that the child can count to 20 using methods described in Chapter 1; that is, the names for the numbers can be said, and these names are associated with appropriate numbers of objects. Now we are introducing the "base-ten" characteristic of our numeration system. The objects are physically grouped to represent numbers from 11 to 19.

Example 2-1   **C-1**
An activity that requires children to group 11 to 19 objects as 1 ten and an appropriate number of ones

Materials: Happy Face pieces: ten-strips and ones (see Material Sheet 3). In-and-Out Shoebox. Make the shoebox by cutting two holes in the top, big enough for a child's hand to go through. Tape in-and-out cards on the two openings. Decorate the box with children's art.

Teacher directions: We have ten-strips and ones pieces in our In-and-Out Shoebox. I want to trade so that I have as few pieces as possible. Help me do that with the 12 singles I have.

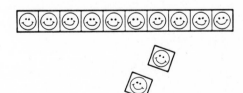

Questions: If I put 10 singles in, what can I take out? (1 ten-strip) How many singles do I have left? (2) So I have 1 ten-strip and 2 singles. Now let's try trading with 14 singles. Help me get as few pieces as possible. After we've traded in our 10 singles, what do we end up with? (1 ten-strip and 4 singles) Here I have 17 singles. What do I do?

This should provide children some familiarity with place-value concepts for 11 to 19. Then, particular attention must be given to going from 19 to 20 by regrouping to 2 tens. Emphasis at similar spots (39 to 40, 99 to 100) is valuable for many children even up through second grade.

Example 2-2   C-1
A concept-development activity that requires children to group from 10 to 99 objects appropriately and to represent their results symbolically

Materials: Happy Face pieces: ten-strips and ones (see Material Sheet 3). An In-and-Out Shoebox, Tens-and-Ones Chart made from heavy paper so that it has pockets to insert cards (see below), and two sets of index cards with the numerals 0 to 9 on them.

Teacher directions: Always use as few pieces as you can. I'll give you [this many] single pieces. Trade them in so you'll have as few pieces as possible.

13   16   18   21   23

Questions: Can you trade in 10 ones for 1 ten-strip? Can you trade in 10 more ones for another ten-strip? How many ten-strips do you have? How many ones do you have? Pick a card that tells how many ten-strips you have. Put it in the place that says "tens." Pick a card that tells how many ones you have. Put it in the place that says "ones." Now write how many Happy Faces you have altogether.

Example 2-3   C-1
A concept-development activity that requires children to compare two-digit numerals with numerals formed by a reversal of the digits

Materials: Happy Face pieces: ten-strips and ones (see Material Sheet 3).

Teacher directions: Always use as few pieces as you can. Show me 21. Now show me 12. Show me 16. Show me 61. Show me 53. Now show me 35. Show me 45. Now show me 54.

Questions: Are [21 and 12] the same? Which is more? How can you tell which is more? Can you tell which number is greater without using your Happy Faces?

It is necessary to include some instructional activities for children that help them develop skills associated with place-value ideas. Example 2-4, in which speed and accuracy are goals, is such an activity. The relevant concepts would have been developed previously.

Example 2-4   S-1
An activity that provides motivational practice involving ideas related to place value

Materials: Bingo cards and chips (or paper squares, beans, bottle caps).

Teacher directions: I'll show you a card like this:

| 2 tens + 3 ones |

23	54	32
45	free	78
16	61	87

If this number is on your card, put a chip on it. The winner is the first one to get three chips in a row.

Questions: Not many; they would detract from the "game."

Example 2-5   C-1
A concept-development activity that requires analysis of two-digit numerals and physical and symbolic representation

Materials: Two sets of index cards with the numerals from 0 to 9 on them (one numeral on each card). Happy Face pieces: ten-strips and ones (see Material Sheet 3). Tens-and-Ones Chart.

Teacher directions: Draw one card from each set. Make the largest number you can (a two-digit numeral). Show it with Happy Faces. Make the smallest number you can. Show it with Happy Faces. Put your cards in the Tens-and-Ones Chart to show your number.

Questions: What is the largest number that is possible? What is the smallest number that is possible? Julie said that her largest number and smallest number were the same. What can we conclude about the cards she drew?

Notice that all five of these activities were designated for Level One. This means that they are most appropriate for children in the range of grades K–2. Each activity involves numbers less than 100. Extensive work should be done in grades 1 and 2 before the introduction of numbers between 100 and 1000. When the children are ready, a slight modification of these activities will make them appropriate for the larger numbers.

■ Place-value activities for later grades

The next activity is a sample one for helping children represent larger numbers. Here Happy Faces are still appropriate, and the major emphasis is—you guessed it—place value.

☞
Example 2-6

C-2

A concept-development activity that requires children to physically represent numbers between 100 and 1000

Materials: Happy Face pieces: hundred-squares, ten-strips, and ones (see Material Sheet 3).

Teacher directions: Show me 231. Show me 542. Show me 300. Show me 207. Show me 630. Show me 111.

Questions: How many hundreds? How many tens? How many ones? Which number is largest? Could you tell without your Happy Faces? How?

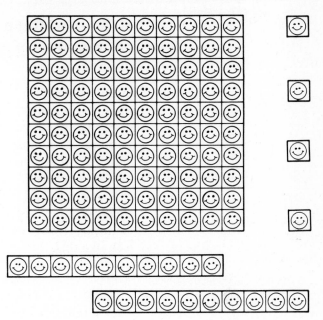

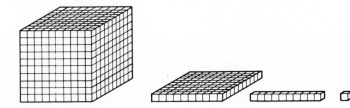

Children need the physical representation of numbers before symbolic representation, although our examples have focused on both types. The extent and type of experiences will depend on the children in your classroom. Your selection should provide for a spiraling of experiences to review and to fix important concepts.

Ice-cream sticks grouped in bundles of ten and held with rubber bands are a variation of the idea in Example 2-6. Some people recommend plastic toothpicks, and commercial materials such as Unifix cubes, or any objects that can be attached in groups of ten, can be used effectively with place-value ideas. Cuisenaire Rods have orange rods for tens and small white cubes for ones. The important thing, of course, is the instruction you give the children to go along with the materials.

For larger numbers at a concrete level, Dienes Blocks are commercially available. The wooden cube that represents one thousand is actually 1000 times the size of the small ones-cube. The hundred-piece contains 100 small cubes, and the ten-rod contains 10 small cubes:

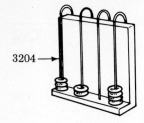

3204 →

A more abstract representation of larger numbers is provided by the abacus, with the poles representing places in the base-ten numeration system.

■ Beyond happy faces: Place-value activities for numbers greater than 1000

Now let's look at some larger numbers. Children need to realize that large numbers exist, and the existence of such numbers is usually verified in their own experiences. Watching television, reading newspapers, or just engaging in everyday conversation often involves the consideration of large numbers. In an instructional setting, the emphasis must again be on place value. With larger numbers we can definitely see the advantage of being able to express a number in exponential form. Although this form of expression should not be expected of children, it's reasonable that teachers should understand its interpretation and appreciate its advantages. For example, if we must write 10^{100} and can't use exponents, the numeral will look like this:

10000000000000000000000000000000000000000000000000000000000000000000000000000000000000000000000000000

This number is called a "googol." Your work with elementary school children won't involve numbers of this size, but you should be prepared to discuss the relative size of numbers and give appropriate examples.

The use of money can be motivating for children in the fifth or sixth grades. For example, you might ask such questions as "Would you rather have a million pennies or a thousand dimes? A 10-centimeter stack of one-dollar bills or a 1-centimeter stack of hundreds?" Money can also be used to emphasize place value: "Suppose you have 3864 dollar bills. If you took them to the bank, how many ten-dollar bills could you get?" (You can't get .4 of a ten-dollar bill, of course, so you'd have 386 tens and your 4 ones.) "How many hundred-dollar bills? How many thousand-dollar bills? If you want only thousands, hundreds, tens, and ones *and* as few bills as possible, what should you get?" (3 thousands, 8 hundreds, 6 tens, and 4 ones) Activities that use play money to pose a variety of questions will broaden children's understanding of place value.

Example 2-7

C-2

**A concept-development activity that requires children
to represent large numbers in several ways**

Materials: Play money: thousands, hundreds, tens,
and ones (use Material Sheet 5).

Teacher directions: Gary is going to be the banker.
What will he give you for your thousand-dollar bill?

Questions:
Banker: All we have is tens today. How many?
Banker: I'll give you all hundreds. How many?
Banker: Nothing but ones for you. How many?

The next activity is just the opposite; that is, children must now exchange small bills for larger ones. The most efficient representation of the
number of bills (with our deliberate restrictions) is the base-ten numeral,
and the children are physically demonstrating it!

Example 2-8   with 😊

**A concept-development activity that requires children to represent
a large number by physically representing its expanded form**

Materials: Play money: thousands, hundreds, tens,
and ones (see Material Sheet 5).

Teacher directions: Suppose you have 4238 one-
dollar bills . . . and no pockets!

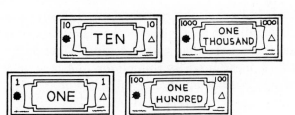

Questions: What's the largest number of hundreds
you could get? What's the largest number of tens you
could get? What's the largest number of thousands
you could get? Trade for thousands and hundreds and
tens and ones. Get as few bills as you can. How many
thousands? Hundreds? Tens? Ones?

The next activity is a concept-development activity that provides some
introduction to decimals through the use of dimes and pennies. It makes
use of familiar information and lays groundwork for further exploration
of the topic.

Example 2-9

C-3

A concept-development activity that introduces decimal fractions and place-value ideas

Materials: Ten story-time cards such as these, and one set of 0–9 cards and coin cards (see Material Sheet 6).

Teacher directions: Don dug one ditch. Daisy did one dish. Patsy picked one peach.

Don Digs Ditches for a Dollar a Ditch $1.00	Daisy Does Dishes for a Dime a Dish $0.10	Patsy Picks Peaches for a Penny a Peach $0.01

Questions:

1. How many peaches must Patsy pick to get as much as Daisy for her one dish? . . . (10)
2. How many dishes must Daisy do to get as much as Don for his one ditch? . . . (10)
3. How many peaches must Patsy pick to get as much as Don for his one ditch? . . . (100)

New directions: Draw a card (select it randomly from the ten cards) that shows the number of ditches

for Don. Draw a card for Daisy. Draw a card for Patsy. (For example, 7 ditches, 4 dishes, 3 peaches.) How much did Don, Daisy, and Patsy earn in all? Replace the cards and draw new ones.

Revise the story so that the story-time bank has only dollars, dimes, and then pennies. Ask questions such as: If Patsy picked 423 peaches, what will the story-time bank exchange her pennies for? Continue variations of the story.

■ Rounding numbers

With many of these activities and other instructional procedures, you should ask questions that allow children to answer to the nearest ten (or hundred or thousand). At the same time that you're asking "How many ten-strips could you get with 42 ones?" ask "Is this closer to 4 tens or 5 tens?" Similarly, if you're asking "With 36 one-dollar bills, how many tens could you get?" also ask "Is this nearer to 3 tens or 4 tens?" The children should verify their own responses with whatever materials are available. A number-line analysis is particularly appropriate for enabling children to decide, for example, that 42 is nearer to 40 than it is to 50:

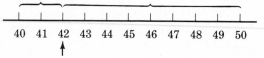

The following activity allows children to practice rounding off numbers less than 100. The square pieces of paper have the numerals 0 through 9 on them. In the container designated "ones," the numeral 5 should be omitted initially. When you believe that the children can accept the arbitrariness of "rounding up" numbers that are halfway between two tens, or of any rule for numbers like this that is given by the particular text you're using, then 5 can be included. This choice, like so many others you'll make, depends on the children in your classroom.

Example 2-10
An activity that provides motivational practice involving rounding off numbers to the nearest ten

S-2

Materials: Lost in Space (see Material Sheet 7).

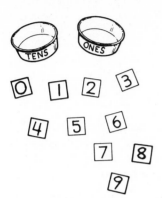

Teacher directions: One of the margarine cups contains little squares of paper telling how many tens you have. The other margarine cup has different-colored squares that tell how many ones you have. Mission Control sends a message that tells you your location. (Draw a number from each cup.) You are LOST IN SPACE, almost out of fuel, and must go to the *nearest* fuel station. The Mission Control message is 4 tens and 3 ones. Which fuel station will you go to? (40. Continue this practice with rounding off until everyone has a chance to participate.)

■ A variety of interpretations for large numbers

The Place-Value Displayer (see Material Sheet 8) is a device that can be used for both concept development and skill development. Children will be working with the standard form for a numeral and several expanded forms of the numeral. This Place Value Displayer is easy to make and use, provides a nice transition to concepts about larger numbers, and can be easily extended.

Rounding off larger numbers can be made easier by considering the different interpretations of a numeral. To round a number to the nearest ten, we decide which two multiples of ten it is between, then choose the one it is closer to. Other cases are similar. For example, 4238 can be rounded off in the following ways:

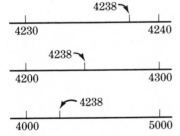

4238 is between 423 tens and 424 tens and is closer to 424 tens. So 4238, rounded to the nearest ten, gives 4240.

4238 is between 42 hundreds and 43 hundreds and is closer to 42 hundreds. So 4238, rounded to the nearest hundred, gives 4200.

Finally, 4238 is between 4 thousands and 5 thousands and is closer to 4 thousands. So 4238, rounded to the nearest thousand, gives 4000.

This variety of interpretations allows children to deal with many mathematical situations they will encounter. One situation involves computational procedures that they will have to learn. Another involves converting metric measures from one unit to another. In some cases, a variety of interpretations allows children to translate numerical data so that a calculator can be used. Remember these ideas as you read the rest of the book!

Example 2-11

C-3

A concept-development activity that provides several interpretations for a numeral

Materials: Place-Value Displayer (see Material Sheet 8). Paper clips to clip parts of the displayer out of the way.

Teacher directions: Insert cards to show 38. Unfold your Place-Value Displayer to answer these questions.

Questions:

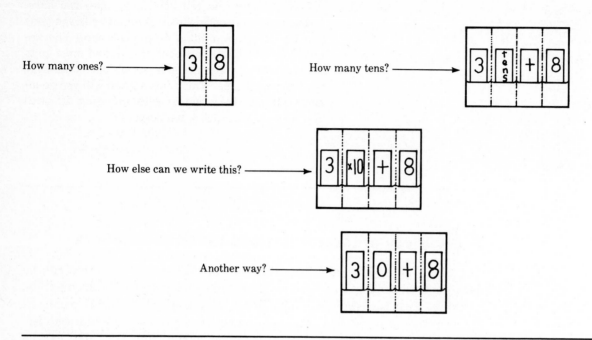

How many ones? → 3 8

How many tens? → 3 | tens | + | 8

How else can we write this? → 3 | ×10 | + | 8

Another way? → 3 | 0 | + | 8

👉

Example 2-12

An activity that provides experience with expanded forms of numerals and emphasizes place value

Materials: Tap-Down Cards (see Material Sheets 9A and 9B), reproduced on heavy paper and cut out like this:

3	0	0.

4	0.

7.

5	0	0	0.

Teacher directions: Draw four cards of different sizes. Stack them so that the longest card is on the bottom up to where the shortest card is on top. The dots line up on the right. Tap down to get your number in a shorter form.

		7.

	4	0.

3	0	0.

5	0	0	0.

So, $5000 + 300 + 40 + 7 = 5347$.

Questions: Can you find other expressions from your cards for 5347? How about $5300 + 47$,
or $5340 + 7$,
or $5000 + 347$,
or $5000 + 300 + 47$?

5	3	4	7.

Reading and writing about large numbers deserves special treatment in the later grades. A motivational topic for exploration involves procedures for naming large numbers. The use of commas or spaces to separate groups of three digits is helpful. In common usage, we find large numbers usually rounded off. An outcome of such exploration by children may well lead to an appreciation of estimation, rounding, and perhaps for expressions using exponents.

☞

Example 2-13   S-3
A skill-development activity that reinforces reading and writing names for large numbers

Materials: Chalkboard, children.

Teacher directions: We're separating the chalkboard into four parts—one for Billy Billion, one for Melissa Million, one for Thornton Thousand, and one for Wanda Ones.

I need four volunteers. Okay. Billy can use one, two, or three digits to write on his section. Everyone else has to use three digits.

Questions: Who can read the entire number they wrote? Let's let them write another one. See if they can make it hard to read. Now let's have four more volunteers.

It's easy to extend this activity so that children can focus on place-value ideas for very large numbers. However, most children (and many adults) have not had experiences that demanded the use of very large numbers. Without these experiences, it's hard to appreciate and understand the magnitude of large numbers.

Using place-value concepts, we can tell that 1 000 000 is 1000 thousands; a million dollars could be exchanged for 1000 thousand-dollar bills, and one million meters is 1000 kilometers. But how big a stack would a million dollars be? How far is a million meters from here? Consider the simple question "How long would it take to count to a million?" This is an interesting and challenging question for children in the fifth or sixth grade. If you could count one number a second nonstop, you'd reach a million after about $11\frac{1}{2}$ days of counting. (*You* figure out how long it would take to count to a billion. Actually, you probably couldn't do one number a second; something like "seven hundred seventy-seven million, seven hundred seventy-seven thousand, seven hundred seventy-seven," or "777 777 777," takes a long time to say!)

The important consideration is this: Before children are introduced to numbers, they need to experience situations that will allow the numbers to make sense. Suggest that children in these later grades find a million of something, such as grains of rice, cereal flakes, leaves, blades of grass, bottle caps, squares on graph paper, telephone numbers, and so on. Let them choose what they'll use to represent a million, and have them verify that it really is a million. You'll be surprised at their creative ideas and methods and at their interest. Of course, your next question could be, "If that's a million, could you show me a billion?"

At this point, the calculations aren't as significant as the ideas, so it's a good place for using a calculator, which provides many opportunities for exploring concepts related to place value. With an inexpensive, standard-type calculator, children are able to deal with positive numbers as large as 99 999 999 or as small as 0.0000001. If the calculator can represent numbers with exponential notations, it's possible to use much larger and much smaller numbers. Children can compare and order these numbers by directly considering place-value aspects of the numerals. Since the computation is so easy to carry out, children can focus on the important question of which operations should be used in the situation. With a calculator, they'll also be forced to extend the base-ten system of numeration for whole numbers to decimals.

The following activities provide an example of the use of calculators in the teaching of place-value concepts.

Example 2-14

S-2

A skill-development activity for place-value ideas using calculators

Materials: Calculators.

Teacher directions: Display 6 8 7 5 4 3 on your calculator. Subtract one number to make the calculator show a 0 in the thousands place, and leave the rest of the digits as they are.

Questions: What can we subtract? (7000) Repeat these directions for other places. Select other whole numbers and continue.

Display 1786.432 on your calculator. What number do we subtract to get a 0 in the hundreds place and keep the other digits? Repeat these directions for other places. Continue with other decimals.

New directions: Play "Keep That Number!" Display 9 8 7 6 5.

Divide by 10; multiply by 10.
Divide by 100; multiply by 100.
Divide by 1000; multiply by 1000.
Divide by 10 000; multiply by 10 000.

Decide what happens.

Example 2-15
A skill-development activity for place value

S-1, 2, 3

Materials: Calculator(s). Cards showing expanded forms of numerals.

Examples:

Level-One Cards

| 20 + 3 | 30 + 5 | 90 + 7 |
| 10 + 0 | 10 + 3 | 80 + 4 |

Level-Two Cards

| 100 + 20 + 4 | 300 + 6 | 200 + 20 |
| (3 × 100) + (2 × 10) + 5 | 500 + 60 + 3 | 900 + 90 + 9 |

Level-Three Cards

| 4000 + 300 + 20 + 1 | 5000 + 5 | 6000 + 20 |
| 7000 + 700 + 6 | 3000 + 200 | (5 × 1000) + (6 × 100) |

Teacher directions: Play "Beat the Calc." I'll draw a card. The person with the calculator has to use the calculator to find the number. If you can write it down before the Calculator Kid can make it appear, you get a point. If not, the calculator gets a point.

Example: First-grade teacher picks $\boxed{10 + 3}$. Calculator Kid punches $\boxed{1}$, $\boxed{0}$, $\boxed{+}$, $\boxed{3}$, $\boxed{=}$, and 13 is displayed. Other children who understand place value can quickly write 13 and get a point for this turn.

Note: The organization of this game will always depend on the number of calculators you have available. With just one calculator, one student can compete with the rest of the class. With enough calculators for each pair of students, one student can compete with one other student. Or, you may have one student competing with several other students.

In problem situations, with or without calculators, attention must be given to practical considerations. If we're concerned with the number of people in our school, we probably want to know the exact number of students. But if we're concerned (for some reason or another) with how many golf balls would fit into the Houston Astrodome, we probably would be satisfied with an approximation.

The computer corner

One of the most neglected concepts in elementary school is place value, particularly when it comes to providing lessons that allow children to associate models with numbers less than 100. But this kind of experience is extremely important since skills related to large whole numbers and computation are based on an understanding of these concepts. Here is a computer program entitled TENS that will provide children with hours of good practice. It is a program that should be used often for short periods of time. Notice that the activity requires children to make records of their *place-value experiences* by writing number sentences.

Example 2-16
Naming tens and ones

C-1

Materials: MGB2 Program TENS on the MGB Diskette.

Teacher directions: This program will help you learn about numbers greater than ten. The computer will show you sets of tens and ones. You must tell the computer the number of tens and the number of ones. Then you must give the number. On a sheet of paper write number sentences for exercises the computer gives you.

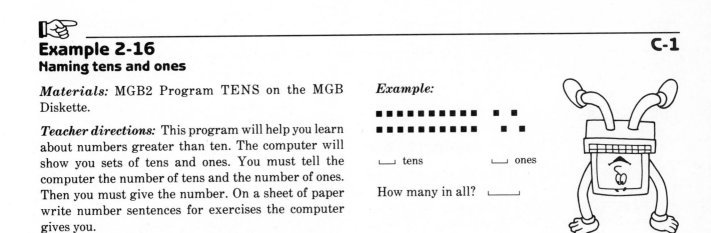

Example:

⬛⬛⬛⬛⬛⬛⬛⬛⬛⬛   ⬛ ⬛
⬛⬛⬛⬛⬛⬛⬛⬛⬛⬛   ⬛ ⬛

⎍ tens   ⎍ ones

How many in all?   ⎍

Let's look at one more program before we get to the computer tip in this section. This program should be used before and while children learn addition facts for greater than ten. The program is called OVER-TEN. Here's what the program does. First, it takes the student's name and date. Then it presents exercises that require students to examine two sets of squares. Each set has less than 10 squares but the number for both sets is greater than 10. The student's task is to answer, "How many more than ten?" The kind of practice this program offers is very beneficial for children at all ability levels. It establishes a pattern of thinking in terms of *tens* and *ones* as you find sums for addition facts greater than ten.

Example 2-17

C-1

A concept-development activity that gets children ready to compute sums greater than ten

Materials: The OVER-TEN program on the MGB Diskette.

■■■■■■■■ ■■■■■■■

$$8 + 7 = 10 + 5$$

Teacher directions: Run the OVER-TEN program. Look at the pictures in each exercise. Find a set of ten squares. Then find how many more than ten there are. Enter your answer for the computer, then write a number sentence to show what you have learned.

Ideas for extention: This activity should be followed up with pencil and paper activities where children group sets to show tens and ones. To make such activities you can use the program MASTER described in Chapter 16.

Run the program a few times, and look for these features of the program: Does it give a progress report? If so, what's the report composed of? Does the program allow students to make mistakes?

■ Computer tips

Not all computer programs should evaluate student progress—some should provide exercises and patiently wait for children to enter correct answers. Of course, these types of programs should provide stimuli that enable children to find answers. In the program TENS, we provide pictures and questions and will not accept incorrect responses from students.

Ideas for enrichment: Other bases

The use of other bases is a challenging topic that involves place-value concepts. Nondecimal numeration systems give students an opportunity to examine structure and explore new ideas. The topic is an excellent one for independent research, and class reports can be worthwhile and enlightening.

Let's take a quick look at base five and see how you can plant a seed for your capable students.

Writing numerals in base five starts out easy enough:

Base Ten	Base Five	
1 =	1	
2 =	2	
3 =	3	
4 =	4	

Now what happens? 5 = 10 We group 1 five and 0 ones.
Again, please? 6 = 11 We group 1 five and 1 one.
Oh, I see. 7 = 12 We group 1 five and 2 ones.

.
.
.

So, for 19 = 34 We group 3 fives and 4 ones.

As our numbers get larger, we continue to group; and grouping is always determined by our base. (In base ten, we have ones, tens, hundreds, etc. In base five, we have ones, fives, twenty-fives, etc.) So, the numeral

$123_{(five)}$ is equal to 1 twenty-five, 2 fives, and 3 ones.

Easier: $123 = (1 \times 25) + (2 \times 5) + (3 \times 1)$.

Determining that important "etc." involves looking at this pattern:

$$5^0 = 1 \quad \text{(by definition)}$$
$$5^1 = 5 \quad (5)$$
$$5^2 = 25 \quad (5 \times 5)$$
$$5^3 = 125 \quad (5 \times 5 \times 5)$$

.
.
.

So, to determine the value (in base ten) of a base five numeral, you find out the value of each place and then multiply and add. Numeration systems with any base have exactly this same structure.

Bright students can appreciate the use of exponents in their exploration. They can handle the generalized idea that place value is determined by the power of the base.

Articles written about other bases usually discuss the reverse procedure as well: how to find a base five numeral for a base ten numeral. Suppose you want to find a base five numeral for a larger (but not too large) number. Try 214. From our chart, we can see that 125 is the largest power of five that we have, and hence we know we have a numeral with four places. Dividing shows us that there is 1 one-hundred-twenty-five in 214, and there is a remainder of 89. It's a good

idea here to write the 1 in the largest place (on the left).

Now, let's look at the 89. The next power of five (on our chart) is 25. Dividing 89 by 25 shows us that there are 3 twenty-fives in 89. Record this 3 in the twenty-fives place. Then find that there is a remainder of 14.

We can tell by looking that this 14 contains 2 fives and 4 ones. So we can fill in our last two places:

$$
1 \quad 3 \quad 2 \quad 4 \quad \text{(five)} \qquad 125\overline{\smash)214} \qquad 25\overline{\smash)89}
$$
$$
\underline{125} \qquad\quad \underline{75}
$$
$$
89 \qquad\qquad 14
$$

So, we are reasonably convinced that
$$214 = 1324_{(five)}$$

As usual, teaching procedures for this topic begin with concrete objects. Chip-trading activities—where children can trade five chips for one of another color worth five times as much—are motivational and introduce the idea that other systems of grouping are possible.

Fives	Ones	*How many?* (base five)
	o	1
	oo	2
	ooo	3
	oooo	4
o		10
o	o	11
o	oo	12
ooo	oooo	34

Five of the chips worth five each will be traded for the next higher value, a chip worth twenty-five. The analogy between pennies, nickels, and now quarters is very useful here. Children can use this to help them trade in either direction. Three nickels and four pennies has the same value as 19, or one dime and nine pennies, and vice versa.

We can see from this that numbers have different names. A clear understanding of that fact can be a foundation for important insights into mathematics. For example, we all know that 7 is an odd number. We just saw, too, that

$$7 = 12_{(five)}$$

So, we have to conclude that $12_{(five)}$ is an odd number. The properties of the numbers do not change, even though we use a different numeration system and different names for the numbers.

Children can appreciate that they have not changed, even though someone uses a different name for them. The same boy in class may be called "Jim," "Jimmy," "James," "James Earl!," and "James Earl Johnson!!" all in the same day.

You can begin any discussion about base five with a story concerning aliens only having five fingers. Since they could only count to five on their fingers, they had to group in fives. Etc.!

Encourage your students to learn more about this. They can become the experts and everyone in class can benefit from their research.

In closing . . .

In this chapter, we've presented some ideas from place value that are appropriate to the elementary-school mathematics program. These ideas are crucial to an understanding of the way in which numbers are represented and to the development of the computational procedures we present in later chapters. We've provided examples of materials that can help children develop these concepts, especially at the lower levels. Teachers in grades 1–3 are encouraged to use the materials to help children develop a firm understanding of the representation of numbers less than 1000. These or similar materials can also be used to justify procedures for computation in addition, subtraction, multiplication, and division. Teachers in grades 4–6 can use these materials to continue the justification of computational procedures, to extend the representation process to numbers larger than 1000, and to help children fully understand the decimal system of numeration through exploration of decimal fractions.

In general, we agree with indications that children have an inadequate understanding of place value. Therefore, we've presented a variety of procedures you can use to remedy this deficiency. Place value *must* be emphasized at each grade level. So, select what is appropriate for your children, and emphasize the base-ten system of numeration!

It's think-tank time

1. Determine which of the activities in this chapter involve higher-order thinking skills and could be used to promote problem solving by elementary school children.
2. What two operations are necessary to interpret numerals in our base-ten numeration system?
3. What is a numeration system?
4. Sketch Happy Face pieces to represent (a) 235, (b) 20, and (c) 104.

5. How many dollar bills could you get for a million pennies?
6. One kilometer is 1000 meters. How many kilometers is 53 000 meters? How many meters is 13 kilometers?
7. Think of representing 862 with as few Happy Face pieces as possible. How many ten-strips are there? For the number 862, how many ten-strips could you get? Exactly how many tens are in 862? What two multiples of ten is

862 between? Which multiple of ten is it closer to?

8. Extend the sample activities in Examples 2-1 through 2-5 to make them applicable to numbers up to 1000.

9. Write the following in expanded form in two ways.
   a. 918 764
   b. 500
   c. 36.782
   d. 106.09
   e. .007
   f. 0

10. Write the following base-ten numerals in base five.

a. 39
b. 125
c. 278

Write the following base-five numerals in base ten.

d. 44$_{(five)}$
e. 432$_{(five)}$
f. 3021$_{(five)}$

11. Make up an original activity for a place-value idea. Decide on your objective(s), construct the activity, and be sure it's appealing!

12. *Bonus:* Select a widely used numeration system that predates our Hindu-Arabic system. Do a bit of historical research and compile a brief summary of your work.

Suggested readings

Beal, Susan Ruth Neuwirth. "Understanding of the Numeration System and Computational Errors in Subtraction." (Doctoral dissertation, University of Chicago, 1984.) *Dissertation Abstracts International,* 44A (1984):2075-2076A.

Brennan, Alison D. "Mr. Ten, You're Too Tall!" *The Arithmetic Teacher,* January 1978, pp. 20-22.

Calvo, Robert C. "Placo: A Number-Place Game." *The Arithmetic Teacher,* May 1968, pp. 465-466.

Easterday, Kenneth E. "Teacher-Made Aids for Teaching Place Value and Estimation." *The Arithmetic Teacher,* January 1978, p. 50.

Flournoy, Frances, Brant, Dorothy, and McCregor, Johnnie. "Pupil Understanding of the Numeration System." *The Arithmetic Teacher,* February 1963, pp. 88-92.

Jackson, Robert Loring. "Numeration Systems: An Experimental Study of Achievement on Selected Objectives of Mathematics Education Resulting from the Study of Different Numeration Systems." (Doctoral dissertation, University of Minnesota, 1965.) *Dissertation Abstracts International,* 26 (1966):5292-5293.

Lichtenberg, Donovan R. "The Use and Misuse of Mathematical Symbolism." *The Arithmetic Teacher,* February 1978, pp. 12-17.

Logan, Henrietta L. "Renaming with a Money Model." *The Arithmetic Teacher,* September 1978, pp. 23-24.

MacRae, Irene R. "A Place Value Game for First Graders." *The Arithmetic Teacher,* November 1957, pp. 217-218.

O'Neil, David, and Jensen, Rosalie S. "Let's Do It, Some Aids for Teaching Place Value." *The Arithmetic Teacher,* November 1981, pp. 6-9.

Parillo, Guy. "Roman Numerals." *The Arithmetic Teacher,* September 1979, p. 45.

Pennington, Mary Jane. "Base Ten Trading Game." *The Arithmetic Teacher,* March 1979, p. 52.

Plunkett, Betty K. "Beginning Ideas about Numbers and Numerals." In Nicholas J. Vigilante (Ed.), *Mathematics in Elementary Education.* New York: Macmillan, 1969.

Reimer, Parrie. "Reading Numerals." *The Arithmetic Teacher,* October 1978, p. 52.

Ronshausen, Nina L. "Introducing Place Value." *The Arithmetic Teacher,* January 1978, pp. 38-41.

Schlesinger, Beth M. "What's in a Number Name?" *The Arithmetic Teacher,* May 1980, p. 35.

Scrivens, Robert William. "A Comparative Study of Different Approaches to Teaching the Hindu-Arabic System to Third-Graders." (Doctoral dissertation, University of Michigan, 1968.) *Dissertation Abstracts International,* 29 (1968):839-840.

Skipper, Slade Welma. "A Study of the Use of Manipulative Materials as Multiple Embodiments for the Study of Numeration Systems by Prospective Elementary Teachers." (Doctoral dissertation, University of Missouri–Columbia, 1972.) *Dissertation Abstracts International,* 34A (1973):1168-1169.

Smith, Robert F. "Diagnosis of Pupil Performance on Place-Value Tasks." *The Arithmetic Teacher,* May 1973, pp. 403-408.

Sulkowski, Toni J. "An Introduction to Place Value." *The Arithmetic Teacher,* September 1978, p. 59.

Thompson, C. S., and Van de Walle, J. "The Power of Ten." *The Arithmetic Teacher,* November 1984, pp. 6-11.

Thompson, Patrick Wilfred. "A Theoretical Framework for Understanding Young Children's Concept of Whole Number Numeration." (Doctoral dissertation, University of Georgia, 1982.) *Dissertation Abstracts International,* 43A (1983):1868A.

Verbeke, Linda M. "An Approach to Teaching Numeration Systems." *The Arithmetic Teacher,* January 1977, pp. 76-78.

Ziesche, Shirley S. "Understanding Place Value." *The Arithmetic Teacher,* December 1970, pp. 683-684.

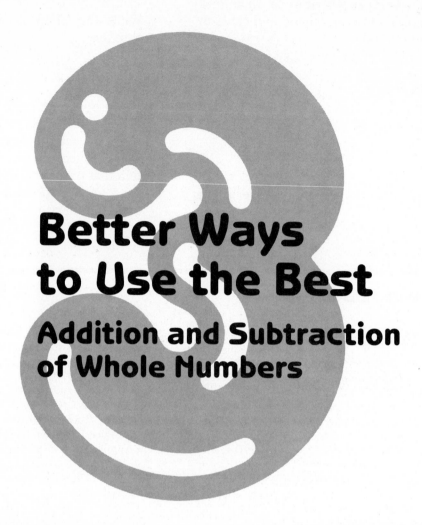

Better Ways to Use the Best

Addition and Subtraction of Whole Numbers

■ *In this chapter, we discuss the two operations of addition and subtraction of whole numbers and their associated computational procedures. Initially, we emphasize interpretations for the operations. Readiness activities using physical and pictorial models are described in conjunction with realistic problems that children may encounter. Then the emphasis shifts to the development of meaningful ways to compute and estimate. Finally, we suggest activities that provide for skill development. Throughout the chapter, we stress the importance of using number sentences for the communication of mathematical ideas.*

This chapter also emphasizes understanding the operations, associating the appropriate operation with a problem situation, and computing efficiently. Teaching strategies for developing each of the operations on whole numbers are similar, and there are important relationships among the operations that must be made explicit. Therefore, this chapter and the following one, which treats multiplication and division of whole numbers, have the same emphases and overlapping strategies. ■

Objectives for the child

1. Combines and separates sets; tells about the results.
2. Matches addition and subtraction situations to appropriate models.
3. Matches addition and subtraction situations to appropriate number sentences.
4. Matches number sentences to appropriate models for addition and subtraction.
5. Constructs number sentences or models for addition and subtraction situations.
6. Recalls the basic addition facts and related subtraction sentences.
7. Illustrates properties for addition using sets of objects, pictorial models, number sentences, or simple computations.
8. Uses properties for addition to reduce the amount of memorizing required to learn the basic addition facts.
9. Uses the associative property of addition and place-value ideas to aid in computing sums such as 8 + 7.
10. Identifies equivalent addition and subtraction statements.
11. Uses place-value models to illustrate and compute in addition and subtraction exercises.
12. Performs addition and subtraction computations in which regrouping *is not* involved.
13. Performs addition and subtraction computations in which regrouping *is* involved.
14. Performs addition and subtraction estimations.

Teacher goals

1. Be able to identify the interpretations for the operations of addition and subtraction.
2. Be able to demonstrate physical and pictorial models for each interpretation of the operations of addition and subtraction.
3. Be able to devise problem situations appropriate for children that reflect the interpretations for the operations of addition and subtraction.
4. Be able to explain why a particular number sentence is appropriate for a given interpretation of an operation.
5. Be able to construct models that demonstrate the steps in the computational procedures for addition and subtraction of whole numbers.
6. Be able to justify the steps in the computational procedures for addition and subtraction of whole numbers.
7. Be able to show relationships that exist between addition and subtraction.
8. Be able to devise motivating practice activities that enable children to compute efficiently.
9. Be able to diagnose computational errors commonly made by children and to devise appropriate remedial activities.
10. Be able to identify and sequence student learning objectives that involve addition and subtraction of whole numbers.
11. Be able to construct activities enabling children to achieve the student learning objectives.

To be mathematically fit (if we can talk about "physical fitness," why not "mathematical fitness?"), a person must know:

when to add, subtract, multiply, or divide
and
how to add, subtract, multiply, or divide.

Pre-number and number concepts developed in the early years provide a framework for the introduction of the operations of whole numbers. An emphasis on concepts related to place value will facilitate the development of the computational procedures.

Children's understanding of the meaning of these operations develops through analysis of the situations they encounter. A child's choice of an appropriate operation depends on experience with a variety of situations. There are several types of situations for each of the four operations. Thus, children must have experiences that enable them to classify each type of situation as an interpretation of the appropriate operation. Moreover, the experiences should be of a kind that children can realistically be expected to encounter. Therefore, teachers must be aware of the different interpretations for the operations and must be prepared to systematically present these interpretations to children. Then, perhaps, we can expect children to know *when* to add (or subtract or multiply or divide).

Computational procedures, called *algorithms*, combine previously learned concepts into succinct steps that result in the standard name for a number; for example, the standard name for 36 + 48 is 84. If the algorithms are developed gradually and meaningfully, they can be a tremendous asset to a person's ability to function. If they are developed too quickly and mechanically, they can be difficult to use and easily forgotten. Many of us who "learned" to compute something like $\sqrt{92643.175}$ can verify that a procedure learned without meaning or justification is frustrating, if not detestable, and isn't retained anyway without extensive practice. Sometimes, even *with* practice it isn't retained.

The algorithms that children learn have been developed to make things easier. Justification for the computational procedures draws heavily on place-value concepts related to the base-ten system of numeration. The vertical arrangement of numerals in an addition algorithm, for example, aligns the corresponding places. Similarly, what many people know as "carrying" or "borrowing" is simply a recording of the regrouping process. As each algorithm is introduced in this chapter, the proper meaning is attached to each step in the process. Explanations for students, provided in this way, will make it easier for them to remember the procedures and to extend them to different situations.

> ESTIMATE
> Since 90 000 = 300 × 300 this is a little more than 300.

Addition of whole numbers

Addition should be introduced with appropriate situations that allow you to give children objects to manipulate. All of the concepts discussed in Chapter 1 are prerequisite to the teaching of addition. Children must be able to associate a number with the appropriate set of objects it represents, and they must be able to count meaningfully. Addition, then, is a more efficient means of counting.

■ Interpretations

Suppose Mary Jane has two apples and her mother gives her three apples. Now how many does she have? In this situation, the child's first response may be to count "one, two, three, four, five." So this interpretation of addition involves *combining* the two distinct sets and finding out how many there are in the new set. A child at a more advanced level may respond "three, four, five," reasoning that, since there are already two, it's possible to begin counting at three. Counting to find the solution to this problem is a legitimate and satisfactory response. In fact, that's how we develop the basic facts of addition.

Counting by beginning with the first element in the second set is a little easier and should be encouraged. It's easier to arrive at 8 for 3 + 5 by saying "four, five, six, seven, eight" than it is by saying "one, two, three, four, five, six, seven, eight." Even at this beginning level of dealing with the operation of addition, notice that knowing 3 + 5 is the same as 5 + 3 would be helpful. A child who knew this would be able to say "six, seven, eight," and that's even quicker and easier than the other two ways.

Teachers working with students at this level of development must have materials that students can handle, move about, and count. In this first interpretation, you should physically combine the two sets—that is, move the objects together and find the number of objects in the new set.

In other cases, you should leave the two sets static. Both situations occur in real-life contexts, and children should have exposure to both. The trees in the front and back yards, for example, provide a good instance for the static interpretation. Certainly, if you have two pennies in one pocket and three pennies in the other pocket, we know that you have 2 + 3 or 5 pennies, whether they're physically put together or not. Actually, most real-life situations are of this type. When you go to the bank to deposit your money, you don't ask to see the other money that you have in your account.

Another interpretation of addition involves increments, or increases on scales. Examples of this type are gaining weight, growing taller, or even having a fever when you're sick. There are no objects to count, but we do need to use addition.

Now that we've briefly discussed "when to add," we will discuss an algorithm. This set of procedures is based on *place value*, and its use depends on a knowledge of the *basic facts of addition*.

■ Learning the basic facts of addition

The basic facts of addition give the standard names for the sums of any two whole numbers less than 10. (For example, 7 + 4 = 11. "7 + 4" indicates the sum of 7 and 4, and "11" is the standard name of this sum. In other words, both of these expressions name the same number.) The basic facts must be developed with children at the concrete level, so physical and pictorial representations are essential. We'll call these representations *models*. The use of objects to demonstrate an addition situation is probably the easiest and soundest model for leading children to the symbolic representation.

How many? 2   How many? 3

How many in all? 2 + 3 or 5

The introduction of the addition symbol "+" as a proper way to answer "How many in all?" is important. Emphasizing this for children helps give them some meaning for addition. Connecting the equals sign "=" with the phrase "is the same as" is another way to help give meaning to the operation of addition. Finally, since 2 + 3 tells how many, and 5 tells how many, we know that

> 2 + 3 is the same as 5
> or
> 2 + 3 = 5 (Say, "Two plus three equals five.")

At this early level, children also should have experience with both forms of the number sentence—that is,

$$2 + 3 = 5 \quad and \quad 5 = 2 + 3$$

This will help them to develop an appropriate understanding of equality.

Workbooks at this level provide pictorial representations for addition sentences. Children should be encouraged to make their own pictures to demonstrate these number sentences by drawing pictures or cutting them from magazines.

Julia Ann
Age 5

$2 + 3 = 5$

The Happy Face pieces can be used to represent the basic facts of addition. Students will already have ten-strips and ones from their place-value activities and can use them this way:

$4 + 3$

■ Sums less than 10

The basic facts that should be dealt with first involve sums less than 10. The numbers should be represented by physical objects. Thus, with young children the expression $4 + 3$ is associated with four things together with three things; they then decide that this is the same as seven things. So,

$$4 + 3 = 7$$

Easier facts in this group are those where we're adding 1, followed, perhaps, by adding doubles, such as $3 + 3 = 6$. These should be taught first; then proceed to harder ones such as $4 + 5 = 9$. The following is an example of a concept-development activity for these basic facts.

Example 3-1   C-1
A concept-development activity for basic addition facts

Materials: Happy Face pieces—ones (Material Sheet 3).

Teacher directions: Show me $4 + 1$. Show me $5 + 2$. Show me $3 + 3$. Show me $2 + 4$.

Questions: How many? Good! So $4 + 1 = 5$. How many? Right! So $7 = 5 + 2$. How many? Yes. So $3 + 3$ is the same as 6. How many? Good! So 6 is the same as $2 + 4$.

Instructional activities with these facts should use the language "is the same as." Thus, children will be saying:

> $4 + 1$ is the same as 5
> and
> 5 is the same as $3 + 2$

The next logical sentence can be "$4 + 1$ is the same as $3 + 2$." This mathematical property (here called *transitivity of equality*) is very important throughout the study of mathematics. It permits many of the substitution-type things we do in logical arguments and supports the reasoning procedures in many formal proofs. At the first grade level, not only can we lay a foundation for further development, but also we can demonstrate significant concepts about numbers.

Children focusing on the number 5 in the next activity can generate each addition expression that is the same as 5. A rubber band in the first groove in the posterboard generates $0 + 5 = 5$, and with the rubber band in the last groove $5 + 0 = 5$ is generated. This activity is also for concept development and involves inexpensive teacher-made materials to help children generate the number sentences.

☞

Example 3-2

A concept-development activity for basic facts (sums less than 10)

Materials: Groovy Boards for numbers 5 through 9 (Material Sheet 10; made from posterboard and stick-on dots). Rubber bands for each board.

Teacher directions: I'll put the rubber band here on my 5 board.

Questions: Who can put the rubber band somewhere else and tell us what it shows? Continue until all six of the facts for 5 have been indicated. Repeat the activity for other numbers at different times.

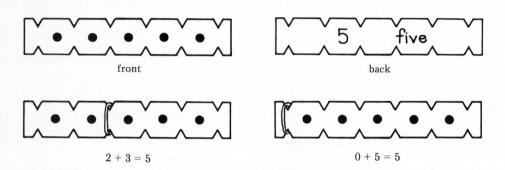

front   back

$2 + 3 = 5$ $0 + 5 = 5$

This shows that $2 + 3$ is the same as 5.   This shows that $0 + 5$ is the same as 5.

The preceding activity can be used to generate a *family of facts* for appropriate numbers. The family of facts for 5 is made up of these number sentences:

$$
\begin{array}{ll}
5 = 5 + 0 & \quad 5 = 2 + 3 \\
5 = 4 + 1 & \quad 5 = 1 + 4 \\
5 = 3 + 2 & \quad 5 = 0 + 5
\end{array}
$$

Eventually, even the children will notice that there's always one more sentence than their number. So, if they start with the number 8, they should be able to come up with nine addition facts in the "family."

■ Sums that equal 10

Because the sums that equal 10 are so important to the place-value features of our base-ten numeration system, they must receive special attention. These sums will also be crucial in the discussion of larger sums and in developing the computational procedures.

Use the following concept-development activity to stress these combinations for 10. They're not necessarily easier to learn, so you'll have to emphasize them.

Example 3-3   C-1
A concept-development activity for basic addition facts

Materials: Happy Face pieces—ten strips (Material Sheet 3). Index cards (optional—regular paper will do fine).

fold

Teacher directions: Fold your ten-strip. Write an addition sentence to show what you have.

10 = 7 + 3

Questions: How many sentences did we get? What are they? Could we make a sentence to show no fold?

Right! 10 = 10 + 0.

■ Sums greater than 10

After many activities with sums less than 10 and sums that equal 10, you'll want to present situations that require children to use place-value concepts. Example 2-2 in the preceding chapter stressed place-value models by having children represent 12, for example, with as few Happy Face pieces as possible (1 ten-strip and 2 ones). The next activity extends the activities already discussed in this chapter by combining them with this idea from Chapter 2.

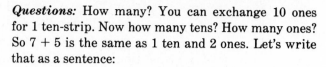

Example 3-4

C-1

A concept-development activity using combinations for sums greater than or equal to 10

Materials: Happy Face pieces—ten-strips and ones (Material Sheet 3). In-and-Out Shoebox constructed like this one, with two holes in the top.

Teacher directions: Show me 7 + 5. Show me 8 + 3. Show me 1 + 9. Show me 6 + 7.

Questions: How many? You can exchange 10 ones for 1 ten-strip. Now how many tens? How many ones? So 7 + 5 is the same as 1 ten and 2 ones. Let's write that as a sentence:

$$7 + 5 = 12 \text{ and } 12 = 7 + 5$$

Use similar questions for each example. Stress the process of exchanging for as few pieces as possible.

After children understand the interpretations for addition and realize that some of the basic facts require place-value ideas and that others do not, you will need to become more systematic. There are 100 basic facts of addition that children need to learn. It would be a formidable task for a 6-year-old to have to memorize these as though they were a set of unrelated facts. Fortunately, they are very much related, and they build on concepts that children should have acquired previously. For the purpose of our analysis, we're presenting the facts in the form of an addition table. The table, in this form, is not appropriate for children until they've had exposure to all of the addition facts and can sensibly discuss the organizational scheme. Children in the second grade seem to handle this quite well.

Table for Basic Facts of Addition

+	0	1	2	3	4	5	6	7	8	9	
0	0	1	2	3	4	5	6	7	8	9	← first row
1	1	2	3	4	5	6	7	8	9	10*	
2	2	3	4	5	6	7	8	9	10*	11	
3	3	4	5	6	7	8	9	10*	11	12	
4	4	5	6	7	8	9	10*	11	12	13	
5	5	6	7	8	9	10*	11	12	13	14	
6	6	7	8	9	10*	11	12	13	14	15	
7	7	8	9	10*	11	12	13	14	15	16	
8	8	9	10*	11	12	13	14	15	16	17	
9	9	10*	11	12	13	14	15	16	17	18	

first column   diagonal

*combination for 10

The first row represents the sums $0 + 0$, $0 + 1$, $0 + 2$, $0 + 3$, and so on. The first column represents the sums $0 + 0$, $1 + 0$, $2 + 0$, $3 + 0$, and so on. The diagonal from upper left to lower right represents the sums $0 + 0$, $1 + 1$, $2 + 2$, $3 + 3$, and so forth.

Now let's look at some properties of addition of whole numbers.

Mathematically, we say:	*For the child, this means:*
1. For any whole number a, $a + 0 = a$; $0 + a = a$. *Zero is the identity element for addition.*	1. Whenever I add 0 to a number, I always get the same number back.
2. For any whole numbers a and b, $a + b = b + a$. *Addition is a commutative operation.*	2. If I know that $5 + 3 = 8$, I don't have to worry about $3 + 5$; it'll be 8, too.
3. For any whole numbers a, b, and c, $(a + b) + c = a + (b + c)$. *Addition is an associative operation.*	3. If I have $7 + (3 + 5)$, I can add 7 and 3 first and then add 5. That's easier than adding 3 and 5 and then 7. Either way, it's the same thing.

And with respect to the addition table:

1. It's easy to fill in the first row and the first column.
2. If I know the part of the table below the diagonal, I can fill in the part of the table above the diagonal.
3. I can think of $7 + 5$ as $7 + (3 + 2)$, and that's $(7 + 3) + 2$, or 12.

$$(7 + 3) + 2$$
$$\searbackslash \quad \nearrow$$
$$12$$

The first two sentences take care of 55 of the basic addition facts. The associative property of addition, the third sentence, helps make some of the other 45 facts easier to recall.

Any sum that is greater than 10 can be found by rewriting one of the numbers so that a combination for 10 is obvious. Then, the expression involving 10 plus another number is simply a place-value expression that can be written directly. The following examples illustrate the importance of the associative property in making this regrouping possible.

Sum	*Rewriting*	*Use of associative property*	*Simplification*	
$9 + 5$	$9 + (1 + 4)$	$(9 + 1) + 4$	$10 + 4$	14
$8 + 7$	$8 + (2 + 5)$	$(8 + 2) + 5$	$10 + 5$	15
$7 + 6$	$7 + (3 + 3)$	$(7 + 3) + 3$	$10 + 3$	13

We are not suggesting that young children use either the symbolism or the language above. We are suggesting that *teachers* use these *ideas*.

Since we assume that children being introduced to addition can already count, we can use that ability to help them learn and remember some of the 45 remaining facts. Since 1 added to any whole number produces the next consecutive whole number, adding 1 is like counting for the child. Thus, it is easy for children to fill in the second row and the second column of the addition table.

There is some support for the idea that doubles (for example, $6 + 6 = 12$) are easier for children to remember. These occur on the diagonal of the addition table, and they should be presented before some of the other facts.

Given 6 + 7, some people decide that this is 13 by thinking of it as 1 more than 6 + 6 or 1 less than 7 + 7. Combinations such as these are sometimes called *near-doubles* and are supposedly easier to remember. Try this idea with children; it may help.

Putting all of this together, we have an addition table that can be analyzed this way:

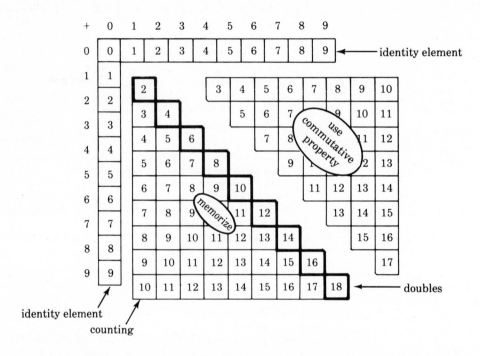

Using these properties of addition and presenting easier facts first is a reasonable way to organize instruction for children.

The combinations for 10 are important because of place value, and using these combinations can help children to determine any fact in the table when the sum is greater than 10. Realizing and using the fact that addition is both commutative and associative and selecting combinations that add to 10 can be an excellent way of facilitating more difficult computation as well. Notice that 10 occurs in the table along a diagonal from bottom left to top right, and the basic addition facts for 10 can be determined for the pairs associated with these squares.

10 ⟶	9, 1	8, 2	7, 3	6, 4	5, 5	4, 6	3, 7	2, 8	1, 9

Stressing combinations for 9, 8, 7, 6, and so on is also important. Using these combinations, combinations for 10, place-value concepts, and the associative property makes it easy to manage sums greater than 10. Let's see how.

Consider 7 + 5. Have the children get 7 objects and 5 objects. They then should actually put 3 of the 5 objects with the 7 objects to make 10.

Now have them complete the sentence $7 + 5 = \square$ by renaming their 1 ten and 2 ones as "12." The following procedure indicates what the child has physically completed.

$7 + 5 = 7 + (3 + 2)$	renaming 5 as $3 + 2$
$7 + (3 + 2) = (7 + 3) + 2$	using associative property of addition
$(7 + 3) + 2 = 10 + 2$	making a combination for 10
$10 + 2 = 12$	using place value

Or how about $8 + 6$? Have children get 8 Happy Face pieces and 6 more Happy Face pieces. Then 2 of the 6 can be combined with the 8 and exchanged for 1 ten-strip. At this point, there will be 1 ten-strip and 4 ones, and this can be written as 14.

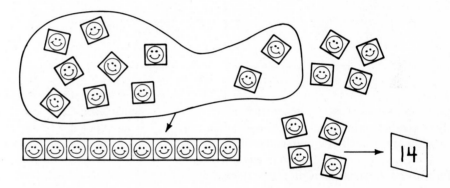

The procedure below demonstrates what the child has done:

$8 + 6 = 8 + (2 + 4)$	renaming 6 as $2 + 4$
$8 + (2 + 4) = (8 + 2) + 4$	using associative property of addition to get 2 with 8
$(8 + 2) + 4 = 10 + 4$	making a combination for 10
$10 + 4 = 14$	using place value

The basic facts can be developed and verified through the use of a *number line*. The number line should start at 0 and have equally spaced intervals between points labeled by consecutive whole numbers. Addition can then be interpreted as moving two specified distances to the right. For example, $6 + 5$ is interpreted this way:

Begin at 0. Move 6 spaces to the right. From there move 5 more spaces. The sum of 6 and 5 is designated by the point where you end up.

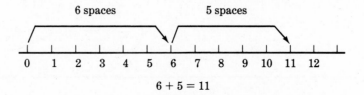

$$6 + 5 = 11$$

You may want to introduce cute elements such as hopping crickets or jumping frogs to make the presentation of this idea even more interesting to children.

So, you can use objects, Happy Face pieces, pictures, and the number line to illustrate the basic facts of addition. Then it will be time for practice activities that help children memorize these facts. Here are some examples of activities for skill development.

Example 3-5   S/C-1
A skill- or concept-development activity for basic facts of addition

Materials: Addition slide rule (see Material Sheet 11).

 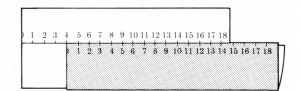

Teacher directions: Use your addition slide rule. Let's find another name for 4 + 5. Slide the 0 on the bottom scale over until it's under the 4 on the top scale. Now find the 5 on the bottom scale. The sum will appear on the top scale, right above the 5.

Questions: Can you find another name for 6 + 3? How about 8 + 5? Let's all find 9 + 4. Are you sure it's working right? How do you know?

Example 3-6   S-1
A skill-development activity for practice with basic addition facts (sums less than 10)

Materials: Karrie Kangaroo and babies with addition expressions on them (see Material Sheet 12).

Teacher directions: Karrie is taking her babies to the zoo. She can only take the ones with 6 today. Help her find which ones to take.

Questions: Not many.

Example 3-7

A concept-development activity for basic addition facts

Materials: Index cards showing addition expressions such as 3 + 2, 4 + 5, 6 + 1, and so on.

Teacher directions: Draw a card. Now tell us a story to fit your card. (Play continues until everyone has a chance to tell an addition story.)

3+5 2+1 6+4 8+9 6+5 4+0

Example 3-8

A skill-development activity for basic addition facts (sums greater than 10)

Materials: Train sets (see Material Sheet 13).

Teacher directions: Pick out an engine. Now match all the cars that go with that engine. Find the caboose that goes with your engine and cars.

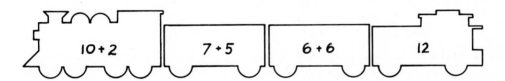

10 + 2   7 + 5   6 + 6   12

These activities can be used with children to develop a "family of facts" for each of the numbers appearing in the body of the addition table.

Example 3-6 uses a figure, Karrie Kangaroo, who should be appealing to children as they find addition expressions for numbers less than ten. For instance, if Karrie's flag has a 6 on it, the child would pick out all the little kangaroos with expressions for 6: 0 + 6, 1 + 5, 2 + 4, 3 + 3, 4 + 2, 5 + 1, and 6 + 0. Other figures that teachers have used include a flower and appropriate petals, a mailbox and letters, dinosaur and eggs, and so forth. The idea is the same, and it's more fun than worksheets!

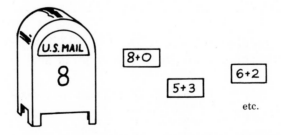

Example 3-8 makes use of the same general idea. However, here the sums are greater than 10. The engine (the most powerful car) contains the place-value representation for the number, and the caboose (the end car) has the compact, or standard, form. Each of the cars has an addition expression for the engine and caboose. Some trains will be longer than others. For example, Engine 10 + 1 and Caboose 11 have Cars 5 + 6, 7 + 4, 8 + 3, 9 + 2, 6 + 5, 4 + 7, 3 + 8, and 2 + 9. Engine 10 + 8 and Caboose 18 have only Car 9 + 9.

These or similar activities can be used to classify all of the basic facts of addition, and they help children to remember these combinations. Immediate recall of the basic facts of addition is essential to the efficient performance of more complicated addition computations. It's also important in the learning of related subtraction ideas.

■ Developing an addition algorithm

Now, with a sufficient understanding of place value and a knowledge of the basic facts (even if they cannot be recalled immediately, they can be reconstructed), the child is ready to add larger numbers. No matter how large the numbers are, these are all that are needed for addition computation. Adding all of the numbers in each of the places involves only the ability to regroup when necessary and knowledge of the basic facts. Similar procedures are appropriate to this development: Happy Face pieces can be used to develop the computational procedures for adding numbers less than 100 where no regrouping is required, as is demonstrated in the following concept-development activity.

Example 3-9   C-1
A concept-development activity for addition with no regrouping required

Materials: Happy Face pieces—ten-strips and ones (Material Sheet 3).

Teacher directions: Let's look at 23 + 31 = □.

Questions: How many ones? 4. How many tens? 5. So, we have 5 tens and 4 ones. Fill in the box.

$23 + 31 = \boxed{54}$

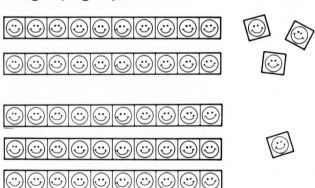

Through these experiences, it should be established that we add first the number of ones and then the number of tens.

$$
\begin{array}{r}
23 = 2 \text{ tens} + 3 \text{ ones} \\
+\,31 = 3 \text{ tens} + 1 \text{ one} \\
\hline
5 \text{ tens} + 4 \text{ ones} = \boxed{54}
\end{array}
$$

The vertical arrangement is sensible at this point, because this computational form can help us keep track of the tens and the ones. Also, since the computation begins on the right, the questions in the activity are consistent with this, starting with "How many *ones?*"

Example 3-9 also can be extended to situations in which regrouping is required. The numbers are represented by a place-value model. Thus, children are "forced" to regroup 10 ones for 1 ten. The teacher and children, together, will be writing down their procedures, recording the exchange, and discussing their results.

Example 3-10
A concept-development activity for addition with regrouping required

C-2

Materials: Happy Face pieces—ten-strips and ones (Material Sheet 3). In-and-Out Shoebox.

Teacher directions: Let's look at $46 + 27 = \square$.

46 →

27 →

Exchange 10 ones
for 1 ten

Questions: How many ones? 13. Can you trade in 10 ones for a ten-strip? Yes. Now how many ones are left? 3, with the ten-strip that you got and the ten-strips that you already had; now how many tens do you have? 7.

Fill in the box:

$46 + 27 = \boxed{73}$

$$\begin{array}{r} 1 \\ 46 \\ + 27 \\ \hline 73 \end{array}$$

☞

Example 3-11
A concept-development activity for addition with regrouping required

C-2

Materials: Play money (Material Sheet 5).

Teacher directions: Let's look at $38 + $14 = □. We're using only ten-dollar bills and one-dollar bills. Select $38, using the smallest number of bills that you can. Now select $14, using the smallest number of bills.

Questions: How many ones? 12. Can you exchange 10 of them for a ten? Yes. Now how many ones do you have? 2. How many tens did you have to begin with? 4 and one more that you exchanged for the ones given you . . .? 5.

Fill in the box:

$$38 + 14 = \boxed{52}$$

$$\begin{array}{r} 1 \\ 38 \\ + 14 \\ \hline 52 \end{array}$$

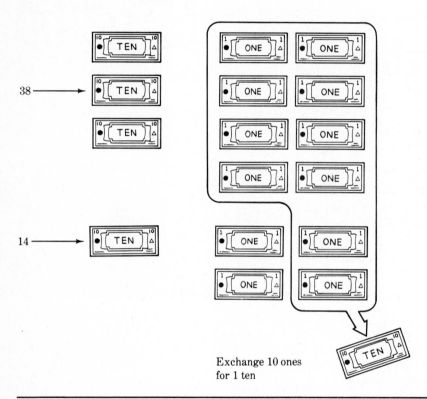

38 ⟶

14 ⟶

Exchange 10 ones
for 1 ten

It may be helpful for children to see the expanded forms that show what is happening in this kind of situation:

$$\begin{array}{l} 46 = 4 \text{ tens} + 6 \text{ ones} \\ + 27 = 2 \text{ tens} + 7 \text{ ones} \\ \hline 6 \text{ tens} + 13 \text{ ones} = 6 \text{ tens} + 1 \text{ ten} + 3 \text{ ones} = 7 \text{ tens} + 3 \text{ ones} = 73 \end{array}$$

or

$$\begin{array}{l} 46 = 40 + 6 \\ + 27 = 20 + 7 \\ \hline 60 + 13 = 60 + 10 + 3 = 70 + 3 = 73 \end{array}$$

They should realize that these are simply further justifications for recording a 1 above the other numerals that tell how many tens.

Physical experiences in conjunction with pictorial representations and charts are extremely important at this level. Our ultimate goal for computations involving the addition of whole numbers is for students to be able to compute sums with reasonable speed and accuracy for numbers of any size. The transfer to addition computations with larger numbers is not a difficult one, provided that children understand the regrouping process and the associated place-value ideas. Some physical representation is necessary for most children, as in the concept-development activity shown in Example 3-12.

Example 3-12   C-2
A concept-development activity for addition with regrouping required

Materials: Happy Face pieces—hundred-squares, ten-strips, and ones (Material Sheet 3).

Teacher directions: Let's look at $142 + 173 = \square$.

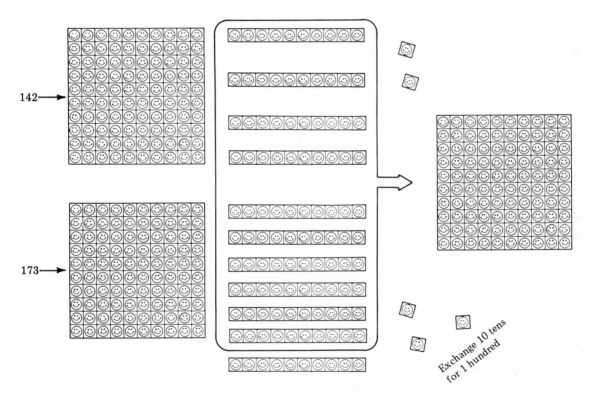

Questions: How many ones? 5. How many tens? 11. Can you exchange 10 tens for 1 hundred? Yes. How many tens do you have left? 1. How many hundreds did you have to begin with? 2. Now with this 1 more, you have . . .? 3.

So, our number sentence is: $142 + 173 = \boxed{315}$

$$\begin{array}{r} 1 \\ 142 \\ + 173 \\ \hline 315 \end{array}$$

The use of expanded notation can be helpful to children at this stage.

$$175 = 1 \text{ hundred } + 7 \text{ tens} + 5 \text{ ones}$$
$$+ 237 = 2 \text{ hundreds} + 3 \text{ tens} + 7 \text{ ones}$$
$$3 \text{ hundreds} + 10 \text{ tens} + 12 \text{ ones}$$
$$= 3 \text{ hundreds} + 1 \text{ hundred} + 1 \text{ ten} + 2 \text{ ones}$$
$$= 4 \text{ hundreds} + 1 \text{ ten} + 2 \text{ ones} = 412$$

$$175 = 100 + 70 + 5$$
$$+ 237 = 200 + 30 + 7$$
$$300 + 100 + 12 = 400 + 10 + 2 = 412$$

An intermediate step toward the shortened form of the algorithm may look like this:

$$\begin{array}{r} 123 \\ 456 \\ + \ 583 \\ \hline 12 \\ 150 \\ 1000 \\ \hline 1162 \end{array}$$

12 $\longrightarrow$	$3 + 6 + 3$ (adding the ones)
150 $\longrightarrow$	$20 + 50 + 80$ (adding the tens)
1000 $\longrightarrow$	$100 + 400 + 500$ (adding the hundreds)

It takes a little more pencil lead, but you can more easily explain where everything comes from.

Finally, the shortened form of the algorithm allows us to note by "1" that we have regrouped from 10 ones to 1 ten and from 10 tens to 1 hundred. Squared paper for computational work can help children keep things straight, especially if they find a sum of more than two numbers:

	1	1	
	1	2	3
	4	5	6
	5	8	3
1	1	6	2

Subtraction of whole numbers

There are several types of situations that can be interpreted as requiring subtraction, and problems that exemplify these various types should be presented to the students so that they can make decisions about when to subtract.

Jim has 5 apples.
He gives away 2.
Now how many does Jim $5 - 2 = \square$
    have?

Laurie has $2.
She wants to buy a $5
    shirt.
How much more does $2 + \square = 5$
    she need?

Jerry has 5 pretzels.
Keith has 2 pretzels.
How many more pretzels $5 = 2 + \square$
    does Jerry have than Keith?

Kim has 5 cars.
2 are red.
The rest are blue. $5 = 2 + \square$
How many are blue?

We have 5 quarts of punch. $5 - 2 = \square$
We drink 2 quarts.
How much is left?

AHA!!
I'LL
SUBTRACT

$5 - 2 = 3$

Say, "5 minus 2 equals 3."

■ Interpretations

The first kind of situation, the *take-away* type, is probably the easiest for the child to interpret because it's easy to represent physically. Children will already be accustomed to associating physical objects and number concepts at this point, and this kind of association should definitely be continued in the development of subtraction.

Let's consider the first situation. Jim had 5, and he gave away 2; now how many does he have?

The child should actually take 5 objects, remove 2 of them, and count to see how many are left. Use whatever physical materials are available (would you believe Happy Face pieces?) to provide many experiences like this for the children. Be sure to associate the appropriate number sentence with the example.

After these experiences involving actual "taking away," children are ready to move to a pictorial representation of the operation by covering up or marking out pictures of objects.

Workbooks also include many exercises that exemplify these ideas, such as:

There are 5.
Mark out 2.
Now how many?

The second type of subtraction situation is sometimes called an *additive* type and asks, "How much must be added to what I already have to obtain a certain amount?" For example, Jane has $2 and she needs $5. This can be represented as:

$a + b = c$, where a and c are known and b is not; thus, $2 + \square = 5$.

Some textbooks call this a *missing addend* problem. Whatever it's called, it does appear often in real-life situations, and children must be led to interpret a sentence such as $2 + \square = 5$ as being logically equivalent to $5 - 2 = \square$. Thus, it is extremely important to focus on the relationship between addition and subtraction.

Again, your initial approach should employ objects to illustrate the question being asked.

You have 6 pennies and want to buy a candy bar that costs 10 pennies. How many more pennies do you need?

$$6 + \square = 10$$

You have 4 Happy Face pieces and want to exchange them for a ten-strip. How many more pieces do you need?

$$4 + \square = 10$$

Pictorial representation of these situations can help children answer questions of this type.

Draw what you need to make the pictures have the same number.

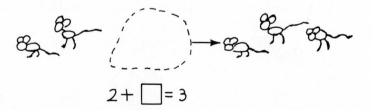

$$2 + \square = 3$$

The third type of subtraction situation, the *comparative* type, involves questions such as "How many more does Jerry have than Keith?"

Jerry ⎯⎯⎯⎯⎯⟶

Keith ⎯⎯⎯⎯⟶

A foundation based on the one-to-one matching test for comparison of two sets can be used here to help children with this application. Teachers who proceed from the question "Which is more?" to "How much more?" provide children with a procedure to solve problems of this type. For actual problem-solving skills to be developed, it is extremely important for teachers to use this language and to talk about the situations with the children.

A fourth type of subtraction situation, the *partitioning type*, involves separating, or partitioning, a set of objects into parts.

5 cars . . . 2 are blue . . .
The others are red . . . How many are red?

Children will start with a given number of objects, then separate, count out, or move a number of objects. Finally, they examine to determine how many objects are left. With the five cars, they've separated them into smaller groups of two and three.

The last type of subtraction situation involves a <u>decrease</u>, but we're not dealing with concrete objects that can be counted. All measurements are like this. Other examples would be loss of weight, being shortened in length, or having the temperature drop when it gets cold. We want children to realize that in these situations, also, we need to subtract.

Again, all five types of situations occur in real life and must be explored by children. Attention must be focused on the basic idea of subtraction as it relates to each of these situations, and the relationship between subtraction and addition must be emphasized.

■ Relating subtraction sentences to the basic facts of addition

The model for development of the basic facts of addition can and should be used to teach subtraction. In the addition model, we combined two distinct sets of objects, A and B, to form another set, C, and found out how many objects were in this new set. If we begin with C, we can separate it into two sets and find out how many objects are in each of the sets. Thus, previously we might have had

$$3 + 2 = 5$$

and now we're interested in

$$5 - 3 = 2$$
$$5 - 2 = 3$$

Notice that this association is a tricky one. It's not as intuitive as it appears, not for young children. So, it takes a deliberate effort on the part of the teacher to help children see that, for any whole numbers a, b, and c:

If $a + b = c$, then $c - b = a$ and $c - a = b$.
If $3 + 2 = 5$, then $5 - 2 = 3$ and $5 - 3 = 2$.

Thus, for every number appearing in the addition table, there are two related subtraction sentences.

$$7 + 3 = 10$$

TWO FREEBIES! ———— $\begin{cases} 10 - 7 = 3 \\ 10 - 3 = 7 \end{cases}$

FREE BEE?

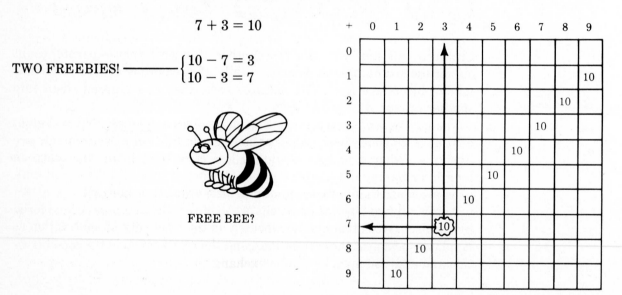

+	0	1	2	3	4	5	6	7	8	9
0										
1										10
2									10	
3								10		
4							10			
5						10				
6					10					
7				10						
8			10							
9		10								

Children need to be told explicitly that if they know their addition facts, the subtraction facts are "thrown in free," and we don't have to study a subtraction table. When you're using models for addition, don't pass up the opportunity to ask questions that call for subtraction ideas as well.

Earlier in this chapter we discussed using counters, objects, and the Groovy Board to generate sentences such as "1 + 4 is the same as 5." At the same time, we can ask first-grade children "1 + *what* is the same as 5?" or "*what* + 4 is the same as 5?" Then, $5 - 1 = 4$ and $5 - 4 = 1$ are reasonable subtraction sentences to discuss. When they're sure of the concepts involved, they can practice for speed and accuracy in the form of immediate recall of basic addition facts.

The Groovy Boards used in Example 3-2 provide an excellent model for relating addition and subtraction ideas, and they can be used with the different interpretations . . . if you ask the right questions!

Example 3-13
C-2

A concept-development activity that allows children to represent a subtraction situation pictorially

Materials: Groovy Boards and Transparent Slider (Material Sheet 10).

Teacher directions: Get your 5-Board and Slider. Fit the Slider onto the board.

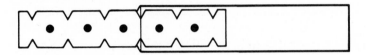

Questions: 1. If I have 5 and cover up 2, how many are left? *(take-away)*

or 2. If I have 2 dots and know there are 5 in all, how many more do I need? *(additive)*

or 3. This is my 5-Board. The 2-Board is this long. How much longer is the 5-Board than the 2-Board? *(comparative)*

or 4. There are 5 in all. 2 are uncovered. How many are covered up? *(partitioning)*

All of these problems are appropriate for $5 - 2 = 3$ and therefore $5 = 2 + 3$.

The activity in Example 3-3 that involved folding a ten-strip of Happy Faces can be used to generate subtraction sentences as well. For example, folding back three of the Happy Faces on the ten-strip fits the sentence $10 - 3 = 7$.

Now let's consider situations in which regrouping is necessary to physically represent the subtraction. In $12 - 5 = \square$, for example, 12 is represented as 1 ten and 2 ones, or, with the Happy Face pieces, by 1 ten-strip and 2 ones. A child cannot physically remove 5 pieces until the ten-strip is exchanged for 10 ones. After this exchange, the child can remove 5 pieces, leaving 7; this demonstrates the necessity of regrouping. Example 3-14 can be used in connection with this development.

Example 3-14   C-1
A concept-development activity to illustrate subtraction computations that require regrouping

Materials: Happy Face pieces—ten-strips and ones
(Material Sheet 3). In-and-Out Shoebox.

Teacher directions: Let's look at $12 - 6 = \square$. Show me 12 with as few pieces as you can. Can you remove 6 pieces? Exchange your ten-strip for 10 ones. Now remove 6 pieces.

Questions: How many do you have left? Then what goes in the box?

The next activity demonstrates how the slide rule can be used to practice and reinforce the subtraction combinations. Other examples in this chapter can be modified to serve the same purpose.

Example 3-15   S/C-1
An activity that allows children to use the slide rule and to write appropriate subtraction sentences for development of skills and concepts

Materials: Slide rule (see Material Sheet 11).

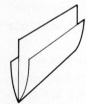
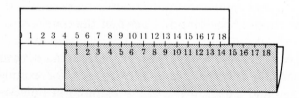

Teacher directions: Let's see if we can use our slide rules to subtract. Can anyone discover how to do it? Right! To find $11 - 7$, we line up the 11 on the top scale with the 7 on the bottom scale. The answer appears over here, right above the 0 on the bottom scale.

Questions: Can you find $13 - 9$? (Etc.) What can you tell me about all of the pairs that are lined up?

Example 3-16
A skill-development activity that requires children to practice subtraction combinations to develop speed and accuracy

Materials: Subtract As You Go (Material Sheet 14). Index cards, showing numerals from 0 through 9. Some plastic chips (or beans, tokens, etc.).

Teacher directions: Draw a card. Subtract the number on your card from every number as you go up the ladder. Say your sentence as you go. If you can do this, you get a chip. The one with the most chips is the winner.

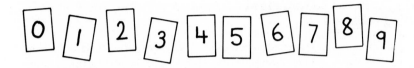

The number line can be used to demonstrate subtraction. For example, to find $8 - 5$: Start at 0. Move 8 spaces to the right. Now move back 5 spaces. You end up at 3. So, $8 - 5 = 3$.

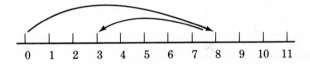

The slide rule, used for both addition and subtraction, is based on the number line. In addition, we ended up with a segment that is as long as the combined length of the two segments. In subtraction, we are comparing one segment with another and finding the difference between them. In this case, the difference between 8 and 5 is 3.

■ Developing a subtraction algorithm

So far, we've been concerned with subtraction sentences related to the basic facts of addition and with the identification of problem situations related to subtraction. With computations involving larger numbers, we need a set of efficient procedures (an algorithm) that will be consistent with strategies used earlier and that relies heavily on place-value ideas.

Easier problems are those in which no regrouping is required in the computation, such as $35 - 12$. The 35 can be represented by 3 tens and 5 ones. Thus, we can subtract 2 ones from 5 ones and have 3 ones. Then we can subtract 1 ten from the 3 tens and have 2 tens.

3 tens + 5 ones	$3(10) + 5$	$30 + 5$	35
1 ten + 2 ones	$1(10) + 2$	$10 + 2$	$- 12$
2 tens + 3 ones = 23	$2(10) + 3 = 23$	$20 + 3 = 23$	23

Notice that the vertical arrangement is sensible at this point, since it allows us to line up the corresponding places and "keep things straight." Also notice that the minus sign "−" is deliberately omitted until the last form is presented. This avoids the need for parentheses around 1 ten + 2 ones, 1(10) + 2, and 10 + 2. It would be confusing to have the sign without the parentheses. Usually, in textbook exercises, the word *subtract* is included in the directions anyway.

For physical models to represent this kind of computation, you could use Happy Face pieces or play money, as well as commercial materials that you may have available.

Example 3-17

C-1

A concept-development activity allowing children to represent subtraction situations with no regrouping required

Materials: Happy Face pieces—ten-strips and ones (Material Sheet 3).

Teacher directions: Represent 47 with Happy Face pieces. Remove 32. Represent 53 with Happy Face pieces. Remove 21. (Etc.)

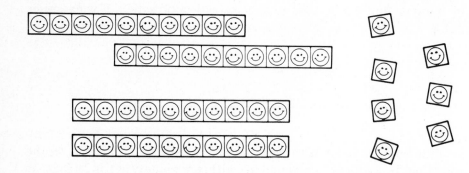

The next logical step concerns situations in which regrouping is required to carry out the computation. For example, consider 34 − 16. If the child has represented 34 by 3 ten-strips and 4 ones, it is physically impossible to remove 6 ones and 1 ten. One of the ten-strips *has* to be exchanged for 10 ones. Then, with 2 tens and 14 ones, it's easy enough to remove 6 ones and 1 ten and record what's left—18. The child is forced to "act out" what you'll be presenting shortly in the form of a set of procedures for subtraction computation. But first you'll provide many experiences like these so that the rules, or procedures, will be meaningful.

Example 3-18

A concept-development activity allowing children to represent subtraction situations with regrouping required

Materials: Happy Face pieces—ten-strips and ones (Material Sheet 3). In-and-Out Shoebox.

Teacher directions: Show me 53, using as few pieces as you can. We want to subtract 35. We can't do it without a trade-in. Exchange 1 ten-strip for 10 ones.

Questions: Now how many ones do you have? Right, 13. How many tens do you have? Right, 4. Can you remove 5 ones? Yes. How many ones are left? Right, 8. Can you remove 3 tens? How many tens now? 1. So 1 ten and 8 ones is 18.

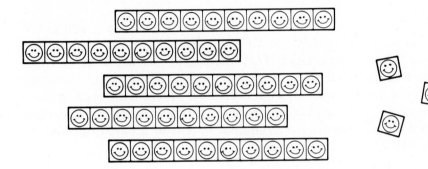

Notice here, as in addition, that questioning procedures in the instructional activities are consistent with the computational steps. We begin with ones and are forced to regroup and then proceed from right to left.

The use of play money is a realistic model to use in subtraction where regrouping is required. Restrict the Place-Value Bank so that it has only tens and ones (later, hundreds and thousands). To illustrate $37 - 18$, have children begin with 3 ten-dollar bills and 7 one-dollar bills. A story such as "owing someone $18" fits nicely. The child is then forced to exchange a ten for 10 ones and pay back 8 ones and 1 ten. Thus, the computational procedure fits the model exactly:

$$
\begin{array}{r}
\overset{2\,1}{\cancel{3}7} \\
-\,18 \\
\hline
19
\end{array}
$$

Exchange 1 ten for 10 ones (leaves 2 tens, 17 ones).
Pay back 8 ones (leaves 9 ones).
Pay back 1 ten (leaves 1 ten).

So, $37 - 18 = 19$.

In children's textbooks, you may see the following forms:

$$
\begin{aligned}
53 &= 5 \text{ tens} + 3 \text{ ones} \\
&= 4 \text{ tens} + 1 \text{ ten} + 3 \text{ ones} \\
&= 4 \text{ tens} + 10 \text{ ones} + 3 \text{ ones} \\
&= 4 \text{ tens} + 13 \text{ ones}
\end{aligned}
$$

$$
\begin{array}{l}
53 = 5 \text{ tens} + 3 \text{ ones} = 4 \text{ tens} + 13 \text{ ones} \\
\underline{35 =} \qquad\qquad\qquad\quad \underline{3 \text{ tens} + \ 5 \text{ ones}} \\
\qquad\qquad\qquad\qquad\quad\ 1 \text{ ten} \ + \ 8 \text{ ones} = 18
\end{array}
$$

$$
\begin{array}{l}
53 = 50 + 3 = 40 + 10 + 3 = 40 + 13 \\
\underline{35 = 30 + 5 =} \qquad\qquad\qquad \underline{30 + \ 5} \\
\qquad\qquad\qquad\qquad\qquad\quad\ 10 + \ 8 = 18
\end{array}
$$

Although these procedures may seem long and drawn-out, they are exactly what the child has to do with the physical models. The shortened form of the algorithm can be explained only in terms of place value, regrouping, and the basic combinations learned earlier. You can hurry things up, but only at the risk of having children not understand—and that risk isn't worth taking!

$$
\begin{array}{r}
4\,1 \\
\cancel{5}3 \\
35 \\
\hline
18
\end{array}
$$

Once the algorithm is developed, the children will need paper-and-pencil practice, game situations that call for accuracy and speed, and situations that call for "mental arithmetic." They need many successful experiences to fix these ideas in their mind. Competition at this level can involve teams or children competing against their individual scores. Alternating these arrangements can help to take into account individual differences among children with respect to the time required to master a particular skill or with respect to ability level in general.

Example 3-19
A practice activity for reinforcing subtraction concepts

S-2

Materials: Index cards showing the digits.

Teacher directions: I'll draw four cards. Use any two of the cards I've drawn, and make the largest number you can. Use the other cards to make the smallest number you can. Subtract your small number from your large number. Let's copy our problem.

Questions: We should all get the same thing, right?

As is the case with addition, if careful attention and sufficient time are given to the physical, pictorial, and symbolic representation of subtraction for numbers less than 100, the transfer to larger numbers will be easier. Consequently, less time will be required in the extension of the algorithm, and there won't be as much need for physical and pictorial representations of the larger numbers. Of course, this decision should be based on your knowledge of the children you're teaching.

■ Subtraction computation involving numbers greater than 100

Notice that the activities using play money are on a more abstract level than those using the Happy Face pieces. The 1 ten-dollar bill *represents* 10 one-dollar bills; 1 one-hundred-dollar bill *represents* 10 ten-dollar bills. In contrast, the ten-strip actually contains 10 ones, and the hundred-square actually contains 10 ten-strips, or 100 ones. With subtraction involving larger numbers, models similar to play money are easier to use.

For example, suppose we want to teach children to subtract 2254 from 5332. We will represent 5332 by 5 thousand-dollar bills, 3 hundred-dollar bills, 3 tens, and 2 ones. The computational form looks like this:

$$\begin{array}{r} 5332 \\ -\,2254 \\ \hline \end{array}$$

We'll begin on the right. The explanation includes the following points:

1. You can't subtract 4 ones from 2 ones.
2. Exchange 1 ten-dollar bill for 10 ones.
3. You now have 12 ones and 2 tens left.
4. Subtract 4 ones from 12 ones to get 8 ones.
5. Proceed to tens, and you can't subtract 5 tens from 2 tens.
6. Exchange a hundred-dollar bill for 10 tens.
7. You now have 2 hundreds left and 12 tens.
8. Subtract 5 tens from 12 tens and get 7 tens.
9. Subtract 2 hundreds from 2 hundreds and get 0 hundreds.
10. Subtract 2 thousands from 5 thousands and get 3 thousands.

When completed, the computation looks like this:

$$\begin{array}{r} \overset{2\ 12\ 1}{5\cancel{3}\cancel{3}2} \\ -\,2254 \\ \hline 3078 \end{array}$$

We can explain every single step! With the extra motivation of "paying back money to a classmate," the algorithm and the story make sense.

The computation for $600 - 248$ deserves special mention because it can be explained in a more efficient manner than the traditional subtraction algorithm. By thinking of 600 as 60 tens, children can then think of 600 as 59 tens and 1 ten or 59 tens and 10 ones, and they're ready to subtract:

$$\begin{array}{r} 600 \\ -\,248 \\ \hline \end{array} \longrightarrow \begin{array}{r} \overset{5\,9\ 1}{\cancel{6}00} \\ 248 \\ \hline 352 \end{array}$$

There are a number of applications—again particularly with money—where this idea is useful.

■ Other computational ideas

In many cases, a method called *equal additions* can be used to make computations easier. The mathematical principle is, "For whole numbers a, b, and c, $a - b = (a + c) - (b + c)$." Children can be convinced that this "works" by looking at the following patterns:

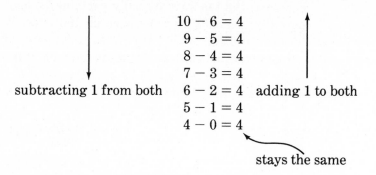

$$10 - 6 = 4$$
$$9 - 5 = 4$$
$$8 - 4 = 4$$
$$7 - 3 = 4$$
subtracting 1 from both $6 - 2 = 4$ adding 1 to both
$$5 - 1 = 4$$
$$4 - 0 = 4$$

stays the same

Now let's look at the following applications.

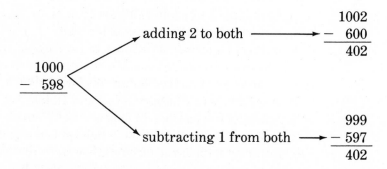

Both computations are easier to carry out than the original, and both are equivalent substitutes.

Direct applications of this idea are possible in situations involving money. It can also be motivational for children to find ways to make hard computations into easy ones—a type of thinking that should be encouraged.

Example 3-20   S-2

A skill-development activity allowing children to use the equal-additions method

Materials: "Hard" computational exercises on cards.

Teacher directions: Make this easy so you won't have to regroup. (Responses will vary, and they should be verified.)

Question: How did you do that?

$$\begin{array}{r} 1002 \\ -97 \\ \hline 905 \end{array}$$

$$\begin{array}{r} 999 \\ -94 \\ \hline 905 \end{array} \qquad \begin{array}{r} 1005 \\ -100 \\ \hline 905 \end{array} \qquad \begin{array}{r} 995 \\ -90 \\ \hline 905 \end{array}$$

There is no doubt about the value of mental arithmetic; computation doesn't always have to involve paper and pencil. We know that practice can increase children's ability to compute mentally, so teachers should provide them with opportunities to practice this skill.

Computational practice with exercises like the ones in Example 3-21 provides good experience for children. The children are made to combine ideas from addition and subtraction, and they should be encouraged to verify their results.

Example 3-21   S-2

Materials: Computation exercises such as the ones below. (Now's the time for a handout sheet!)

In these first three exercises, there's only one way to fill in the boxes; in the next three, there are more ways than one to fill them in.

```
         □56                                  43□      □56
2□32     8□2      9□5              5□8      3□8     5□63
+□38□    +32□     −20□             +□23     +42□    −□4□□
────     ────     ────             ────     ────    ────
4705     □498     □82              1□□8     □7□1     13□2
```

By the end of children's experiences in elementary school, it is entirely reasonable to expect mastery of the algorithms of addition and subtraction of whole numbers. Thus, activities in the fifth and sixth grades that deal with these topics will focus on practice and review. Children at all levels should be encouraged to estimate results before computing as a check for reasonableness of response. For computation with larger numbers, estimation involves rounding procedures similar to these:

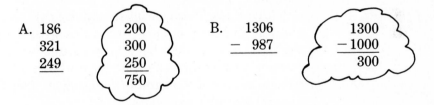

A. 186   200   B.   1306   1300
   321   300   − 987   −1000
   249   250   ────   ────
   ───   ───   300
            750

The following activity combines estimation and calculators with a motivational gimmick that kids love.

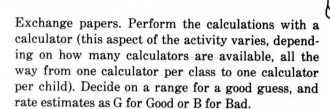

Example 3-22
A skill-development activity for estimation

S-3

Materials: Calculators (hidden at first). Worksheets with a whole lot of messy-looking addition and subtraction exercises on them.

Teacher directions: You've been good today, so we'll do some math. (Passes out worksheets, one for every two people.) Do not do these exercises! I repeat, do *not* do these exercises! I'm giving you five minutes, working with one other person quietly, to estimate what the answers are. Use your pencil *only* to write your estimate or "good guess."

Exchange papers. Perform the calculations with a calculator (this aspect of the activity varies, depending on how many calculators are available, all the way from one calculator per class to one calculator per child). Decide on a range for a good guess, and rate estimates as G for Good or B for Bad.

Question: Who got the most G's? (Discuss estimating procedures.)

Using mental arithmetic

After the algorithms for addition and subtraction have been developed and practiced, *and* you're sure your students understand the meaning of the two operations, it's time to encourage speed and accuracy. The very best way to perform a computation is in your head—you'll always have it with you.

Mitten math: This continued activity can be started with the story that children in Minnesota are very quick with the basic facts. The reason is that it is so cold they have to wear mittens all day and can't use their fingers to count. So, "It's time for mitten math" will encourage your children to get out of the habit of counting on their fingers and will lead them to more desirable behaviors.

Unleaded arithmetic: No pencils allowed here. We have to teach children to do this and we have to convince them that thinking is an appropriate response to exercises in arithmetic. Again, your head is the best thing to use.

The exercises that were discussed in this chapter can be done mentally by many of your students. Show them some strategies to use and encourage them to try without a pencil.

Exercises	*Think*
23 + 31	20 + 30 and 3 + 1. That's 50 + 4, or 54.
46 + 27	40 + 20 and 6 + 7. That's 60 + 13, or 73.

Exercises	*Think*
$142 + 173$	$100 + 100$ is 200. $40 + 70$ is 110. That makes 310, so far. $2 + 3 = 5$. So now I have 315.

$$\begin{array}{r} 5269 \\ + 7426 \\ \hline \end{array}$$

If I take 1 from the bottom number and add it to the top number, that makes this pretty easy.

$$\begin{array}{r} 5270 \\ + 7425 \\ \hline 12695 \end{array}$$

$47 - 32$	$40 - 30$ is 10, and $7 - 2$ is 5. That's 15.
$53 - 21$	$50 - 20$ is 30, and $3 - 1$ is 2. That's 32.
$53 - 35$	I guess I need a different plan. Let's see. 53 is $40 + 13$. Now, $40 - 30$ is 10, and $13 - 5$ is 8. That's 18.

$$\begin{array}{r} 1002 \\ - 97 \\ \hline \end{array}$$

I can add 3 to the bottom number and get 100. Then add 3 to the top number, too, and get 1005. That makes this one a lot nicer. $1005 - 100$ is 905.

Some children will be able to devise their own personal plans for mental arithmetic, a valuable consequence. The resulting positive attitudes toward arithmetic should convince you that the idea is worthwhile.

■ When to teach what

Procedures for teaching concepts related to addition and subtraction of whole numbers depend on the size of the numbers and the amount of regrouping required. Young children in the first and second grades should have extensive experience with the basic facts of addition and related subtraction facts. Many instructional activities at this level will focus on concept development.

As place-value concepts are being developed, the algorithms for addition and subtraction can be presented simultaneously, first with numbers less than 100. Basic facts will continue to be reviewed. In the middle grades, then, activities will stress both skill and concept development. As new ideas are presented, the use of models is extremely important—at any level. Finally, in the later grades, children should demonstrate their knowledge of basic facts and place-value ideas by efficient performance of the algorithms for addition and subtraction of whole numbers.

At all levels, teachers need to emphasize the different interpretations for the operations, thereby emphasizing problem solving. Another strand that is appropriately found in all grades involves estimation techniques. Throughout the elementary school, the use of calculators can be integrated into instruction.

 ━━━━━━━━━━━━━━━━━━━━━━━━━━━━━

The computer corner

In this computer corner, we present a program entitled MASTER that was designed especially for you, the teacher. With it, you can design student activities based on individual needs. The program will do two things for you: it will print place-value models, or it will print models with two sets having less than ten objects. In both cases, the models are randomly generated, so there is an endless supply. The program uses these models to prepare lessons. It will ask you to select one of the model types—place value or sets less than ten—and to enter the lesson's title and directions. Then, one, two, three, the program will print a nice, tidy lesson with eight exercises. Below is part of one such generated lesson. By entering different directions, you can acquire lessons for different instructional objectives.

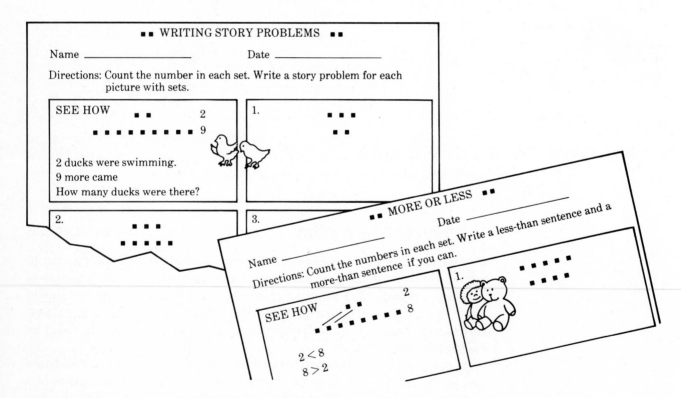

■ Computer tips

There's a lot more to instructional computing than using computers as tutors and teaching children to write programs. As the program described above demonstrates, the computer can help the teacher construct good learning materials for regular or special needs.

Ideas for enrichment: Integers

The topic of integers is a challenging one that can be used with some children in elementary school. This set contains the whole numbers and their additive inverses:

$$\ldots -3, -2, -1, 0, 1, 2, 3, \ldots$$

Just as there is no largest whole number, there is also no *largest* integer, *and* there is no *smallest* integer. Children who have some familiarity with a number-line model for whole numbers can accept this model for integers:

$$\longleftarrow \quad -4 \; -3 \; -2 \; -1 \quad 0 \quad 1 \quad 2 \quad 3 \quad 4 \quad \longrightarrow$$

A money story can be used to introduce these "new" numbers. One effective approach involves the idea of checks representing positive amounts and bills representing negative amounts. Other resources will also provide a variety of suggested teaching procedures and aids, and you can encourage some of your students to pursue this topic on an independent study basis.

Addition of integers can be taught similarly to the addition of whole numbers. Let's look at a few examples.

You have a check for $3.
You get another check for $2.
Now you have $5 worth. $(+3) + (2) = +5$

You have a check for $3.
Someone gives you a bill for $1.
Now you are worth $2. $(+3) + (-1) = +2$

You have a check for $3.
Someone gives you a bill for $3.
Now you're not worth anything. $(+3) + (-3) = 0$

You have a bill for $3.
Someone gives you a check for $2.
Now you are $1 in debt. $(-3) + (+2) = -1$

You have a bill for $3.
Someone gives you another bill for $3.
Now you are $6 in debt. $(-3) + (-3) = -6$

These ideas can also be illustrated on the number line. There are numerous applications of integers that are within the grasp of elementary school students. Some examples of familiar uses are temperatures that fall below zero, elevations above and below sea level, and scorekeeping procedures in games that allow you to go "in the hole." This last use is particularly motivational.

In closing . . .

This chapter describes concepts and skills related to addition and subtraction of whole numbers. Elementary school children are expected to master these skills because they are prerequisite to other mathematical ideas, including the development of multiplication and division. A variety of interpretations for the operations are presented. The chapter includes instructional activities for both concept and skill development. The models for the computational procedures embody the place-value aspects of the algorithms.

It's think-tank time

1. Determine which of the activities in this chapter involve higher-order thinking skills and could be used to promote problem solving by elementary school children.

2. What skills and concepts are prerequisite to the development of addition?

3. Why is the writing of numerals in vertical form unnecessary with the basic facts of addition?

4. How can the fact that addition is a commutative operation help a child who needs to find 3 + 9?

5. a. Make up a problem situation for a child that involves a basic fact of addition and in which the sets would be physically moved together.
   b. Make up a problem situation for a child that involves a basic fact of addition and in which the sets would remain static.

6. Explain and sketch how you can use the 6 Groovy Board:
   a. to generate the family of facts for 6.
   b. to generate the related subtraction facts for the addition facts.
   c. to convince children that addition is a commutative operation. (You won't use the word *commutative* with them.)

7. Sketch Happy Face pieces to demonstrate that 8 + 5 is the same as 13.

8. List the basic facts that could be called "near-doubles." Where do these occur in the addition table?

9. Consider the sentence 8 + 6 = □. Show how you could help a child fill in the box by renaming 6 to emphasize place value and to make the computation easier. Using the same idea, rename 8.

10. Sketch a number line to demonstrate 7 + 5 = 12.

11. Explain how the slide rule, used for addition, is equivalent to the number-line demonstration for addition of whole numbers.

12. Sketch Happy Face pieces to demonstrate the computation for 27 + 38.

13. Sketch play money to demonstrate the computation for 187 + 235.

14. What is an algorithm?

15. The activity using the train sets (Example 3-8) is designed in such a way that children can find a family of facts for the numbers 11, 12, 13, 14, 15, 16, 17, and 18. List the expressions that should go on the engine, caboose, and cars for each number.

16. You can mentally compute 196 + 87 by thinking of 87 as 4 + 83. (What's the sum of 196 and 4? Easy? Easy!) Use this idea to compute the following:
   a. 298 + 167   b. 599 + 381   c. 989 + 215
   Show how you did it.

17. What skills and concepts are prerequisite to the development of subtraction?

18. Write a word problem suitable for children in first, second, or third grade to represent each of the following types of subtraction situations:
   a. take-away   b. additive   c. comparative   d. partitioning   e. decreasing

19. Use a number line to demonstrate the following:
   a. 12 − 3 = 9   b. 7 − 4 = 3   c. 6 − 6 = 0

20. Write three number sentences demonstrating that subtraction is not a commutative operation.

21. Sketch Happy Face pieces to demonstrate the following.

    a. $13 - 6 = 7$ b.
    $$\begin{array}{r} 31 \\ -14 \\ \hline 17 \end{array}$$
    c.
    $$\begin{array}{r} 106 \\ -94 \\ \hline 12 \end{array}$$

22. Explain each step in the procedures for computing the following.
    $$\begin{array}{r} 126 \\ -78 \end{array}$$

23. Use the method of equal additions to compute the following.

    a.
    $$\begin{array}{r} 1500 \\ -997 \end{array}$$
    b.
    $$\begin{array}{r} 87 \\ -29 \end{array}$$
    c.
    $$\begin{array}{r} 1026 \\ -917 \end{array}$$
    d.
    $$\begin{array}{r} 1000 \\ -186 \end{array}$$
    e.
    $$\begin{array}{r} 100 \\ -54 \end{array}$$

24. Explain how you would help children estimate the results of these computations.

    a.
    $$\begin{array}{r} 348 \\ +796 \end{array}$$
    b.
    $$\begin{array}{r} 904 \\ -299 \end{array}$$

25. Fill in the boxes:

    a. $3 + (-3) = \square$ b. $5 + (-7) = \square$
    c. $(-6) + 2 = \square$ d. $(-4) + (-3) = \square$

26. Use the program called MASTER to plan at least five different lessons. Use the Objectives for Children found at the beginning of this chapter as a resource for choosing learning objectives for the lessons. Choose lesson directions so that the resulting lessons will be activities that ready the child to learn the chosen objectives. Print at least two of your lessons.

Suggested readings

Baroody, Arthur J. "Mastery of Basic Number Combinations: Internalization of Relationships or Facts?" *Journal for Research in Mathematics Education*, March 1985, pp. 83–98.

Blume, Glendon Wilbur. "Kindergarten and First Grade Children's Strategies for Solving Addition and Subtraction Problems in Abstract and Verbal Problem Contexts." (Doctoral dissertation, University of Wisconsin–Madison, 1981.) *Dissertation Abstracts International, 42A* (1982): 3482A.

Bolduc, Elroy J., Jr. "A Factorial Study of the Effect of Three Variables on the Ability of First-Grade Children to Solve Arithmetic Addition Problems." (Doctoral dissertation, University of Tennessee, 1969.) *Dissertation Abstracts International, 30A* (1970):3358.

Brownell, William A., and Moser, Harold E. "Meaningful vs. Mechanical Learning: A Study in Grade III Subtraction." *Duke University Studies in Education*, 8 (1949): 1–207.

Bruni, James V., and Silverman, Helene J. "More Indoor Games to Motivate Computational Skills." *The Arithmetic Teacher*, May 1977, pp. 354–365.

Burton, Grace M. "Teaching the Most Basic Basic." *The Arithmetic Teacher*, September 1984, pp. 20–25.

Coxford, Arthur Frank, Jr. "The Effects of Two Instructional Approaches on the Learning of Addition and Subtraction Concepts in Grade One." (Doctoral dissertation, University of Michigan, 1965.) *Dissertation Abstracts International*, 26 (1966):6543–6544.

Fisher, Nancy Capozzolo. "Practical Paper Models for Number Concepts." *The Arithmetic Teacher*, December 1973, pp. 630–633.

Flournoy, Mary Frances. "The Effectiveness of Instruction in Mental Arithmetic." *Elementary School Journal*, November 1954, pp. 148–153.

Gibb, E. Glenadine. "Children's Thinking in the Process of Subtraction." *Journal of Experimental Education*, September 1956, pp. 71–78.

Goff, Gerald K., and Troxel, Vernon E. "Number Chains: Drill and Discovery." *The Arithmetic Teacher*, December 1979, pp. 35–37.

Grouws, Douglas A. "Open Sentences: Some Structural Consideration from Research." *The Arithmetic Teacher*, November 1972, pp. 595–599.

Hall, Donald E., and Hall, Cynthia T. "The Odometer in the Addition Algorithm." *The Arithmetic Teacher*, January 1977, pp. 18–21.

Holz, Alan W. "A Slide Rule for Elementary School." *The Arithmetic Teacher*, May 1973, pp. 353–359.

King, Julia A. "Missing Addends: A Case of Reading Comprehension." *The Arithmetic Teacher*, September 1982, pp. 44–45.

Lazerick, Beth E. "Mastering Basic Facts of Addition: An Alternate Strategy." *The Arithmetic Teacher*, March 1981, pp. 20–24.

Leutzinger, Larry P., and Nelson, Glenn. "Using Addition Facts to Learn Subtraction Facts." *The Arithmetic Teacher*, December 1979, pp. 8–15.

Madell, Rob. "Children's Natural Processes." *The Arithmetic Teacher*, March 1985, pp. 20–22.

Osborne, Alan Reid. "The Effects of Two Instructional Approaches on the Understanding of Subtraction by Grade Two Pupils." (Doctoral dissertation, University of Michigan, 1966.) *Dissertation Abstracts International, 28A* (1968):158.

Secada, Walter G., Fuson, Karen C., and Halls, James W. "The Transition from Counting-All to Counting-On in Addition." *Journal for Research in Mathematics Education,* January 1983, pp. 47–57.

Sherrill, James M. "Egg Cartons Again?!" *The Arithmetic Teacher*, January 1973, pp. 13–16.

Sherrill, James M. "Subtraction: Decomposition versus Equal Addends." *The Arithmetic Teacher*, September 1979, pp. 16–17.

Steinburg, Ruth. "A Teaching Experiment of the Learning of Addition and Subtraction Facts." (Doctoral dissertation, University of Wisconsin–Madison, 1983.) *Dissertation Abstracts International, 44A* (1984):3313A.

Thompson, Charles S., and Van de Walle, John. "Let's Do It, Transition Boards: Moving from Materials to Symbols in Subtraction." *The Arithmetic Teacher*, January 1981, pp. 4–9.

Thompson, Charles S., and Van de Walle, John. "Let's Do It, Modeling Subtraction Situations." *The Arithmetic Teacher*, October 1984, pp. 8–12.

Tucker, Benny F. "Give and Take: Getting Ready to Regroup." *The Arithmetic Teacher*, April 1981, pp. 24–26.

Vest, Floyd. "Introducing Additional Concrete Models of Operations." *The Arithmetic Teacher*, April 1978, pp. 44–46.

Weaver, J. Fred. "Some Factors Associated with Pupil Performance Levels on Simple Open Addition and Subtraction Sentences." *The Arithmetic Teacher*, November 1971, pp. 513–519.

Weaver, J. Fred, and Suydam, Marilyn N. *Meaningful Instruction in Mathematics Education.* Columbus, Oh: ERIC Information Analysis Center for Science, Mathematics and Environmental Education, 1972.

Better Ways to Use the Best
Multiplication and Division

■ *In this chapter, we present multiplication and division of whole numbers and associated computational procedures. Initially, we emphasize interpretations for the operations. Readiness activities employing physical and pictorial models are then described in conjunction with realistic problems that children may encounter. The emphasis then shifts to the development of meaningful ways to compute and estimate. Finally, we suggest activities that provide for skill development. Throughout the chapter, we stress the importance of using number sentences for the communication of mathematical ideas.*

The content and organization of the chapter are developed from the belief that children should become competent problem solvers. Understanding each operation, associating the appropriate operation with a problem situation, and computing efficiently are prerequisite abilities for competent problem solving. Thus, we stress the development of each of these abilities and show how their attainment is related. ■

Objectives for the child

1. Counts up to 100 by twos, threes, fours, fives, and so on.
2. Matches multiplication and division situations to appropriate models.
3. Matches multiplication and division situations to appropriate number sentences.
4. Matches number sentences to appropriate models for multiplication and division.
5. Constructs number sentences or models for multiplication and division situations.
6. Recalls the basic multiplication facts and identifies missing factors in a multiplication sentence that involves a basic fact.
7. Illustrates properties for multiplication and division by using sets of objects, pictorial models, number sentences, or simple computations.
8. Uses properties of multiplication to reduce the amount of memorizing required to learn the basic multiplication facts.
9. Uses the associative property of multiplication and place-value ideas to compute products of multiples of 10.
10. Uses place-value ideas and the distributive property of multiplication over addition to compute the product of two whole numbers.
11. Associates a rectangular model with a computed product.
12. Computes products whose factors are less than 1000.
13. Computes the quotient of two whole numbers when the divisor is less than 100. (The divisor may be larger than 100 when the exercise is a simple one.)
14. Divides by powers of 10, 100, 1000, and so on.
15. Divides by multiples of 10 or 100.
16. Estimates products and quotients.

Teacher goals

1. Be able to identify the interpretations for the operations.
2. Be able to demonstrate physical and pictorial models for each interpretation of the operations.
3. Be able to devise problem situations appropriate for children that reflect the interpretations for the operations.
4. Be able to explain why a particular number sentence is appropriate for a given interpretation of an operation.
5. Be able to construct models that demonstrate the steps in the computational procedures for whole numbers.
6. Be able to justify the steps in the computational procedures for whole numbers.
7. Be able to identify and sequence student learning objectives that involve operations on whole numbers.
8. Be able to devise motivating practice activities that enable children to compute efficiently.
9. Be able to construct activities enabling children to achieve student learning objectives.

Multiplication of whole numbers

What's the name of that old song—"Anything You Can Do, I Can Do Better"? If we look at history, it seems that a better title would be "Anything We Can Do, We Can Do Better." Think about it. There hasn't been a plan, scheme, machine, or method that when once mastered, humans didn't try to improve. The development of arithmetic has been no different. No sooner had we devised addition as a shorter way to count than we set about looking for quicker and easier ways to add. And that's one way to interpret multiplication.

■ Interpretations

Suppose you work as a pickle packer, and you pack 17 pickles per package. If at the end of the day you've packed 296 packages, how many pickles have you packed? One way to find out is to compute the sum of 296 seventeens. Don't laugh! People who used earlier numeration systems had to do just that, although it's doubtful they were worried about pickles. Of course, you interpreted this as a multiplication problem. Here's another problem in which it's nice to have methods more efficient than addition. Suppose you want to help Norma find the total number of trading stamps she has. She has 8407 rows and each row has 874 stamps. (This could get sticky.) You'd tell Norma to multiply, wouldn't you?

Let's look at one more problem. An ice-cream parlor sells 28 flavors of ice cream and 22 flavors of topping. If it takes one flavor of ice cream and one flavor of topping to make a sundae, how many different kinds of sundaes can be made? One way to find the answer would be to do the following.

1. Use all 28 flavors of the ice cream with the first topping and make sundaes; then use all 28 flavors of the ice cream with the second topping and make sundaes. Continue until all 28 flavors of the ice cream have been used with each of the 22 toppings.
2. Count all the sundaes.
3. Now take a Sunday to rest!

Of course, an easier way to solve the problem is to compute 28×22.

Although the most practical way to solve each of these problems is to multiply, each represents a different type of *interpretation*. In general, problems like the pickle problem will be called the *additive type*, those like the trading-stamp problem will be called the *row-by-column type*, and those like the ice-cream-sundae problem will be called the *combination type*. Now, to see how we arrived at these names for the different types of multiplication problems, let's look at problems involving smaller numbers.

Additive-type multiplication problems involve a given number of sets, each set having the same number of objects. The goal is to find the total number of objects. These problems can be represented using objects, pictures of objects, or number sentences.

Problem: 4 packages of cupcakes.
          3 cupcakes in each package.
          How many cupcakes in all?

Picture
diagram:

Number
sentences: $3 + 3 + 3 + 3 = 12$ $4 \times 3 = 12$

The row-by-column type of multiplication problem involves a given number of rows, each row having the same number of objects. Again, the goal is to find the total number of objects. These problems can be illustrated using objects, pictures of objects, or rectangular models made from squared paper (see Material Sheet 17C).

Problem: 4 rows of cars.
          3 cars in each row.
          How many cars in all?

Picture
diagram:

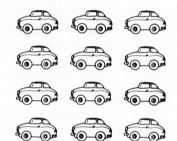

Rectangular region
diagram:

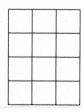

Number
sentence: $4 \times 3 = 12$

Combination-type multiplication problems are those in which there are two sets of objects, and one object from the first set is to be combined with one object from the second set. The goal is to find the total number of possible pairs. These problems can be illustrated using a tree diagram.

Problem: 3 kinds of sandwiches—hot dog, hamburger, cheese.
          2 kinds of drinks—cola, uncola.
          How many different lunches of a sandwich and a drink?

Tree
diagram:

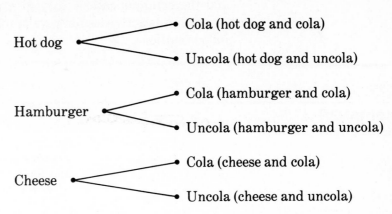

Number
sentence: $3 \times 2 = 6$

Why all this fuss about the different interpretations? The answer is simple. Eventually, we want children to be able to distinguish problem situations that require multiplication from those that don't (we did the same thing for addition and subtraction). The ability to relate a problem to one of these three interpretations helps children to decide when multiplication is appropriate.

Are we interested in having children name or even discriminate among the different problem types? Not necessarily. The interpretations aren't exclusive. A case could be made for placing any problem requiring multiplication into any of the categories. The different types are described only to provide a guide for teachers so that they can expose children to the range of problems requiring multiplication.

When do we begin to teach multiplication? Early! In the latter part of the first grade, children should carry out activities in which they make up sets having the same number of objects and then combine the sets to find the total number of objects.

After number concepts and addition and subtraction ideas have been developed, children should learn to find sums such as $2 + 2 + 2 + 2$ or $3 + 3 + 3 + 3$ and should learn to count by twos, threes, and so on. These kinds of activities emphasize the additive approach to multiplication and prepare children for solving simple problems of this type.

The development of specific multiplication ideas probably should begin sometime in the third grade. The additive type should be developed first, inasmuch as this interpretation is a natural extension of children's previous experiences with addition. As children gain security in solving simple additive-type problems, the row-by-column interpretation should be introduced.

A great deal goes into teaching these two interpretations of multiplication. The development of each one should involve *realistic* or "fun" *problem situations, manipulation of materials, models,* and *number sentences.* Ultimately, we want children to see how a story problem, its solution, and a mathematical communication—namely, a number sentence—are connected. We also want children to feel sure that 3×5 *really does equal* 15.

As children gain experience, the materials they use to solve the problems and the communications they give can become more abstract. Here are some sample activities for developing the additive and the row-by-column interpretations.

Example 4-1   C-2
Additive-type problems

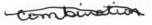
Combination

Materials: Simple story problems, beans, margarine tubs.

Sample problem: 3 baskets; 2 eggs in each basket. How many eggs in all?

Teacher directions: Solve the problem. Use the beans and the margarine tubs to help you. Draw a picture to show what you did.

Example 4-2   C-2
Additive-type problems

Materials: Three sets of cards—cards with story problems, cards showing objects in sets having the same number, and cards with multiplication sentences.

Teacher directions: Match the cards so that the right story problem, the right picture, and the right number sentence are together in a stack.

Sample cards:

3 rugs. 2 bugs on each. How many bugs?	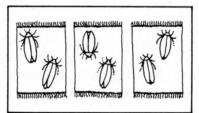	$3 \times 2 = 6$

Example 4-3
Row-by-column-type problems

Materials: Simple story problems. Happy Face pieces. A card with blank spaces for numerals. Numerals on small cards.

Example: 3 rows.
4 children in each row.
How many children?

Teacher directions: Solve the problems. Use Happy Face pieces to help you. Then make a multiplication sentence to show what you find out.

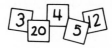

Example 4-4
Row-by-column-type problems

C-2

Materials: Pictures with sets of objects arranged in rows and columns. Cards with number sentences.

Example:

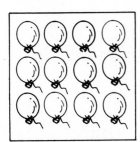

Teacher directions: Match each picture with a number sentence.

$3 \times 4 = 12$

So far, we've discussed strategies and activities for teaching the additive and row-by-column interpretations for multiplication. Similar methods could be suggested for teaching the combination type, but this type of interpretation is more difficult for children to learn and probably should not be developed until the fifth or sixth grade.

■ Learning the basic facts of multiplication

Learning interpretations for multiplication is necessary for knowing when to use multiplication. Being able to represent problem situations with models is also important. But to learn efficient computational procedures, children need to memorize the basic multiplication facts. The amount of time that this task takes for children varies considerably. It is not necessary to wait until all of the children have learned all of the facts in order to proceed with the development of the multiplication algorithm. You will continue to review as you go. Some children may still have to look up basic multiplication facts in a table, and the inefficiency of this should motivate them to memorize the facts.

There are 100 basic facts, each involving the product of two whole numbers less than 10; for example, $7 \times 9 = 63$. The 7 and the 9 in this expression are called *factors*. The number 63 is called a *product*. The expression itself (7×9) is called a *product expression*, or simply a *product*. (After all, 7×9 equals 63.)

Table for basic facts of multiplication

×	0	1	2	3	4	5	6	7	8	9	←factor
0	0	0	0	0	0	0	0	0	0	0	
1	0	1	2	3	4	5	6	7	8	9	
2	0	2	4	6	8	10	12	14	16	18	
3	0	3	6	9	12	15	18	21	24	27	
4	0	4	8	12	16	20	24	28	32	36	
5	0	5	10	15	20	25	30	35	40	45	
6	0	6	12	18	24	30	36	42	48	54	
factor→ 7	0	7	14	21	28	35	42	49	56	63	←product
8	0	8	16	24	32	40	48	56	64	72	
9	0	9	18	27	36	45	54	63	72	81	

As children learn the interpretations, they will undoubtedly learn many of the facts. The order in which the facts are learned, however, might lack organization. The teacher must devise activities that enable children to fill in gaps and to organize what they've learned. Having them construct a multiplication table will help. Of course, the entire table should not be worked on at one time. The idea is to design several small activities that will eventually lead to its completion. These activities should involve (1) the use of manipulative aids for discovering facts and (2) the use of properties of multiplication for eliminating unnecessary memorization. After children complete the table, it's time for them to practice. Games and drill activities should be carefully planned so that immediate recall of basic facts becomes a reality.

Throughout this section, we stress the use of concrete and pictorial aids for illustrating multiplication ideas. These materials should always be used until the interpretations are understood. Then, rectangular models become a very important tool for illustrating computation ideas.

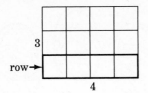

A rectangular model is a rectangular region that is subdivided into squares. For example, the rectangular region at the left is subdivided into 3 rows with 4 squares in each row. Since there are 12 squares in all, the model illustrates that $3 \times 4 = 12$.

The following is an activity that uses rectangular models to aid in the memorization of basic facts. By counting squares, children can discover these basic facts while being exposed to a model for multiplication.

Example 4-5
Using an aid to find the basic facts

C-2

Materials: Fact Finder (see Material Sheet 15).

Teacher directions: Put the cover of your Fact Finder over the grid to show a rectangle with squares. This rectangle will represent a multiplication expression. By counting the squares, you can find the product. Use your Fact Finder to find the missing products in these number sentences: $3 \times 9 = \square$, $4 \times 8 = \square$, $6 \times 2 = \square$, etc.

Example: This shows 3 rows with 7 squares, or a total of 21 squares. So, $3 \times 7 = 21$.

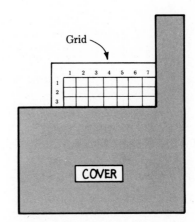

■ Properties of multiplication

Knowing the properties of multiplication can help children memorize the basic facts. Let's see how.

Some special numbers. Appropriate experiences with sets of objects easily convince children that the following relationships hold:

$$n \times 1 = n \quad \text{and} \quad 1 \times n = n \qquad \text{for all whole numbers } n.$$

Because of these relationships, we say that 1 is the *identity element for multiplication.* You can use problems like the following to illustrate this idea.

1. 3 baskets. 1 egg in each basket.
   How many eggs?

$3 \times 1 = 3$

→ number of eggs in all

→ number of eggs in each basket

→ number of baskets

2. 1 apple. 2 worms in each apple.
   How many worms?

$1 \times 2 = 2$

→ number of worms in all

→ number of worms in each apple

→ number of apples

It is also useful to know that these relationships hold:

$n \times 0 = 0$ and $0 \times n = 0$ for all whole numbers n.

Using the language of video games, zero is a "zapper" in multiplication.

Convincing children of these relationships is more difficult than it might seem. Sentences such as $3 \times 0 = 0$ and $0 \times 3 = 0$ must result from problem situations that make sense to children; otherwise, the number sentences are meaningless. Illustrating that $3 \times 0 = 0$ will be easier than illustrating that $0 \times 3 = 0$. However, the problems that you use to illustrate either of these number sentences should follow the presentation of similar problems in which the products are not 0. Here are a few examples.

1. Example for illustrating $n \times 0 = 0$.

Questions: How many circles?
             How many beans in each circle?
             How many beans in all?

$4 \times 3 = 12$

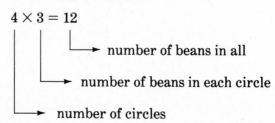

→ number of beans in all

→ number of beans in each circle

→ number of circles

Now remove the beans from the circles.

◯   ◯   ◯   ◯

Questions: How many circles?
How many beans in each circle?
How many beans in all?

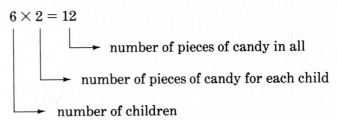

$4 \times 0 = 0$
→ number of beans in all
→ number of beans in each circle
→ number of circles

2. Example for illustrating $0 \times n = 0$.

Children who go into **Mr. Mac's** store get 2 pieces of candy apiece. On Monday, 6 children went in. How many pieces of candy were given away?

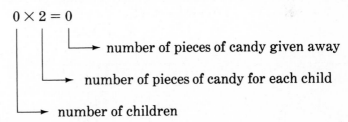

$6 \times 2 = 12$
→ number of pieces of candy in all
→ number of pieces of candy for each child
→ number of children

On Tuesday, 0 children went in. How many pieces of candy were given away?

$0 \times 2 = 0$
→ number of pieces of candy given away
→ number of pieces of candy for each child
→ number of children

The commutative property for multiplication. When children are familiar with the multiplication sentences $3 \times 5 = 15$ and $5 \times 3 = 15$, you can use rectangular models to verify that each sentence is true—that is, that 3×5 and 5×3 are equal.

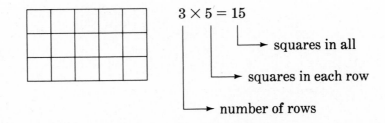

$3 \times 5 = 15$
→ squares in all
→ squares in each row
→ number of rows

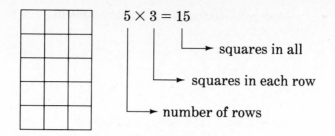

$5 \times 3 = 15$

→ squares in all

→ squares in each row

→ number of rows

Each model has the same number of squares. In fact, the first model can simply be turned to produce the second one. Then how are the product expressions different? The only difference is the order in which the factors occur, and changing this order does not affect the product. Here's how we express this idea mathematically:

For all whole numbers a and b, $a \times b = b \times a$.

We say that multiplication of whole numbers is *commutative*. This property of multiplication is important for many reasons. One is that it reduces the amount of memorization required to learn the basic facts. If children learn $7 \times 3 = 21$, they can use this fact and the commutative property to find 3×7. Children should have experience discovering that the commutative property works for multiplication. They might then use their discoveries to make "life with basic facts" a little easier.

So far in this section we've presented three ideas:

1. $n \times 1 = n$ and $1 \times n = n$
2. $n \times 0 = 0$ and $0 \times n = 0$
3. $a \times b = b \times a$

How do these ideas affect the number of basic facts children must memorize? We can examine the multiplication table to find out. Using the properties we've discussed, we can complete a good bit of the table, leaving only 36 facts for children to memorize.

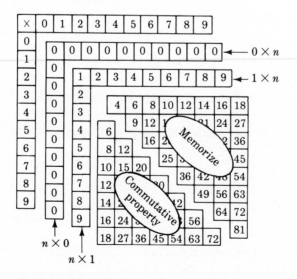

The distributive property of multiplication over addition. This rectangular model shows that 8 × 7 = 56

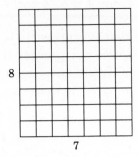

Of course, 8 × 7 is the same as 8 × (5 + 2), so 8 × (5 + 2) = 56. You can easily illustrate this fact with the same rectangular model by simply darkening a line segment.

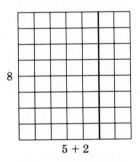

If we cut the rectangular model along the darkened line segment, the result is two rectangular models. One illustrates that 8 × 5 = 40; the other illustrates that 8 × 2 = 16.

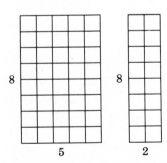

Altogether there are still 56 squares. These models collectively illustrate that

$$8 \times 7 = 8 \times (5 + 2) = (8 \times 5) + (8 \times 2) = 40 + 16 = 56$$

We can demonstrate this relationship for all whole numbers. For example:

$$3 \times 6 = 3 \times (5 + 1) = (3 \times 5) + (3 \times 1) = 15 + 3 = 18$$
$$7 \times 9 = 7 \times (3 + 6) = (7 \times 3) + (7 \times 6) = 21 + 42 = 63$$

These examples illustrate the *distributive property of multiplication over addition* for whole numbers. Mathematically, we say:

For all whole numbers a, b, and c, $a \times (b + c) = (a \times b) + (a \times c)$.

Children will find later that this property is helpful in learning to compute products when the factors are greater than 10. It is also helpful in learning the facts. Children should have experiences discovering that this property "works," and then use it to help learn the facts. Here are some sample activities.

Example 4-6 *row and column*   C-2
Verifying the distributive property of multiplication over addition

Materials: Ditto sheet showing rectangular models and number sentences.

Example:

Teacher directions: Study the A picture, then find □. Study the B picture, then find △. Compare □ and △. See if you can make a discovery.

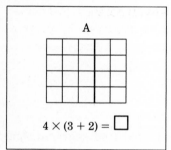

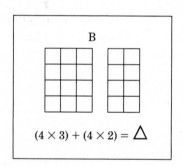

Example 4-7   C-2
Using the distributive property to find the facts for 8

Teacher directions: Use the facts for 3 and the facts for 5 to find the facts for 8.

Example: $4 \times 8 = 4 \times (5 + 3) = (4 \times 5) + (4 \times 3) = 20 + 12 = 32$. So, $4 \times 8 = 32$.

The associative property for multiplication. One more property remains to be discussed, and it's an extremely important one for developing computational procedures involving large numbers. This property is called the *associative property of multiplication*. Mathematically, we say:

For all whole numbers a, b, and c, $a \times (b \times c) = (a \times b) \times c$.

Suppose we have the product expression $3 \times 2 \times 4$. The associative property says we can find (2×4) and then find 3×8:

or we can find 3×2 and then find 6×4:

■ Activities for skill development

Memorizing the basic facts doesn't have to be dull. Creative teachers can devise games and other activities that will enable children to practice the facts while saving the teacher time, money, and energy. A game board or a deck of cards, for example, can be used repeatedly, whereas a ditto sheet is used only once and then discarded. Here are some activities for practicing facts. The directions are a little complicated, but third-graders can learn to play and enjoy every minute of it.

Example 4-8

S-2

Materials: A set of 32 cards labeled with product expressions—facts for 2, 3, 4, and 5 (don't repeat facts such as 3×5 when a card is made for 5×3).

Directions: Play "Do You Dare?" This is a game for up to four players. Cards are shuffled and placed face down. First player draws three cards but looks at only two. He or she then guesses whether the product shown on the third card is less than, between, or greater than the products shown on the first two cards. (In each case the standard name for the product must be stated.) If player guesses correctly, the cards go in his or her "point" stack. Otherwise, the cards are placed in a "dead" stack. Other players play accordingly. Each new play involves three new cards. The game is over when all the cards in the original pile have been played. The winner is the player with the most cards in his or her "point" stack.

Example: Player draws three cards and turns up two, keeping one face down.

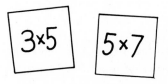

Player guesses "between." Player turns third card over and states product for each card: "15, 35, and 32." The player places all three cards in his or her "point" stack because the guess is correct (32 is between 15 and 35).

Example 4-9
The 5-second multiplication club

S-2

Materials: Chips labeled with numerals 0 through 9. Several cards, each labeled with 5 numbers between 0 through 9 inclusively. Clock that indicates seconds.

Teacher directions: Select a card and a chip without looking at the numerals. On the signal, give the products obtained by multiplying the number shown on the chip by each number shown on the card. If you can give products correctly for three cards, using no more than 5 seconds for each, then you will become a member of the *5-second multiplication club.*

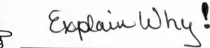

[handwritten: Self Compotition / Idividual goal / Important to Know]

[handwritten: Explain Why!]

[handwritten: organize]

Example 4-10

S-2

Materials: Ten fingers.

Teacher directions: You can use your fingers to multiply numbers between 5 and 10. Here's how to compute 6×8. Subtract 5 from 6 and 5 from 8. You get 1 and 3. Fold down all but 1 finger on one hand and all but 3 fingers on the other hand. Find the sum of the numbers shown by the extended fingers on the two hands. This sum indicates tens. ($1 + 3 = 4$, so we have 4 tens.) Now find the product of the numbers shown by the folded-down fingers of the two hands. These are ones. ($4 \times 2 = 8$, so we have 8 ones.) So, using *finger multiplication*, we find that 6×8 is 48. (This method of computing was once used by poor people because they didn't have pencils and paper, and they didn't memorize their facts as we do.) Now use finger multiplication to find these products: 6×6, 6×7, 8×9, etc. (Note that in 6×6 you get 2 tens plus 16 ones, or 36; also, in 6×7, you get 3 tens plus 12 ones, or 42.)

■ Products of multiples of 10

We need to develop one more idea before we can deal with products like 32×496. We need to know how to compute products like 300×4000 that involve multiples of 10. These multiples are:

10, 20, 30, . . . , 90
100, 200, 300, . . . , 900
1000, 2000, 3000, . . . , 9000
etc.

Let's break the problem into parts. First, we deal with the product of a number between 0 and 9 inclusively and a power of 10—that is, 10, 100, 1000, and so on. Hold on there! This is just place value!

3×10 is 3 tens, or 30
3×100 is 3 hundreds, or 300
3×1000 is 3 thousands, or 3000

How can we compute products like 3×40? Simple—use those concepts we took pages to develop earlier:

$3 \times 40 = 3 \times (4 \times 10)$	We use place-value ideas to express 40 as 4×10
$3 \times (4 \times 10) = (3 \times 4) \times 10$	We use associativity to get the 3 and 4 together.
$(3 \times 4) \times 10 = 12 \times 10$	Here we use a basic fact.
$12 \times 10 = 120$	And here we use place value again!
So . . . $3 \times 40 = 120$	

(Since we know that $3 \times 40 = 120$, we also know that $40 \times 3 = 120$. Why?)

Now we can compute many kinds of products without going through each step!

$6 \times 40 = (6 \times 4) \times 10 = 240$
$7 \times 900 = (7 \times 9) \times 100 = 6300$
$2000 \times 3 = 1000 \times (2 \times 3) = 6000$
$4 \times 500 = (4 \times 5) \times 100$
$20 \times 30 = . . .$ OOPS!

Now we're ready for another kind of product. We'll find a way to compute 20×30.

$20 \times 30 = (20 \times 3) \times 10$	We learned in the previous example that $2 \times 30 = (2 \times 3) \times 10$ So, 20×30 must be $(20 \times 3) \times 10$
$(20 \times 3) \times 10 = [(2 \times 3) \times 10] \times 10$	We learned in the previous example that $20 \times 3 = (2 \times 3) \times 10$
$[(2 \times 3) \times 10] \times 10 = (2 \times 3) \times (10 \times 10)$	We can use associativity to get the 10s together.
$20 \times 30 = (2 \times 3) \times 100 = 600$	You tell why!

Using play money or Happy Face pieces to illustrate these procedures for computing will convince children of the conclusions. For example, Happy Face pieces can be used to make displays like the following:

3 rows with 20, or →
(3 × 2) tens, or
6 tens, or
60

20 rows with 30 in each row, or
(2 × 10) rows with (3 × 10) in each row, or
(2 × 3) sets with (10 × 10) in each set, or
6 sets of 100, or
600

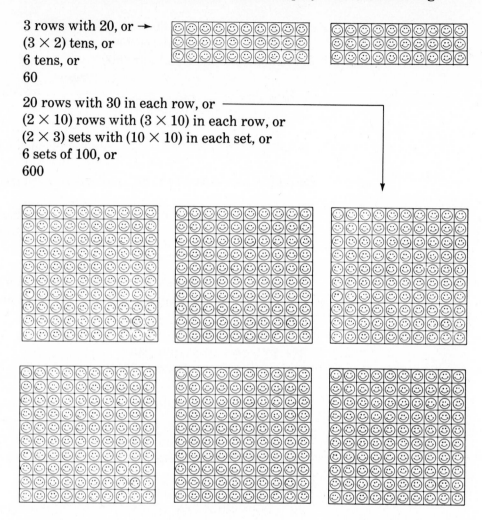

There's no stopping us now! We can do all kinds of computations, and we don't have to go through all the steps (since most of the work can be done in our heads). So can children, and if the work has been associated with concrete and pictorial models, it's even easier.

$$70 \times 60 = (7 \times 6) \times 100 = 4200$$
$$900 \times 60 = (9 \times 6) \times 1000 = 54\ 000$$

The Tap-Down Cards (Material Sheets 9A and 9B) can be used effectively here. Children can select two cards and find the product.

The important thing to realize is that our knowledge of properties for multiplication, place-value ideas, and basic facts allows us to use this procedure.

■ Developing a multiplication algorithm

Now we're ready to find ways to compute products such as 42 × 496. First let's analyze what we know:

1. interpretations
2. basic facts
3. ways of illustrating basic facts with rectangular models
4. properties
5. ways of computing products involving multiples of 10.

To compute the products, we will use *only* what we already know; we won't invent anything new. Let's start with a simple exercise.

To compute 3 × 17, we illustrate 3 × 17 by means of a rectangular model having 3 rows, with 17 squares in each row.

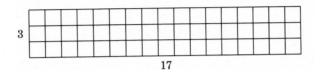

We can separate this model into two rectangular models, A and B.

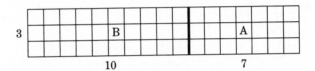

B is a rectangular model that illustrates 3 × 10, since it has 3 rows with 10 squares in each. Similarly, A illustrates 3 × 7. So:

$$3 \times 17 = 3 \times (10 + 7) = (3 \times 10) + (3 \times 7)$$

Of course, we already knew this because multiplication is distributive over addition. We can summarize all of this in a compact computational form.

$$
\begin{array}{r}
17 \\
\times\ 3 \\
\hline
21 \\
+30 \\
\hline
51
\end{array}
\quad
\begin{array}{l}
3 \times\ 7 \\
3 \times 10 \\
3 \times 17
\end{array}
\quad
\begin{array}{l}
\text{model A} \\
\text{model B} \\
\text{entire model}
\end{array}
$$

So, 3 × 17 = 51.

We could also use Happy Face pieces to illustrate this product.

Here's a tougher one: compute 17×23. First, we illustrate 17×23 with a rectangular model. It has 17 rows of 23 squares each. This can be thought of as $(10 + 7)$ rows with $(20 + 3)$ squares in each row. We use this idea to separate the model into four parts. Part A illustrates 7×3, B illustrates 7×20, C illustrates 10×3, and D illustrates 10×20.

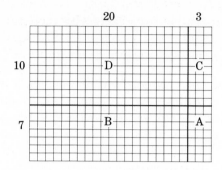

Collectively, the parts illustrate that

$$17 \times 23 = (7 \times 3) + (7 \times 20) + (10 \times 3) + (10 \times 20)$$

We summarize in the "long" computational form:

$$
\begin{array}{r}
23 \\
\times\ 17 \\
\hline
21 \\
140 \\
30 \\
+200 \\
\hline
391
\end{array}
\qquad
\begin{array}{l}
7 \times\ 3 \quad\ \text{A} \\
7 \times 20 \quad \text{B} \\
10 \times\ 3 \quad \text{C} \\
10 \times 20 \quad \text{D} \\
17 \times 23 \quad \text{the entire model}
\end{array}
$$

We can also use the "short" form:

$$
\begin{array}{r}
23 \\
\times\ 17 \\
\hline
161 \\
+230 \\
\hline
391
\end{array}
$$

$7 \times 23 = 7 \times (3 + 20) = (7 \times 3) + (7 \times 20)$ A and B

$10 \times 23 = 10 \times (3 + 20) = (10 \times 3) + (10 \times 20)$ C and D

Now we can extend the procedure to handle products like 42×396. Try it. Of course, you won't use squared paper models, unless your wall needs papering! But if the children are mature enough, you can use sketches, although it must be explained that the sketches will not be drawn to scale.

42×396

	300	90	6
40	40×300	40×90	40×6
2	2×300	2×90	2×6

As you can see, learning to compute products of whole numbers is a complicated process. First, children must acquire background concepts for learning computation procedures. They must then have experiences fitting previously learned concepts together to find quick and efficient ways to compute. To represent products, children should construct models such as those given in this section. They should have experience separating the models into "parts" and relating these parts to their computations. The long form of the computation should be practiced extensively before the short form is introduced, and simple exercises should be practiced before harder ones. Here are some sample activities for developing computational procedures for multiplication.

Example 4-11 C-3

Materials: Index cards showing rectangular parts for products. Index cards showing computations for products (use Material Sheet 17A).

Example:

Rectangular model

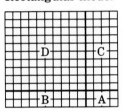

Computation

$$
\begin{array}{r}
14 \\
\times\ 12 \\
\hline
8 \\
20 \\
40 \\
+100 \\
\hline
168
\end{array}
$$

Teacher directions: Make pairs of cards by matching the appropriate computation with the appropriate rectangular model.

Example 4-12 C-3

Materials: Calculator and cards labeled with numerals. Each numeral should represent a multiple of 10, 100, or 1000 and should have two nonzero digits.

Teacher directions: With the calculator and a little common sense, you can find products easily for some large numbers. Draw two cards; suppose they're these:

420   8700

Think "42 tens × 87 hundreds." Use the calculator to find 42 × 87. Use your head to find 10 × 100.

Write 3654 thousands
or
3 654 000

Now draw two more cards; etc.

■ When to teach what

Obviously, children can't learn overnight all they need to know about multiplying whole numbers. Actually, it takes a number of years. Somewhere between the third and the eighth grades, however, most children can learn to compute efficiently, use number sentences appropriately, and solve problems. So a good question is *"When do you teach what to whom?"*

Of course, children differ in ability and experience, but there is a progression of ideas that most children can accommodate.

In the latter part of the second grade, children can learn to do "skip counting." Counting by twos, fives, and tens seems to come easy. Counting by threes can be accomplished with a little more effort. It's time to learn the meaning of multiplication and to memorize the basic facts in the third and fourth grades, but computation can be introduced before all the facts are memorized. However, exercises that are chosen should reflect the facts that children are learning.

At first, computations should involve at least one factor that is less than 10; then computation with factors greater than 10 can be introduced. The greatest emphasis on multiplication computation should occur in the fifth and sixth grades. It is at these levels that some degree of mastery can be expected. Teachers can also expect to find many children who can perform steps in a computation but who are still wrestling with the basic facts. In the seventh and eighth grades, the emphasis is on review. No matter what the level, however, the teacher has an obligation to teach the writing of number sentences and the solving of problems. When children are capable of rounding numbers, the obligation extends to teaching how to estimate products.

These goals are not always achieved easily. Good teachers are constantly asking themselves questions such as these:

Do these children have the prerequisite skills?
Do they understand important place-value concepts, such as how to regroup?
Can they compute sums?
Do they know the basic facts?
Can they compute products involving multiples of ten?
Do they know the interpretations for multiplication?

Good teachers also make sure that children understand why steps in an algorithm are performed as they are. Good teachers illustrate and explain, and then give children opportunities to illustrate and explain. In the final analysis, the results are worth the efforts of both teachers and students.

Division of whole numbers

Because division is so closely related to multiplication, the development of division does not have to start from scratch. Earlier, we learned that multiplication can be interpreted as finding the total number of objects, c, where there are a sets of objects with b objects in each set.

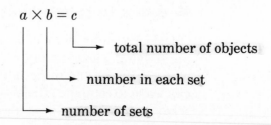

Or, multiplication can be interpreted as finding the total number of objects, *c*, when there are *a* rows with *b* objects in each row. In other words, we are "thinking multiplication" when we have *a* and *b* and try to find *c*. Suppose we know *c*, the total number, and *b*, the number in each set, and want to find *a*, the number of sets. In this kind of situation, we must "think division." For example, suppose you want to put 63 oranges into bags so that each bag will have 9 oranges. How many bags will you need? Here we know the number of objects in each set and the total number of objects. What we need to find is the number of sets.

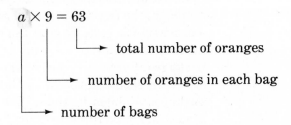

$$a \times 9 = 63$$

total number of oranges

number of oranges in each bag

number of bags

Likewise, we must "think division" in this situation: There are 9 bags with the same number of oranges in each and 63 oranges in all. How many oranges are there in each bag?

$$9 \times b = 63$$

total number of oranges

number of oranges in each bag

number of bags

In each of these situations, we are given the *product* and *one of the factors*. The other factor is missing. Problems like the first are called *subtractive-type* problems; ones like the second are called *distributive-type* problems.

You can see that division ideas are dependent on multiplication. In fact, any separation of the two operations is artificial. We separate the two only so that we can devise useful computational procedures. Even so, we continually rely on what we know about multiplication to help us with division.

Readiness activities for division can begin as early as the first grade, but formal instruction should not begin until shortly after children have learned the interpretations for multiplication. When you are introducing and teaching basic facts of multiplication, you should be relating these concepts to division. Real-life situations should be used to present ideas, and concrete or pictorial materials should be used to represent the problem situation.

■ Interpretations

Research indicates that the *subtractive interpretation* is the easiest for children to grasp. Here we know the total number of objects and the number of objects in each set. We need to find the number of sets. Consider the following examples.

30 crackers.	56 hot dogs.	49 days.
6 in a pack.	8 in a pack.	7 in a week.
How many packs?	How many packs?	How many weeks?

To solve problems like these, children could use **counters** (such as beans, tiles, or buttons) and **containers** (such as muffin cups or margarine tubs). They should use these materials to work through the problem. For example, in the first problem we give a child 30 tiles, each tile representing a cracker. We indicate that the child can use as many margarine tubs as needed. The margarine tubs represent the packages. The child makes packages of crackers until no more sets with 6 crackers can be made. Sets of 6 tiles each are removed and put into margarine tubs until no more packages of 6 can be made. When the packaging is completed, children should draw pictures to show "what happened" and write number sentences that represent the situation.

Picture

Number sentence

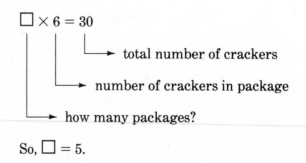

So, $\square = 5$.

Of course, had there been 32 crackers, the picture and number sentence would look like the following:

Picture

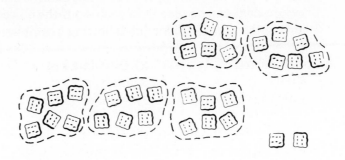

Number sentence

$$(\square \times 6) + 2 = 32$$

After children have grasped this interpretation for division, they're not always eager to manipulate concrete objects—they realize the job can be done more efficiently with numbers. At this point, they should learn to solve subtraction-type problems using repeated subtraction and then streamline their computation by using multiplication facts.

Example: 22 sticks of gum. 5 sticks in a pack. How many packs?

Repeated subtraction:

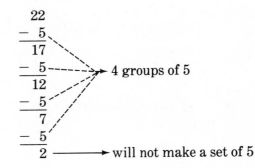

Number sentence:

$$(4 \times 5) + 2 = 22$$

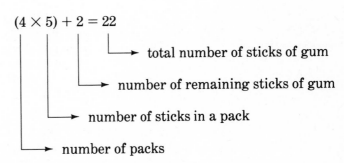

total number of sticks of gum

number of remaining sticks of gum

number of sticks in a pack

number of packs

Subtraction using a multiplication fact as an aid:

$$\begin{array}{r} 22 \\ -20 \\ \hline 2 \end{array}$$ ⟶ 4 sets of 5

⟶ will not make a set of 5

So, $(4 \times 5) + 2 = 22$.

In *distributive-type* division problems, we know the total number of objects and the number of sets. The goal is to find the greatest number of objects that can be placed in each set if the objects are distributed equally among the sets. Consider the following examples.

42 marbles.	27 chocolate kisses.	35 pennies.
6 boys.	9 children.	7 girls.
How many for each?	How many for each?	How many for each?

The strategy for solving this type of problem is different from that used when solving a subtraction-type problem. Instead of removing sets of equal number from the total number of objects, the total number of objects is distributed equally among a given number of sets until there aren't enough objects to "go around again." Of course, objects are sometimes left over.

Let's use this strategy to solve the first problem given above. We give children 42 beans and 6 margarine tubs. The beans represent marbles, the tubs, boys. The children are directed to distribute the beans among the margarine tubs, keeping the number of beans in the tubs equal, until there aren't enough beans to "go around again." When the distributing is completed, children should draw pictures to "show what happened" and write a number sentence to describe their findings.

Picture

Number sentence

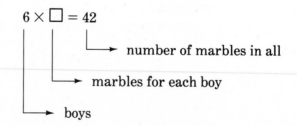

$6 \times \square = 42$

→ number of marbles in all

→ marbles for each boy

→ boys

Here again, children will realize that multiplication can help. The problem of separating 42 objects into 6 sets with the same number of objects, where each set is as large as possible, can be solved by using the fact that $6 \times 7 = 42$. Separating 45 objects into 6 sets is also related to this multiplication fact. The relationship is $(6 \times 7) + 3 = 45$; that is, 7 is the greatest number of objects that can be put into each set, and there will be 3 objects left over. Children may not arrive at this last conclusion immediately; they may have to try cases—for example, 6×5, 6×6, 6×7, 6×8, and so on; 6×5 and 6×6 are too small, 6×8 is too large, but 6×7 is just right.

After ample experiences with both subtractive and distributive interpretations, children should find a relationship between the two. This relationship will be important later when they begin to apply formal computational procedures in problem situations. Here is the relationship:

Suppose we want to replace $\square$ in the sentence $3 \times \square = 18$. The way in which the meaning of the sentence is developed indicates that it is a distributive-type problem; that is, if 18 were distributed into 3 sets, how many would be in each set? Could it be interpreted as a subtractive-type problem? Of course! Whether we distribute 18 into 3 sets with the same number or find the greatest number of sets having 3 that can be formed from a set of 18, the results are the same: 6! In other words, after knowledge of the interpretations is firm, we can use either method to find the replacement for $\square$ in sentences such as $a \times \square = b$ or $\square \times a = b$.

Whatever the interpretation, children should work extensively with problem situations in which the "remainder" is 0 before being introduced to remainders greater than 0. They should also carry out many readiness activities for learning to write number sentences and for learning to perform formal computations.

Example 4-13

C-2

Materials: Ditto showing a table like the one below.
String, scissors, and meter stick.

Cut a piece of string this long.	Cut as many pieces as possible this long.	How many new small pieces do you have?	How much string is left?
72 cm 51 cm etc.	9 cm 6 cm etc.		

Teacher directions: Cut the string as indicated by this table. Use what you find to fill in the blanks. Be sure to measure carefully. Write number sentences after you have completed the table.

☞
Example 4-14

Materials: Large lima beans labeled with the numbers 1 to 50. A game board with a track labeled with numbers 1 to 9. A pair of dice.

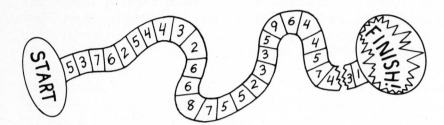

Directions: Play the "Division Raceway Game" (a game for two to four players). Players put playing pieces on "Start." Each player does these things:

1. Rolls dice and moves the indicated number of places.
2. Picks a bean from the margarine tub without looking.
3. Finds the greatest multiple of the number shown on the landing spot contained in the number shown

on the bean. If correct, the player stays on the new landing spot. Otherwise, the player moves back to the previous landing spot.
4. The winner is the first player to land on "finish."

Example: Player lands on $\boxed{5}$, picks the bean labeled ㊽, and states that the greatest number of 5s (whole number) in 49 is 9—that is $(9 \times 5) + 4 = 49$.

Not all quotients are whole numbers. Until now, we have avoided the use of the sign ÷. However, for communication purposes, we'll occasionally use this sign. We've avoided it so far because we believe that the development of division should grow from multiplication ideas. Also, the use of this sign with children who have not studied fractions can lead to confusion, because there is no whole-number replacement for □ in a sentence such as $11 \div 5 = \square$. Notice also that there is no whole-number replacement for □ in a sentence such as $\square \times 3 = 22$. But this situation can be handled conveniently using operations with which children are familiar—multiplication and addition: $\boxed{7} \times 3 + \triangle = 22$. There is no convenient way, however, to handle $11 \div 5 = \square$ because $11 \div 5 = 2.2$, and 2.2 is not a whole number. Children who are prematurely introduced to this sign sometimes get the notion that, for example, $11 \div 5 = 21 \div 10$, especially if they've seen notation that $11 \div 5 = 2R1$ and $21 \div 10 = 2R1$ (2R1 is not a number). The computations are similar:

$$
\begin{array}{r} 2 \\ 5\,\overline{)11} \\ -10 \\ \hline 1 \end{array}
\qquad
\begin{array}{r} 2 \\ 10\,\overline{)21} \\ -20 \\ \hline 1 \end{array}
$$

But $11 \div 5$ cannot equal $21 \div 10$, since $11 \div 5 = 2.2$ and $21 \div 10 = 2.1$. To avoid this kind of confusion, the sign should not be used until children can find quotients that aren't whole numbers. Until then, children can compute using the standard form and can write number sentences in terms of multiplication.

$$3 \overline{)43} \qquad\qquad (3 \times \square) + \triangle = 43$$

standard form   number sentence expressing
for computing   division idea

The use of R to designate the remainder is totally unnecessary.

Division by zero. For each of the interpretations, division has been studied in terms of sentences like these:

$$3 \times \square = 9$$
$$\square \times 4 = 20$$

In both of these sentences, we're trying to find a missing factor for a given product. Suppose we encounter a sentence in which one of the factors is 0 and the product given is not 0. For example:

$$0 \times \square = 10 \quad \text{or} \quad \square \times 0 = 10$$

Can we find a replacement for $\square$? That is, can we find a number to multiply by 0 and get 10? 0 won't work, 1 won't work, and 2, 3, and so on also won't work. Since any number multiplied by 0 is 0, we could never get 10. We've already seen that $0 \times \square = 10$ and $\square \times 0 = 10$ could each be rewritten as $10 \div 0 = \square$. But, since there's no possible replacement for $\square$, these sentences are meaningless!

Now suppose that 0 is a factor and that the product indicated is also 0. For example:

$$0 \times \square = 0 \quad \text{or} \quad \square \times 0 = 0$$

What replacements can we find for $\square$? 0 is all right, since $0 \times 0 = 0$. 2 is all right, since $0 \times 2 = 0$. But 3 is all right, too, since $0 \times 3 = 0$. So is 4, or 57, or 312! In this case, $\square$ has more than one replacement and the corresponding division sentence, $0 \div 0 = \square$, is meaningless.

We found that, in the two possible cases that can be interpreted as division by zero, the results are inconsistent with the way operations are defined. That is, in the first case we could find no solution, and in the second case we found more than one solution. Both results are unacceptable because when an operation is indicated, the result must be one and only one number.

These ideas are subtle and should be explained to children carefully. Whenever arguments cannot be made convincingly, they should be avoided.

Now that all four operations involving basic facts have been discussed, it's time to summarize how nicely they all fit together.

For every addition fact, there are two related subtraction facts.
For every multiplication fact, there are two related division facts (except where 0 is involved.)
Addition is an associative and commutative operation.
Multiplication is an associative and commutative operation.
Multiplication can be thought of as repeated addition.
Division can be thought of as repeated subtraction.

These ideas can provide some much-needed structure for children. Considering the basic facts of addition, the basic fact of multiplication and the corresponding subtraction and division facts, there are 390 number sentences. HOWEVER, they are *very much* related. It's up to you to help your children see the relationship.

Pointing out the characteristics of 0 and 1 can be a unifying strategy. These very important numbers deserve special attention.

	Addition	*Subtraction*	*Multiplication*	*Division*
0	The identity element. $a + 0 = a$ $0 + a = a$	$a - 0 = a$	The "Zapper." $a \times 0 = 0$ $0 \times a = 0$	Impossible to divide by 0. $0 \div a = 0$
1	Adding 1 gives the next whole number.	Subtracting 1 gives the preceding whole number. $(a > 0)$	The identity element. $a \times 1 = a$ $1 \times a = a$	$a \div 1 = a$

■ Developing a division algorithm

Knowing the interpretations for division doesn't mean that you can compute quotients. Sure, you can easily find how many groups of 5 are contained in a set of 26 objects, but finding how many groups of 74 are contained in a set of 4283 is quite another matter. You need a systematic computational procedure, an algorithm. The long division algorithm relies heavily on the *basic facts of multiplication* and *place-value ideas*. Before the teacher introduces this procedure, certain concepts from Chapter 2 should be reviewed with children. Their ability to interpret a numeral in a variety of ways is crucial. For example, they must be able to express 136 as 1 hundred, 3 tens, and 6 ones, or as 13 tens and 6 ones, or as 136 ones. Keep this in mind as you read.

To follow this procedure, you'll want to make a place-value chart and play money and follow along. Using these aids, you'll see how easily the distributive procedure can be demonstrated.

Let's compute $6\overline{)136}$. First we represent 136, using play money in denominations of hundreds, tens, and ones.

Model

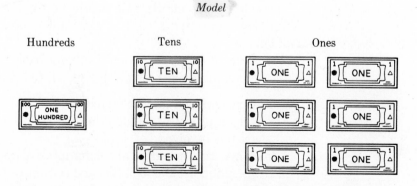

Hundreds   Tens   Ones

We'll distribute the bills of each denomination into 6 sets—each having the same number—if we can. We'll make the sets as large as possible. Bills of a denomination that are "left over" will be exchanged for the appropriate number of bills of the next-smaller denomination. In this case, we start with the hundreds bills.

Our 1 hundred cannot be distributed into 6 sets having hundreds, so we represent the hundred as 10 tens.

Model

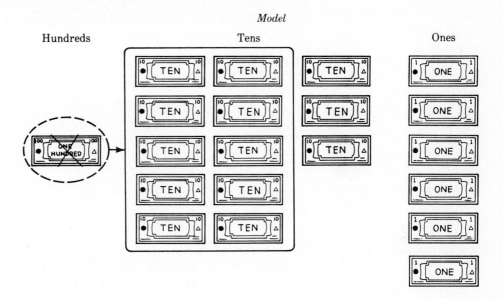

Computation

6 ⟌136

We can clearly "see" that 136 has 13 tens.

Now distribute the tens bills into 6 sets having the same number. Make the sets as large as possible. Represent "left-over" tens as ones. Record how many tens are in each set.

Model

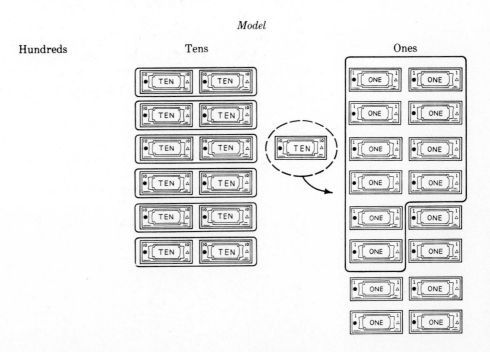

Computation

$$\begin{array}{r} 2 \\ 6\overline{\smash{)}136} \\ 12 \\ \hline 16 \end{array}$$

2 → 2 in the tens place

12 → 6 sets with 2 tens

16 → exchanged 1 ten for 10 ones
already had 6 ones

Our model shows 16 ones. Distribute the ones bills into 6 sets. Record the number in each set. Indicate the number left over.

Model

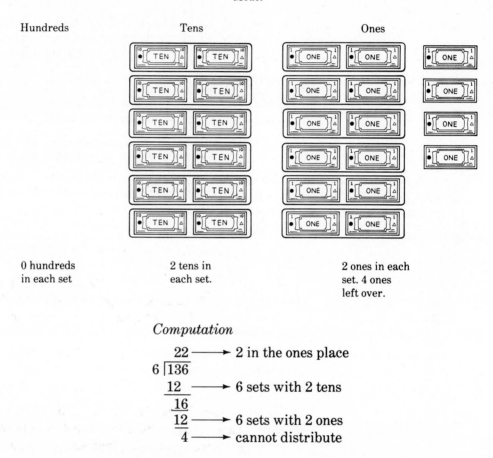

Hundreds   Tens   Ones

0 hundreds   2 tens in   2 ones in each
in each set   each set.   set. 4 ones
                                                      left over.

Computation

$$\begin{array}{r} 22 \\ 6\overline{\smash{)}136} \\ 12 \\ \hline 16 \\ 12 \\ \hline 4 \end{array}$$

22 → 2 in the ones place

12 → 6 sets with 2 tens

12 → 6 sets with 2 ones

4 → cannot distribute

Here's the number sentence that shows the results of our work.

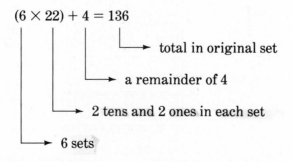

$(6 \times 22) + 4 = 136$

→ total in original set

→ a remainder of 4

→ 2 tens and 2 ones in each set

→ 6 sets

Here's a more difficult computation. This time we won't use play money.

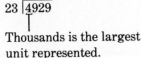

Thousands is the largest unit represented.

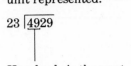

Hundreds is the next largest unit represented.

1. Set up a computation form and find the largest unit represented.

2. We can't distribute 4 thousands into 23 sets without changing them to hundreds, so we use the next largest unit that is represented.

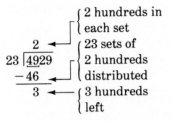

3. Distribute 49 hundreds into 23 sets having the same number of hundreds. Make the sets as large as possible. (Think "23 times what number is close to 49?") Each set will have 2 hundreds. Altogether, 46 hundreds are distributed, so there are 3 hundreds and 29 ones left.

4. The next largest unit that can be distributed is tens. Exchange 3 hundreds for 30 tens. We have 2 tens already. Now there are 32 tens. Distribute the 32 tens into 23 sets having the same number of tens. Make the sets as large as possible. Each set gets 1 ten. Altogether, 23 tens are distributed, so there are 9 tens left.

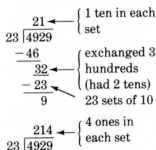

5. The next largest unit that can be distributed is ones. Exchange the 9 tens for 90 ones. We have 9 ones already. Now there are 99 ones. Distribute the 99 ones into 23 sets each having the same number of ones. Make the sets as large as possible. (Think "23 times what number is close to 99?") Each set gets 4 ones. There are 7 ones left that cannot be distributed.

The last step shows that 4929 can be distributed among 23 groups so that there are 2 hundreds, 1 ten, and 4 ones in each group. The remainder shows that there are 7 ones that are not distributed. (After all, 7 is not enough to "go around" again!)

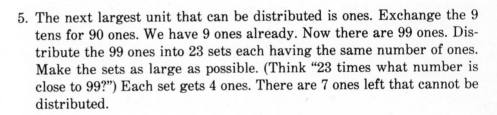

Here's a number sentence that shows the results of our work:

$$(23 \times 214) + 7 = 4929$$

What is the quotient to the nearest whole number?

Play money, Happy Face pieces, and so forth can be used to demonstrate the distributive method of computing quotients. Here is a sample activity for developing this procedure.

Example 4-15
Using play money to divide

C-3

Materials: Play money (see Material Sheet 5).

Teacher directions: You're rich! You have 354 dollars in these denominations: 3 one-hundred-dollar bills, 5 ten-dollar bills, and 4 one-dollar bills. You're going to buy Christmas presents for three friends. What is the most you can spend for each friend if each

is to get a present of the same value? To find out, separate the money into three piles. You may have to exchange larger bills for smaller ones. You can only use hundreds, tens, and ones. Give a division computation to illustrate what you did.

Children should also have exploratory experiences using division concepts with the calculator. Here is an activity that illustrates this idea.

Example 4-16
Using the calculator

C-3

Materials: Calculator. Cards indicating division computations.

42)‾2596‾   4)‾235‾

Teacher directions: Every division computation is related to a number sentence involving multiplication and addition. For example, 42)‾2596‾ is related to ($\square \times 42) + \bigcirc = 2596$. For each card, write a number sentence; then use the calculator to find $\square$ and $\bigcirc$. See if you can come up with an easy method for doing this!

Example 4-17
Problem solving

C-3

Materials: ASMD cards. Four cards like these for each child:

Teacher directions: I'll read a number story. Decide what to do.

 If addition is the answer, hold up the A card.
 If subtraction is the answer, hold up the S card.
 If multiplication is the answer, hold up the M card.
 If division is the answer, hold up the D card.

You've solved the problem.

The preceding activity is an excellent one that provides children with practice in deciding which is the appropriate operation. This is thought by many to be the most crucial aspect of problem solving. Story problems should be taken randomly from an appropriate book; any of the problems in this chapter or in Chapter 3 would be fine. A list of similar problems is well worth making. After some practice with simple or one-step story problems, you can add new problems that require several operations. Children are a good source. Let them make up their own stories and enjoy the results.

One type of activity that can be used with many topics involves competition with the calculator. Children without calculators are encouraged to try to find solutions for arithmetic expressions more quickly than the children who are using calculators to find solutions. The choice of suitable expressions is crucial to the success of this activity. (An expression like 69×47 obviously wouldn't be fair—which shows why calculators are useful!) Examples of suitable expressions are provided for three levels. The psychological advantages of this type of activity include immediate feedback, reinforcement of correct responses and extinction of incorrect responses, repetitive practice, a high probability of success, competition that is nonthreatening, and increased confidence. It also allows the children to give their machines an important message: "Some things you can do *we can do better!*"

Example 4-18

S-1, 2, 3

A skill-development activity for any topic you choose

Materials: Calculator(s). Cards showing number expressions.

Examples:

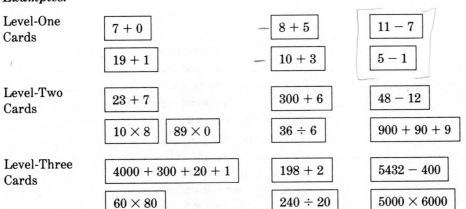

Level-One Cards

| $7 + 0$ | $8 + 5$ | $11 - 7$ |
| $19 + 1$ | $10 + 3$ | $5 - 1$ |

Level-Two Cards

| $23 + 7$ | $300 + 6$ | $48 - 12$ |
| 10×8 89×0 | $36 \div 6$ | $900 + 90 + 9$ |

Level-Three Cards

| $4000 + 300 + 20 + 1$ | $198 + 2$ | $5432 - 400$ |
| 60×80 | $240 \div 20$ | 5000×6000 |

Teacher directions: Play "Beat the Calc." I'll draw a card. The person with the calculator has to use the calculator to find the number. If you can write it down before the Calculator Kid can make it appear, you get a point. If not, the calculator gets a point.

Example: First-grade teacher picks $\boxed{10 + 3}$. Calculator Kid punches $\boxed{1}$, $\boxed{0}$, $\boxed{+}$, $\boxed{3}$, $\boxed{=}$, and 13 is displayed. Other children who understand place

value can quickly write 13 and get a point for this turn.

Note: The organization of this game will always depend on the number of calculators available. With just one calculator, one student can compete against the rest of the class. With enough calculators for each pair of students, one student can compete with another. Or, you may have one student competing with several other students.

→

Estimation and whole numbers

In this chapter and in Chapter 3, we emphasize the interpretations and computational procedures for addition, subtraction, multiplication, and division of whole numbers. Often, however, it's more expedient to *estimate* sums, differences, products, and quotients than it is to compute them. For this reason, children should begin early to develop skill in estimating. As they learn computational procedures for an operation, they should also learn to make corresponding estimates. In fact, it's a good idea for children to estimate answers in a set of practice exercises before computing. In this way, they can judge the reasonableness of the computed value.

Place-value ideas and rounding techniques are the primary tools used for estimating. The idea is to round off numbers to convenient multiples of 10, 100, 1000, and so on, so that computation with the resulting numbers can be done mentally. This usually means "fixing up" the computation so that no regrouping is required. Study these examples:

A. *Computation*	*Estimate*	B. *Computation*	*Estimate*
149	150	361	360
+482	+500	−126	−130
631	650	235	230

C. *Computation*	*Estimate*	D. *Computation*	*Estimate*
482	500	$\begin{array}{r}19\\33\overline{)629}\\-330\\\hline 299\\-297\\\hline 2\end{array}$	$\begin{array}{r}20\\30\overline{)600}\end{array}$
× 37	× 40		
3374	20000		
1446			
17834			

In example A, 149 is rounded off to the nearest ten, whereas 482 is rounded to the nearest hundred. Had 482 been rounded to the nearest ten, the computation would still have been difficult to do mentally. Examples C and D are handled similarly, but in Example B both numbers are rounded to the nearest ten.

Some estimates are easier to make than others because of the numbers involved. In example D, for instance, we can use another method to find an estimate. Since $100 \div 33$ is almost 3 and since 629 is about 600, we can see that there are about (6×3) 33s in 629; that is, $629 \div 33$ is about 18. Likewise, finding estimates that involve divisors close to 10, 20, 25, 50, and so on or 100, 250, 500, and so on is fairly easy. Products are easy to estimate when each factor is rounded to a number that involves only one nonzero digit. Sums and differences are easy to estimate when numbers are rounded off so that regrouping is eliminated. But there are no fixed rules for estimating. The goal is to find "quick answers" that are close enough to actual results to be helpful. Each problem situation dictates what is "close enough" and therefore dictates how the numbers are to be handled. Thus, the best way to develop estimation skills is to have many realistic experiences requiring the use of these skills.

→ Using mental arithmetic

Children should be encouraged to memorize the basic facts because they are so useful and because not doing so results in a handicap to functioning fully. A child can appreciate this line of reasoning. You can help by dealing with the easier ones first. Facts involving 0 are a snap for us all. Facts using 1 are a real breeze. Then 2s are easy because they're just like doubles in addition. Facts with 5 are already familiar and the pattern is great. Children can discover some interesting things about the 9s. If you know the 2s, you can just double them to get 4s. Then double 4s to get the 8s. You can add the 1s and 2s to get the 3s. You can double the 3s to get the 6s. You can add the 5s and 2s to get the 7s, the worst ones of all. Hey, that's all there are!

A very important property that is extremely useful in multiplying numbers greater than 10 is the distributive property of multiplication over addition. It is also helpful in mental arithmetic. Other important and useful ideas involve 0, multiples of 10, and any ideas you like that are sound.

Exercises	*Think*
7×13	That's 7×10 and 7×3. That's $70 + 21$ or 91.
9×101	That's 9×100 and 9×1. That's 900 and 9 or 909.
9468×0	What an easy one! 0.
60×700	6×7 is 42 and 10×100 is 1000. That's 42 000.
$684 \div 2$	That's 600 and 80 and 4 divided by 2. That's 300 and 40 and 2 or 342.
$480 \div 60$	Let's see. What do I multiply 60 by to get 480? The answer is 8.

These are a few suggestions for the typical textbook exercises you will see. Watch for more exciting ideas in other sources. Ask the students to find and share plans or strategies. This is a motivational topic for most children, and the payoff is worth the time and trouble.

The following idea is too good to hide in a chart:

Let's multiply 5 by some large number. Try 4286. In my head . . . the answer is . . . 21430. (Ham it up!) How did I do that? Let's do one more. $5 \times 24668 = $ (Ta daa!) 123340.

These examples are the easiest kind because all of the digits are even. We multiply by 10 by annexing a 0. Then we can divide by 2 quickly. (Multiplying by 10 and dividing by 2 is the same as multiplying by 5.) Try this, you'll like it.

■ When to teach what

Learning to compute quotients can be difficult for children. In the elementary grades, the emphasis should be on the meaning of division. Computational procedures that are developed should proceed slowly and should follow the development of corresponding multiplication concepts. When children in the third or fourth grades have learned the interpretations and basic facts for multiplication, they can be introduced to the interpretations for division and to simple computations such as $3\overline{)12}$, $3\overline{)13}$, $6\overline{)42}$, and $6\overline{)43}$. Likewise, when children in the fourth and fifth grades extend their computational skills to include products where one factor is less than 100 and one factor is 100 or greater, they can learn to compute quotients where the divisor is less than 100. Computing quotients with divisors greater than 100 can be introduced in the fifth and sixth grades, but exercises should remain simple. Emphasis should be placed on learning why steps in the algorithm are performed as they are.

Concrete aids such as Happy Faces or play money should be manipulated as teachers illustrate procedures and as children perform their computations. Special teaching attention should be provided where computing involves regrouping in the dividend or where digits of zero are involved in the divisor, dividend, or quotient. These situations pose the greatest difficulty for children. Mastery of whole-number division skills should occur sometime in the seventh or eighth grade. Throughout the development of division ideas, writing number sentences, estimating, and problem solving should be stressed.

The computer corner

In this corner, we present three programs. The first is a tutorial that allows children to practice their multiplication facts while viewing good models. This program, called ARRAY, tracks student errors and provides a printed progress report that lists specific errors. A more detailed description of ARRAY is given in Chapter 16.

The second program is called AUTHOR. This program allows teachers to customize lessons. Run the program a couple times to see the one we have constructed, then prepare one of your own. Before you change the existing program, however, you should copy AUTHOR onto another diskette. See program documentation for more details.

WITH AUTHOR, WRITING YOUR OWN LESSONS IS A SNAP!

The third program is called FOURS. If you run the program, you will see that it generates multiples of four. Below is a listing of the IBM version of this program. Entering it yourself is good practice. To do so you will have to load BASICA first. This programming language should have been copied to the MGB Diskette when you first received it. After you load BASICA, enter each line of the program exactly as it is given in the book.

```
10  '**********Fours-MGB**********
20  WIDTH 40 'Set width of screen
30  COLOR 14,1,5 'Set color
40  KEY OFF 'Turn function keys off
50  CLS 'CLEAR THE SCREEN
60  PRINT:PRINT 'Give explanation
70  PRINT''   I can count by fours.''
80  PRINT''   Count with me.''
90  PRINT''   See if you can beat me!''
91  FOR P=1 TO 2000:NEXT P 'Pause
92  CLS
100 FOR I=0 TO 100 STEP 4 'Enter loop to generate multiples
110 LOCATE 12,20,0
120 PRINT I 'Print multiple
130 FOR P=1 TO 950:NEXT P 'Pause
140 CLS
150 LOCATE 8,12,0
160 PRINT ''Another 4 coming up!''
170  FOR M=1000 TO 2000 STEP 30' Sound Loop
180    SOUND M,.001
190  NEXT M
200 SOUND 100,0 'Set sound to normal
210 FOR P=1 TO 500:NEXT P 'Pause
220 CLS
230 LOCATE 12,20,0
240 NEXT ' Enter loop for another multiple
250 END
```

Check your program carefully, run it, and then change it so that it will generate multiples for other numbers.

■ Computer tips

Programs that tutor or provide drill and practice should also provide printed progress reports. These reports should not only give scores, but should also indicate the types of errors that children are making so that teachers can take the appropriate remedial action.

Tutorials that can be modfied by teachers are more valuable than those that can't be modified. They allow teachers to tailor lessons to specific learner needs. Furthermore, children are not as likely to tire as quickly when using a program that can be changed.

Many kinds of instructional programs can be found in periodicals and books that teachers can use directly and even modify to obtain a library of free software.

Ideas for enrichment: Finger multiplication

Finger reckoning is one topic that can provide a motivational challenge to some of your children. Sages of long ago explored methods of complex computation using their fingers. Example 4-10 demonstrates one technique for such finger multiplication. Another popular method can be used for multiplying by 9.

Capable students should be encouraged to explore these and other methods. They also can be expected to understand, after some research, *why* these methods work.

The following explanations involve a little bit of algebra. You will be expected to use variables for numbers and to use the distributive property appropriately in the proofs. These ideas are entirely within the grasp of interested and talented elementary school children.

The illustration shows a way to use your fingers to find 3×9. Hold down the third finger. You have 2 fingers up on the left: that's how many tens there are. You have 7 fingers to the right of your third finger: that's how many ones there are. 2 tens and 7 ones $= 27$.

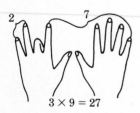

$3 \times 9 = 27$

For the general case, suppose you are multiplying by n. You'll have $n - 1$ fingers on the left and $10 - n$ fingers on the right. (In this picture n is 3, $n - 1$ is 2, and $10 - n$ is 7.)

So, we have $(n - 1)$ tens and $(10 - n)$ ones. Our product is $(n - 1) \times 10 + (10 - n) \times 1$.

This simplifies to $10\,n - 10 + 10 - n$ which is equal to $9\,n$.

Therefore, we have proved that this procedure always works.

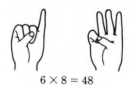

$6 \times 8 = 48$

The next illustration shows a way to use your fingers to find 6×8. Subtract 5 from 6 and hold up that many fingers on your left hand. Subtract 5 from 8 and hold up that many fingers on your right hand: that's how many tens there are in all. Now multiply 4 times 2: that's how many ones there are. You have 4 tens and 8 ones. That's 48.

For the general case, suppose you are multiplying two numbers, a and b.

Subtract 5 from both of them and you have
$a - 5$ and $b - 5$.
This is how many tens you have: $(a - 5) + (b - 5)$.
The number of fingers that are down on one hand is $5 - (a - 5)$.
The number of fingers that are down on the other is $5 - (b - 5)$.
This is how many ones you have:
$$[5 - (a - 5)] \times [5 - (b - 5)].$$
So, we have
$[(a - 5) + (b - 5)]$ tens and $[5 - (a - 5)] \times [5 - (b - 5)]$ ones.
That is,
$$[(a - 5) + (b - 5)] \times 10 + [5 - (a - 5)] \times [5 - (b - 5)].$$

This simplifies to
$$[(a + b - 10) \times 10] + [(10 - a) \times (10 - b)].$$
This is
$$10a + 10b - 100 + 100 - 10a - 10b + a \times b.$$
This is equal to $a \times b$.
Therefore, we have proved that this procedure works.

The topic provides an opportunity to challenge your most academically talented students. The entire class can participate in the "how it works" aspect, particularly in the multiplying-by-9 method.

In closing . . .

This chapter describes multiplication and division concepts and skills that most children should acquire in the elementary-school mathematics program. These concepts and skills are prerequisite to the development of other mathematical ideas that the child will encounter and are necessary for the solution of practical problems.

Children should be able to interpret situations and to associate them with appropriate operations. They should be able to decide when sufficient information is given and when it isn't. Finally, it's important that they be able to compute and estimate efficiently.

By now you should be convinced that this is no small undertaking for either the child or the teacher. But creative, energetic teachers with the proper background can make these goals a reality.

It's think-tank time

1. Determine which of the activities in this chapter involve higher-order thinking skills and could be used to promote problem solving by elementary school children.
2. Write word problems that have these interpretations at a level appropriate for third-or fourth-graders:
   a. additive interpretation for multiplication
   b. row-by-column interpretation for multiplication
   c. subtractive interpretation for division
   d. distributive interpretation for division
3. Draw a rectangular model and a model involving sets of pictured objects to illustrate each of these multiplication sentences. Label the models appropriately.
   a. $3 \times 4 = 12$ b. $5 \times 5 = 25$
   c. $6 \times 2 = 12$ d. $7 \times 1 = 7$
4. Children claim that the multiplication facts for 2 and 5 are easy to learn. Why do you think this is so? Study this chart, which shows factors and products for the facts about 9. See if you can find a pattern that will make the learning of these facts easy.

Factors	1, 9	2, 9	3, 9	4, 9	5, 9	6, 9	7, 9	8, 9	9, 9
Product	9	18	27	36	45	54	63	72	81

5. Here is a chart giving equivalent multiplication and division sentences. Complete it. Do not replace □.

×	$3 \times \square = 12$			
÷		$16 \div 2 = \square$	$\square = 27 \div 3$	$c \div b = a$

6. "Algorithms for Operations with Whole Numbers," Chapter 4 of the National Council of Teachers of Mathematics' *Twenty-ninth Yearbook* (p. 157–166), discusses the subtractive algorithm for division. Read these pages, then use the subtractive method to compute these quotients: $9\overline{)5734}$, $17\overline{)2965}$.
7. Prepare materials for Example 4-12. Carry out this activity.

8. Explain how you would convince children that $3 \times 40 = (3 \times 4) \times 10$. Use play money or Happy Faces (Material Sheets 4 and 7).

9. Compute 13×18 using the "long form." Draw and label the parts of a rectangular model to illustrate your work.

10. Compute both of the following. Explain your steps. Then write a number sentence that expresses the results. What are these quotients to the nearest whole number?
    a. $6\overline{)366}$ b. $72\overline{)1259}$

11. Which is greater—a hundred billions or a billion hundreds?

12. Give the prerequisites for learning to compute (a) products and (b) quotients.

13. Explain why division by 0 is meaningless. Discuss both these situations: $0 \div 0$ and $5 \div 0$.

14. How is the development of division of whole numbers related to multiplication?

15. When would you introduce the sign $\div$? Explain your reasoning.

16. It can be shown that $a \times (b - c) = (a \times b) - (a \times c)$ for all whole numbers a, b, and c. We can use this fact to help us compute. For example:
    $3 \times 98 = (3 \times 100) - (3 \times 2) = 300 - 6$
    $= 294$
    Use this property to compute these products:
    a. 4×97 b. 6×49 c. 7×66 d. 3×1196

17. Decide whether each statement is always true, never true, or sometimes true if a, b, c, and d are whole numbers.
    a. $a \times (b + c) = (a + b) \times (a + c)$
    b. If $c \div a = b$, then $a = b \times c$.

c. $a \div 0 = 0$
d. $0 \div a = 0$
e. $a \times 1 = 1$
f. $a \times (b \times c) = (a \times b) \times c$
g. $a \times b = b \times a$
h. $a \times b \times c = (a \times b) \times (a \times c)$
i. $a \div a = 1$
j. $a \times a = 1$
k. $a \div b = b \div a$
l. If $a - (b - c) = d$, then $(a - b) - c = d$.

18. Estimate. Give brief explanations of the methods you use.
    a. $\begin{array}{r} 348 \\ +796 \end{array}$ b. $\begin{array}{r} 904 \\ -299 \end{array}$ c. $\begin{array}{r} 649 \\ \times 91 \end{array}$ d. $49\overline{)389}$

19. Change the program FOURS so that it generates other multiples, such as those of 5, 7, and so on.

20. Find at least two other programs in magazines or books that will run on the computer you are using. Enter and run the programs. See if you can modify them in some helpful way.

Suggested readings

Baretta-Lorton, Robert. *Mathematics: A Way of Thinking.* Menlo Park, Calif.: Addison-Wesley, 1977, p. 241.

Connelly, Ralph, and Heddens, James. "Remainders That Shouldn't Remain." *The Arithmetic Teacher*, October 1971, p. 379.

Cook, Kathy J., and Dossey, John A. "Basic Fact Thinking Strategies for Multiplication-Revisited." *Journal for Research in Mathematics Education*, May 1982, pp. 163–171.

Dawson, Dan T., and Ruddell, Arden K. "An Experimental Approach to the Division Idea." *The Arithmetic Teacher*, February 1955, pp. 6–9.

Dilley, Clyde A. "A Comparison of Two Methods of Teaching Long Division." (Doctoral dissertation, University of Illinois, 1970.) *Dissertation Abstracts International, 31A* (1970):2248.

Driscoll, Mark J. "Estimation and Mental Arithmetic." In *Research within Reach, Elementary School Mathematics.* St. Louis, Mo.: R & D Interpretive Services, Inc., 1981.

Grafft, William. "A Study of Behavioral Performances within the Structure of Multiplication." *The Arithmetic Teacher*, April 1970, pp. 335–337.

Gray, Roland F. "An Experiment in the Teaching of Introductory Multiplication." *The Arithmetic Teacher*, March 1965, pp. 199–203.

Grouws, Douglas A. "Open Sentences: Some Instructional Considerations from Research." *The Arithmetic Teacher*, November 1972, pp. 595–599.

Hall, William D. "Using Arrays for Teaching Multiplication." *The Arithmetic Teacher*, November 1981, pp. 20–21.

Hall, William D. "Division with Base-Ten Blocks." *The Arithmetic Teacher*, November 1983, pp. 21–23.

Hill, Edwin Henry. "Study of Third, Fourth, Fifth, and Sixth Grade Children's Preferences and Performances on Partition and Measurement Division Problems." (Doctoral dissertation, State University of Iowa, 1952.) *Dissertation Abstracts International, 12* (1952):703.

Hughes, Frank George, "A Comparison of Two Methods of Teaching Multidigit Multiplication." (Doctoral dissertation, University of Tennessee, 1973.) *Dissertation Abstracts International, 34A* (1973):2460–2461.

Irons, Calvin J. "The Division Algorithm: Using an Alternative Approach." *The Arithmetic Teacher*, January 1981, pp. 46–48.

Kratzer, Richard O., and Willoughby, Stephen S. "A Comparison of Initially Teaching Division Employing the Distributive and Greenwood Algorithms with the Aid of a Manipulative Material." *Journal for Research in Mathematics Education*, November 1973, pp. 197–204.

Kurtz, Ray. "Fourth-Grade Division: How Much Is Retained in Grade Five?" *The Arithmetic Teacher*, January 1973, pp. 66–71.

Laing, Robert A., and Meyer, Ruth Ann. "Transitional Division Algorithm." *The Arithmetic Teacher*, March 1982, pp. 10–12.

Patrick, Sarah. "Expanded Division." *The Arithmetic Teacher*, November 1982, pp. 44–45.

Reys, Robert E. "Division and Zero—An Area of Needed Research." *The Arithmetic Teacher*, February 1974, pp. 153–156.

Robold, Alice I. "Grid Arrays for Multiplication." *The Arithmetic Teacher*, January 1983, pp. 14–17.

Salama, Hassan Ali. "The Effect of the Place-Value Method of Teaching Long Division upon the Teaching Ability of Prospective Elementary Teachers." (Doctoral dissertation, Florida State University, 1981.) *Dissertation Abstracts International, 42A* (1981):2547–2548A.

Schrankler, William Jean. "A Study of the Effectiveness of Four Methods for Teaching Multiplication of Whole Numbers in Grade Four." (Doctoral dissertation, University of Minnesota, 1966.) *Dissertation Abstracts International, 27A* (1967):4055.

Sigda, Edward Joseph. "The Development and Evaluation of a Method of Teaching Basic Multiplication Combinations, Array Translation and Operation Identification with Third Grade Students." (Doctoral dissertation, Temple University, 1983.) *Dissertation Abstracts International, 44A* (1984):1717A.

Stuart, M., and Bestgen, B. "Productive Pieces: Exploring Multiplication on the Overhead." *The Arithmetic Teacher*, January 1982, pp. 22–23.

Van de Walle, John, and Thompson, Charles S. "Let's Do It, Partitioning Sets for Number Concepts, Place Value and Long Division." *The Arithmetic Teacher*, January 1985, pp. 6–11.

Vigilante, Nicholas J. "Access to Multiplication Facts." *The Arithmetic Teacher*, September 1978, pp. 42–44.

Some Theory about Numbers

Factors, Multiples, Primes, and Composites

■ *This chapter presents some ideas from number theory that can be useful in developing operations on rational numbers. The classification of whole numbers greater than 1 as either prime or composite and the subsequent prime factorization of composite numbers are useful background concepts related to this development.*

A second major focus of the chapter is based on the belief that topics from number theory can be interesting and challenging to children. Thus, we encourage teachers to promote a spirit of inquiry

by having children explore a variety of ways to represent numbers and by having them make conjectures based on patterns they have observed. Although the emphasis, at this stage, is on motivation, the suggested activities also abound in mathemetical ideas. Children can successfully participate in these instructional activities even if they have not mastered the computational procedures for whole-number operations. In short, number theory can be fun for both teachers and children. ■

Objectives for the child

1. Uses rectangular regions to represent the factorization of whole numbers.
2. Determines the factors of whole numbers.
3. Writes multiplication sentences to represent the factorization of whole numbers.
4. Identifies and determines multiples of whole numbers.
5. Classifies whole numbers as even or odd.
6. Uses divisibility rules to determine factors of whole numbers greater then 100.
7. Identifies prime numbers and composite numbers.
8. Determines the prime factorizations of composite numbers.
9. Determines the greatest common factor of two whole numbers.
10. Determines the least common multiple of two or more whole numbers.

Teacher goals

1. Be able to associate rectangular regions used in the development of multiplication with concepts related to (a) factors and multiples and (b) prime numbers and composite numbers.
2. Be able to justify the procedure involved in using prime factorization to determine the greatest common factors and least common multiples of whole numbers.
3. Be able to devise interesting activities for reinforcing concepts related to number theory.
4. Be able to maintain a classroom atmosphere that encourages questions, conjectures, and experiments.

> $2 \times 3 \times 7 \times 11 \times 19$ is divisible by 14. Is it divisible by 21? By 38? By 77? What are some other numbers it's obviously divisible by?
>
> 3 000 000 is equal to $3 \times 2^6 \times 5^6$. Can you factor 3 000 000 as fast as you can write "3 000 000"? Sure you can!

We saw in earlier chapters that we could obtain a lot of information about a number just by looking at the numeral that represents it. Now we'll see that a little bit of number theory can go a long way in providing additional important information about numbers. Number theory offers most children an opportunity for successful experiences, and children enjoy being able to answer "difficult" questions. So, having just been through several chapters on computation, let's turn to an area of study in which computation takes a back seat.

Much of the terminology used in this chapter should be used in the development of multiplication and division of whole numbers. Thus, even though this chapter develops ideas that are prerequisite to operations on rational numbers usually presented in the intermediate grades, much of the language should be familiar to children at approximately the third-grade level.

Factors and multiples

Any whole number other than 1 can be represented by at least one product expression involving other whole numbers. For example, $12 = 2 \times 6$ or 6×2, $12 = 3 \times 4$ or 4×3, and $12 = 1 \times 12$ or 12×1. Therefore, the numbers 1, 2, 3, 4, 6, and 12 are called *factors* of 12. How about the number 18? $18 = 1 \times 18$ or 18×1, $18 = 2 \times 9$ or 9×2, and $18 = 3 \times 6$ or 6×3. So, 1, 2, 3, 6, 9, and 18 are factors of 18. In general, a whole number m is a factor of a whole number n if there is some whole number that we can multiply m by to get n.

Now consider 13: $13 = 1 \times 13$ or 13×1. So, 1 and 13 are factors of 13. There are only two ways to express 13 as a product of two whole numbers; that is, 13 has exactly two factors. Other numbers having exactly two factors can be listed by children.

Another concept that children should become familiar with concerns *multiples* of whole numbers. If a whole number m is a factor of n, then we say that n is a multiple of m. For example, 4 is a factor of 12, so 12 is a multiple of 4. Children will already have had some contact with this idea from their experience with basic multiplication facts, although they may not have heard it expressed in just this way. They'll quickly agree that 4, 8, 12, 16, 20, 24, 28, 32, and 36 are multiples of 4. These numbers appear in the row and column for 4 in the body of the multiplication table. In fact, any row or column in the table is a list of multiples of some number. A multiplication table could be extended indefinitely and, similarly, a list of multiples of a number can be extended. So, every number has an infinite set of multiples.

It takes a little convincing to get children to agree that 0 is a multiple of every number. Your discussion should lead children to conclude that anytime we multiply 4 by a number we get a multiple of 4. If we multiply 4 by 0 we get 0, so 0 is a multiple of 4, and, by the same reasoning, it is a multiple of every other number. Also, 0 appears in the body of the multiplication table in the row and column of every other number. Thus, through this reasoning also, 0 is a multiple of every number.

■ Even numbers and odd numbers

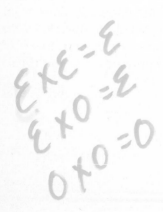

Numbers that are multiples of 2 are called *even* numbers. If a whole number is not even, then it's an *odd* number. Children can be introduced to these ideas very early and can prove informally that the sum of two even numbers is an even number, the sum of an odd number and an even number is an odd number, and the sum of two odd numbers is an even number. They can examine similar situations involving multiplication, make predictions about products, and give reasons for their predictions. Discussions related to addition can center on combining pairs of real objects or combining pairs in pictures of objects. Discussions related to multiplication can involve rectangular arrays and an analysis of the number of pairs of rows or columns. The three activities that follow are designed to develop and reinforce concepts related to even and odd numbers.

Example 5-1

A concept-development activity for even and odd numbers

C-1

Materials: None.

Teacher directions: Let's see whether we can tell, without counting, whether we have an even number of people in the room. Hold hands with one person. Is there anyone not holding hands?

No? Then there's an even number of people, since we can make pairs with no one left over.

or

Yes? Then there's an odd number of people, since we made pairs and had one left over.

Example 5-2

A concept-development activity for addition of even and odd numbers

C-2

Materials: Strips of dots (heavy paper and stick-on circles).

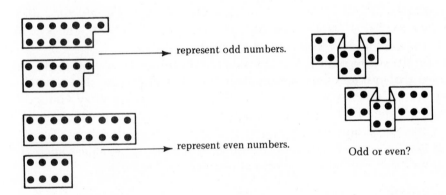

represent odd numbers.

represent even numbers.

Odd or even?

Teacher directions:

A. Put strips for two even numbers together. Do you get a strip that represents an odd number or an even number?

B. Put strips for an even number and an odd number together. Do you get a strip that represents an odd number or an even number?

C. How about when you put together strips for an odd number and an even number?

D. Put strips for two odd numbers together. Do you get a strip that represents an odd number or an even number? Explain why it works this way.

Example 5-3

A concept-development activity for addition of odd and even numbers

C-2

Materials: Pencil and paper.

Teacher directions: Let's make a chart to show what happened. For example:

Questions: Do you think it will always work out this way? Why?

	Strip	Number	Strip	Number	New Strip	Number	Even or Odd?
A	⠿	6 (even)	⠿	4 (even)	⠿⠿	6 + 4	even
B	⠿	2 (even)	⠿	3 (odd)	⠿⠿	2 + 3	odd
C	⠿	3 (odd)	⠿	2 (even)	⠿⠿	3 + 2	odd
D	⠿	5 (odd)	⠿	3 (odd)	⠿⠿	5 + 3	even

■ Classifying on the basis of number of factors

We've seen that 12 has exactly six factors, 18 has exactly six factors, and 13 has exactly two factors. For any nonzero whole number, the set of factors is finite, and the factors can be listed.

Because the concepts of prime numbers and composite numbers will be developed through an examination of the number of factors that a whole number has, it's important to focus on the question "How many factors?" Children can then become ready to accept a classification of whole numbers on this basis and to accept the definitions that will follow. To get them ready for these ideas, they'll need preparatory experiences such as those provided in the following concept-development activities.

Example 5-4

A concept-development activity involving factors of numbers

Materials: Dot paper (Material Sheet 16). Colored pencils (optional).

Teacher directions: Using the dot paper, draw block figures (rectangles) to represent numbers. For example, I can draw four figures for 8:

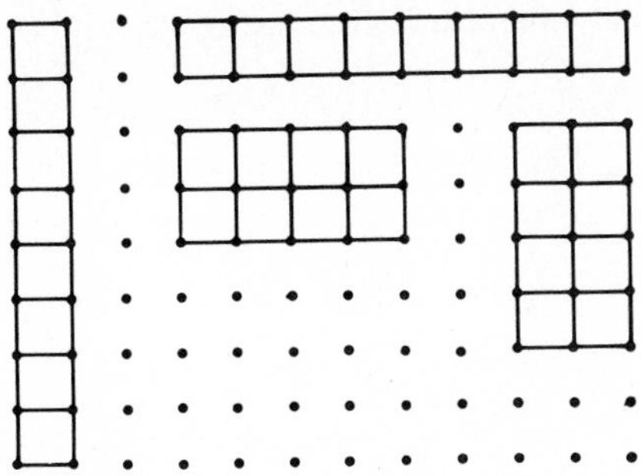

But I couldn't use these:

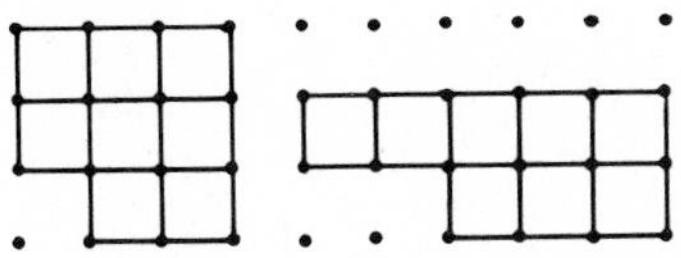

Questions: What do you suppose the following figures have to do with the factors of 8?

1 is a factor of 8. Here's a 1 × 8 figure.————————→
Write a multiplication sentence: 8 = 1 × 8.

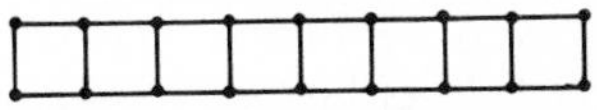

2 is a factor of 8. Here's a 2 × 4 figure.————————→
Write a multiplication sentence: 8 = 2 × 4.

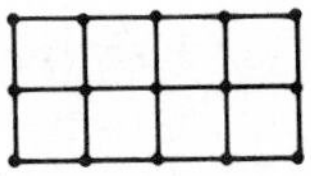

4 is a factor of 8. Here's a 4 × 2 figure.————————→
Write a multiplication sentence: 8 = 4 × 2.

8 is a factor of 8. Here's a 8 × 1 figure.————————→
Write a multiplication sentence: 8 = 8 × 1.

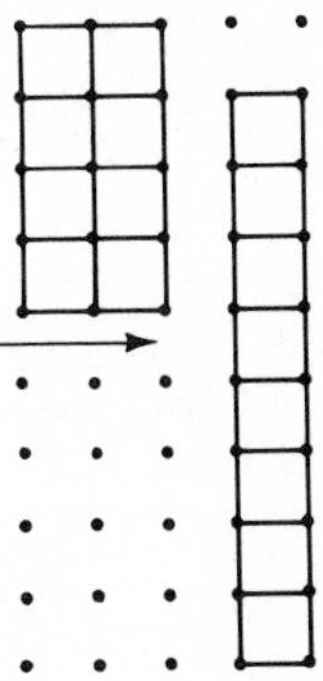

How many factors? Four. How many figures? Four.
How many multiplication sentences? Four.

New teacher directions: Draw figures for each of these numbers: 4, 6, 9, 10, and 12. Tell how many figures you can draw for each number. (Each number can be represented by three or more figures.) Now draw figures for each of these numbers: 2, 3, 5, 7, and 11. (Each will have exactly two figures.)

Questions: How many different figures can you draw for the number 1? Study what you've done. Use the figures to tell one way in which these numbers are different: 1, 7, and 12.

☞

Example 5-5
A concept-development activity involving factors

C-2

Materials: A teacher-made ditto of a chart such as the one below.

Teacher directions: Let's fill in the chart together. We can get most of it by remembering the multiplication facts.

Number	Factors	How many factors?
1	1	1
②	1, 2	2
③	1, 3	2
☆ 4	1, 2, 4	3
⑤	1, 5	2
☆ 6	1, 2, 3, 6	4
.	.	.
.	.	.
.	.	.
☆25	1, 5, 25	3

Some numbers have more than two factors. Put a star by these. Some numbers have exactly two factors. Circle these. Are there any numbers that don't have either a circle or a star? How many such numbers are there?

In the preceding activity, children focus on whether a number is prime or composite. But before they can determine how many factors a number has, they need a systematic way to find all of the factors. Have them try a number that has a lot of factors, such as 36 or 48.

Always start with 1. Find its partner: 36. Write

1   36

Try 2. That works, with a partner of 18. Now fill in these two, in order.

1,   2,   18,   36

Try 3. It works, with a partner of 12. Put these two in. So far, we have:

1,   2,   3,   12,   18,   36

Try 4. It works, with 9. Put these two in the right place.
Try 5. It won't work since 36 doesn't end in 0 or 5.
Try 6. It is a factor of 36, and it doesn't have a partner.
This is as far as we have to go. (Why?) Now our list looks like this:

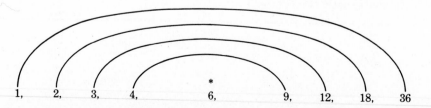

Children find the "rainbow" appealing. They need to realize, however, that one of the numbers doesn't have a partner because 36 is a perfect square. Other such numbers are 25 and 100. Perfect squares are the only numbers with an odd number of factors.

Primes and composites

Now we're ready to define some terms. We can see from the preceding discussion that whole numbers greater than 0 can be classified according to the number of factors they have. Whole numbers having exactly two factors are called *primes;* whole numbers having more than two factors are called *composites*. Since 1 has only one factor, it is neither prime nor composite.

So, using our chart from Example 5-5, we can label the circled numbers as "prime" and the starred numbers as "composite." Thus, the first prime number is 2; other primes are 3, 5, 7, 11, 13, 17, 19, and 23. This is as far as our chart goes, but the set of prime numbers is infinite. Composite numbers in the chart are 4, 6, 8, 9, 10, 12, 14, 15, 16, 18, 20, 21, 22, 24, and 25. The set of composite numbers is also infinite.

The following activity is designed to provide practice for children in classifying whole numbers as either prime or composite and also to develop concepts related to patterns among the prime numbers.

Example 5-6
A skill-development activity with primes and composites

skill concept

S/C-3

Materials: A teacher-made ditto containing the numbers listed below. Colored pencils or crayons.

Teacher directions: We're going to find the prime numbers less than 100 by using a method that's more than 2000 years old! It was invented by the mathematician Eratosthenes.* If you use colored pencils or crayons, you'll see some interesting patterns.

1	2	3	4	5	6
7	8	9	10	11	12
13	14	15	16	17	18
19	20	21	22	23	24
25	26	27	28	29	30
31	32	33	34	35	36
37	38	39	40	41	42
43	44	45	46	47	48
49	50	51	52	53	54
55	56	57	58	59	60
61	62	63	64	65	66
67	68	69	70	71	72
73	74	75	76	77	78
79	80	81	82	83	84
85	86	87	88	89	90
91	92	93	94	95	96
97	98	99	100		

What's the first prime number on the list? It's 2. (Why isn't it 1?) So, circle 2. Now choose a colored pencil to mark out all of the other numbers that have 2 as a factor. What columns are they in?

Now circle the next prime number on the list. It's 3—right! Choose another colored pencil to mark out all of the numbers that have 3 as a factor. What column are they in? Why didn't you have to mark out any in the sixth column?

(Continue this questioning procedure through the number 7.)

Now circle the remaining numbers on the list and look at them.

Question: What patterns do you see?

*er´ə-tos-t̪hə-nēz´

Another important idea that should be emphasized with children is that any composite number can be obtained by multiplying prime numbers together. For instance, if our composite number is 21, we can obtain it by finding the product of the prime numbers 3 and 7. Accordingly, 3×7 is called the *prime factorization* of 21. If we express a number as a product using only prime factors, we have determined the prime factorization of the number.

But before we take a further look at determining the prime factorization of a number, it will be helpful to examine some systematic ways of deciding whether a number is a factor of another number.

■ Finding factors of whole numbers

Anyone who knows the basic facts of multiplication knows some of the factors for some whole numbers, but he or she doesn't necessarily know *all* the factors for just the whole numbers that occur in the multiplication table. For example, 54 occurs in the multiplication table as 6×9 and 9×6. But 54 is also equal to 3×18, 2×27, and 1×54.

There are some relatively easy ways to decide whether one number is a factor of another. These methods are sometimes called *divisibility rules*, because they can be used to tell whether one number is divisible by another. Generally speaking, the statement "a number m is a factor of a number n" means the same thing as "n is divisible by m." For example, "2 is a factor of 14" means the same thing as "14 is divisible by 2." (Children will already be familiar with the term *factor*, however, so there is no need to introduce them to this terminology.)

Here are some divisibility rules. Remember, they're designed to make things easier!

Rules for Finding Factors

1 is a factor of all numbers; thus, $7 \times 1 = 7$, $4 \times 1 = 4$, $89 \times 1 = 89$, and so on.

2 is a factor of a number if and only if the last digit of the numeral is 0, 2, 4, 6, or 8. So, 12 478 has 2 as a factor.

3 is a factor of a number if and only if 3 is a factor of the sum of the digits in the numeral. So, 2112 has 3 as a factor, since $2 + 1 + 1 + 2 = 6$, and 3 is a factor of 6.

4 is a factor of a number if and only if the number represented by the last two digits has 4 as a factor. So 916 has 4 as a factor, since 4 is a factor of 16.

5 is a factor of a number if and only if the last digit of the numeral is either 0 or 5. So 625 has 5 as a factor.

6 is a factor of a number if and only if the last digit of the numeral is either 0, 2, 4, 6, or 8, *and* the sum of the digits has 3 as a factor. In other words, the rules for 2 and 3 have to apply. So 4230 has 6 as a factor, since it ends in 0 and $4 + 2 + 3 + 0 = 9$, which has 3 as a factor.

7 is a factor if the number can be divided by 7. There are other ways, but this is probably the easiest.

8 is a factor of a number if and only if the number represented by the last three digits has 8 as a factor. So 7104 has 8 as a factor, since 104 has 8 as a factor.

9 is a factor of a number if and only if 9 is a factor of the sum of the digits in the numeral. So, 1872 has 9 as a factor, since $1 + 8 + 7 + 2 = 18$, and 9 is a factor of 18.

10 is a factor of a number if and only if the last digit of the numeral is 0. So 1240 has 10 as a factor.

There are sound mathematical reasons to justify these rules. Some of these can be understood by children if teachers provide appropriate questions to guide them. Several of the patterns can be discovered easily, and no child should be discouraged from asking "Why does that work?" For an inquiring child, this topic can be challenging and enriching.

Now let's return to prime factorization. The rules for finding factors can be helpful, and they also provide the basis for an easier approach to determining the *prime* factorization of a number.

Let's find the prime factorization of 90. Finding two factors that are "close together" is usually an efficient way to begin. In this case (from place value or from the rule about 10 as a factor), we can see without computation that 90 is 9×10. Now, from a knowledge of the basic facts of multiplication, we can think of 9 as 3×3 and 10 as 2×5, and that gives us the prime factorization of 90: $3 \times 3 \times 2 \times 5$.

Children seem to enjoy "factor trees" for representing this procedure. Our tree for 90 is shown below.

A "standard" method that always works is to start with 2, use it as a factor as often as possible, and then try 3, then 5, and so on through the prime numbers until all of the factors are prime factors. In this case, our factor tree for 90 would look like the one below.

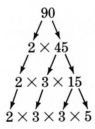

This procedure can be tiresome, however, and it often provides more chances for error.

Regardless of the method used, the results will be the same, because a number can be expressed as a product of prime numbers in only one way. It doesn't matter that the order of the factors is different. There are the same number of twos, threes, and fives in the expressions whether we write $3 \times 3 \times 2 \times 5$ or $2 \times 3 \times 3 \times 5$. There will always be the same primes, and the same number of those primes, in the last row of the factor trees. Usually, however, in writing the prime factorization of a number, we list the factors from smallest to largest. Thus, we write $90 = 2 \times 3 \times 3 \times 5$, or $2 \times 3^2 \times 5$.

Try one more. How about 500? This time we won't be too concerned if the factors aren't about the same size. It should be obvious (from place value again) that 500 is 5×100. Now for the easy part: 100 is 10×10, and $10 = 2 \times 5$; so, for every factor of 10, we have to have a 2×5. Then, almost without thinking and certainly without computing, we can express 500 as

$$5 \times \underbrace{2 \times 5}_{10} \times \underbrace{2 \times 5}_{10}$$

Try this on 700 or 9000. Now you should be able to write the prime factorization of 3 000 000 quicker than you can say "Supercalifragilistic-expialidocious!"

Example 5-7 is a skill-development activity designed to help reinforce ideas related to prime factorization of numbers. Examples 5-8 and 5-9 are related concept-development activities. Through these activities, children can learn much about the topic in a motivational setting.

Example 5-7

S-3

A skill-development activity for factoring numbers

Materials: 48 index cards labeled as follows.

2	3	5	7

You need: ten 2s
eight 3s
four 5s
two 7s

35	25	21	14	4	6
8	9	27	10	12	12
16	16	18	15	24	32
40	42	48	56	54	12

(24 prime cards)   (24 composite cards)

Teacher directions: This is a game for two or more players. Shuffle and deal all the cards. Each player looks at his or her cards and forms "Factor Stacks." The product of the factors must be on top of the stack. The object is to make the stacks as big as possible and to make as many stacks as possible.

Scoring: A player gets 5 points for each stack of 3 cards, 10 points for each stack of 4 cards, 20 points for each stack of 5 cards, 40 points for each stack of 6 cards.

Example: Suppose a player gets this hand:

2	2	3	5	5	9	25	18	42	12	6	54

The cards could be stacked like this:

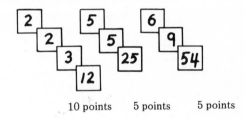

10 points   5 points   5 points

This player would get 20 points.

Note: The scoring makes it a definite advantage to try to use the prime factorization of a number.

Question: What are the possible 6-card stacks in the game?

Example 5-8

C-3

A concept-development activity involving prime factorization

Materials: A box containing chips (or buttons, beans, etc.) of four colors. Red chips represent 2, blue chips 3, green chips 5, and yellow chips 7.

Teacher directions: Draw four chips from the box. Multiply the numbers represented by the colors you have, and tell me the number. I can tell you what colors your chips are.

Example: A student draws four chips and says "60." The prime factorization of 60 is $2 \times 2 \times 3 \times 5$. So, the student has 2 red chips, 1 blue chip, and 1 green chip.

New teacher directions: Try to figure out the method. (Have students become the "teacher.")

☞

Example 5-9   C-3
A concept-development activity involving prime factorization

Materials: A teacher-made ditto of a chart like the one below (to be filled in with the class).

Teacher directions: Here is a chart to record information from the activity we did with the colored chips. Let's fill it in together.

Draw	Red (2)	Blue (3)	Green (5)	Yellow (7)	Product
1	x x	x	x		60
2	x	x x	x		90
3	x	x x		x	126
Etc.					

Questions: What's the largest product we could get? What's the smallest product we could get? If we put in some purple chips, what number would we want them to represent? Etc.

■ Greatest common factor

A phrase that means exactly what it says in mathematics is *greatest common factor.* This can be interpreted literally to mean the largest factor that two numbers have in common.

The greatest common factor (GCF) of two numbers doesn't have to be a prime number. As a matter of fact, we are seldom interested in the greatest prime factor of a number. Let's start with the number 21:

$$21 = 3 \times 7$$

3 is a prime factor of 21.
7 is a prime factor (the greatest) of 21.

But the greatest or largest number that is a factor of 21 is "3×7."

Now let's see how we can use prime factorization to find the greatest common factor of two numbers. Here are some questions.

Consider the number $2 \times 3 \times 5 \times 7 \times 11$. (Don't bother with computation! This is a perfectly good number just as it is, in its prime-factored form.)

Questions	Answers
Is 2 a factor of that number?	yes
Is 3 a factor of that number?	yes
Is 5 a factor of that number?	?
Is 7 a factor of that number?	?
Is 11 a factor of that number?	?
Is 19 a factor of that number?	no
Is 23 a factor of that number?	?
Is 2×3 a factor of that number?	yes
Is 3×5 a factor of that number?	?
Is 7×11 a factor of that number?	?
Is 2×5 a factor of that number?	yes

Questions	*Answers*
What would you multiply 2×5 by to get the number?	$3 \times 7 \times 11$
What would you multiply $3 \times 5 \times 7$ by to get the number?	?
What's the smallest factor of that number?	1
Why isn't 1 in that expression?	?
What's the largest factor of that number?	$2 \times 3 \times 5 \times 7 \times 11$

The reason that 1, the smallest factor, is not included in that prime factorization is that 1 is not a prime number. The largest factor of any number is the number itself. For a number with many prime factors, there are a large number of possible combinations for other factors.

Now consider the following two numbers:

$$2 \times 3 \times 5 \times 11 \quad \text{and} \quad 2 \times 5 \times 7 \times 13$$

Questions	*Answers*
Is 2 a factor of both of these numbers?	yes
Is 5 a factor of both of these numbers?	yes
Is 7 a factor of both of these numbers?	no
Is 2×5 a factor of both of these numbers?	yes
What's the largest number that is a factor of both of these?	2×5

So, the *greatest common factor* of $2 \times 3 \times 5 \times 11$ and $2 \times 5 \times 7 \times 13$ is 2×5. Finding the greatest common factor is pretty easy—if we start with the prime factorizations of the numbers. If we begin with two numbers like 36 and 90, what will we do to determine the greatest common factor? Right! First, find their prime factorizations.

$$36 = 2 \times 2 \times 3 \times 3 \quad \text{and} \quad 90 = 2 \times 3 \times 3 \times 5$$

Now it should be a matter of inspection: the GCF is $2 \times 3 \times 3$, which is 18.

Suppose our two numbers are 12 and 25; 12 is the same as $2 \times 2 \times 3$, and 25 is 5×5. There are no common prime factors for 12 and 25. Remembering that 1 is a factor of every number, however, helps us decide that when the numbers have no common prime factors their greatest common factor is 1. So the greatest common factor of 12 and 25 is 1.

To extend these ideas to three numbers, we again interpret the phrase literally and find the largest factor common to all three numbers. For example, consider 12, 18, and 24.

$$12 = 2 \times \underbrace{2 \times 3} \quad 18 = \underbrace{2 \times 3} \times 3 \quad 24 = 2 \times 2 \times \underbrace{2 \times 3}$$

The expression 2×3 is the greatest combination that appears in each of these. So, the greatest common factor of 12, 18, and 24 is 2×3, or 6.

The application of this idea will be evident to children when they begin the study of fractions. (You should be able to rewrite $\frac{36}{90}$ as $\frac{2}{5}$, using the fact that $2 \times 3 \times 3$, or 18, is the greatest common factor of 36 and 90. Wait for Chapter 6.)

■ Least common multiple

Another idea from number theory that has application to children's study of fractions is that of the *least common multiple* (LCM) of two whole numbers. Like the greatest common factor, *least common multiple* can be interpreted literally. It refers to the whole number that is the smallest nonzero multiple that two numbers have in common. Remember that we've already decided that 0 is a multiple of every number, so 0 would be the least common multiple of any two whole numbers. But since we'll eventually be using the least common multiple as the denominator of fractions, we're going to exclude 0 as a possibility (there will be more about this in Chapter 6). Again, let's begin with some problems. Consider $3 \times 5 \times 7 \times 11 \times 13$.

Questions	Answers
Is that number a multiple of 3?	yes
Is that number a multiple of 7?	yes
Is that number a multiple of 11×13?	yes
Is that number a multiple of 5×7?	?
Is that number a multiple of $3 \times 5 \times 13$?	?
Is that number a multiple of 3×3?	no (why?)
Is that number a multiple of $2 \times 3 \times 5 \times 7$?	?
What other factor would it need in order to be a multiple of $2 \times 3 \times 5 \times 7$?	2
What other factor(s) would it need in order to be a multiple of $3 \times 3 \times 5$?	?

For one number to be a multiple of another, it must contain all of the prime factors of that number.

Now consider the problem of finding a multiple of both of the following numbers:

$$2 \times 3 \times 5 \quad \text{and} \quad 3 \times 3 \times 5 \times 7$$

Questions	Answers
What factors have to be included to have a multiple of $2 \times 3 \times 5$?	one 2, one 3, and one 5
What factors have to be included to have a multiple of $3 \times 3 \times 5 \times 7$?	two 3s, one 5, and one 7
What's the smallest number with all of these factors?	$2 \times 3 \times 3 \times 5 \times 7$

So with $2 \times 3 \times 5$, we need another 3 and a 7 to have a multiple of $3 \times 3 \times 5 \times 7$. Thus, the least common multiple of $2 \times 3 \times 5$ and $3 \times 3 \times 5 \times 7$ is $2 \times 3 \times 3 \times 5 \times 7$, or 630. Again, if we're given the prime factorizations, it's pretty easy. If not, then we must find the prime factorizations. For example, find the least common multiple of 24 and 16:

$$24 = 2 \times 2 \times 2 \times 3$$
$$16 = 2 \times 2 \times 2 \times 2$$

By inspection, the least common multiple is 48. Notice that the greatest common factor is 8, the part that is common in this. $2 \times \boxed{2 \times 2 \times 2} \times 3$.

$$24$$
$$\overbrace{\hspace{3cm}}$$
$$2 \times 2 \times 2 \times 2 \times 3, \text{ or } 48$$
$$\underbrace{\hspace{2cm}}$$
$$16$$

This is the smallest nonzero number that is a multiple of both 24 and 16, which would be a very helpful thing to know if we were trying to add numbers like $\frac{5}{24}$ and $\frac{7}{16}$ (again, wait until Chapter 6).

These ideas can be extended to finding the least common multiple of more than two whole numbers, since the procedure is essentially the same. All that we need to do to ensure that one number is a multiple of another is to see that it contains all of the factors of that number. The *least* multiple that is common to some set of numbers will have all of the factors of those numbers and no other factors. For example, consider the procedure for finding the least common multiple of 12, 15, and 18:

$$12 = 2 \times 2 \times 3$$
$$15 = 3 \times 5$$
$$18 = 2 \times 3 \times 3$$

The least common multiple is:

$$18$$
$$\overbrace{\hspace{2cm}}$$
$$2 \times 2 \times 3 \times 3 \times 5, \text{ or } 180.$$
$$\underbrace{\hspace{1.5cm}}_{12} \quad \underbrace{\hspace{1cm}}_{15}$$

We can also see that to get this number, 12 must be multiplied by 15, or 18 must be multiplied by 10. So, by examining the prime factorizations of 12, 15, and 18, we found the least common multiple of the three numbers to be $2 \times 2 \times 3 \times 3 \times 5$, or 180. Since $12 = 2 \times 2 \times 3$, to obtain 180, 12 must be multiplied by 3×5. Since $15 = 3 \times 5$, to obtain 180, 15 must be multiplied by $2 \times 2 \times 3$. And since $18 = 2 \times 3 \times 3$, to obtain 180, 18 must be multiplied by 2×5. So, we've not only found the LCM by this procedure, but we've also discovered the other factors we need to combine with the three numbers to generate this least common multiple.

Activities such as the following can pull together for children some of the ideas in this chapter and reinforce concepts related to number theory. Remember, this topic should be fun!

Example 5-10
A skill-development activity for ideas related to number theory

S-3

Materials: None.

Teacher directions: Play "What's My Number?" I'm thinking of a number. You can ask me questions that can be answered by "yes" or "no." Girls against the boys. Each team alternates asking a question. If you guess a certain number and you're wrong, your team loses two turns. Some good questions are: Is it less than 100? Is it prime? Is it even? Is it a multiple of 3?

Example 5-11 allows children to make use of the divisibility rules for finding factors and to use calculators to check their theories and conclusions as well as taking care of the tedious computations for them.

☞

Example 5-11
An activity that involves finding factors for numbers

S-3

Materials: Chart to be filled in. Calculators (optional but very helpful).

Teacher directions: Use the rules for finding factors to complete the following table.

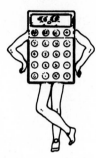

Number	*1*	*2*	*3*	*4*	*5*	*6*	*7*	*8*	*9*	*10*
				Factors						
540	x	x	x	x	x	x			x	x
9246	x	x	x			x				
185										
151										
8021										
1316										
1225										
125										
1002										
126										
136										
700										
543672										
1800										

I've completed the first two rows for you. Don't use the calculator unless you *need* it.

Questions: What will the calculator display look like if the number you divide by *is* a factor?
What will it look like if it *isn't?*
Which columns do you absolutely not need to use a calculator for? Why? (1, then 10, are most obvious.)
How do we tell if 2 is a factor?
(Repeat this question for 3 through 9.)

If I let you use your calculator for only one column, which one would you choose? (Yea! 7.)

Encourage discussion both before children begin filling in the chart and after they're finished. Once they begin filling in the chart, speed them up!

Example 5-12
A skill-development activity for factoring

S-3

Materials: 10×10 grid for each pair of players. Different-colored pencils for each player. Beans (or chips) labeled from 1 to 10.

1	2	3	4	5	6	7	8	9	10
1	2	3	4	5	6	7	8	9	10
11	12	13	14	15	16	17	18	19	20
21	22	23	24	25	26	27	28	29	30
31	32	33	34	35	36	37	38	39	40
41	42	43	44	45	46	47	48	49	50
51	52	53	54	55	56	57	58	59	60
61	62	63	64	65	66	67	68	69	70
71	72	73	74	75	76	77	78	79	80
81	82	83	84	85	86	87	88	89	90
91	92	93	94	95	96	97	98	99	100

Teacher directions: Get into pairs to play. List the numbers 1 to 100 like this on your grid.

To first player: Select a bean. Now look in the column for that number. The number on the bean tells you only which column you can choose from. Now select any number from that column. Color that square. Now color the remaining (uncolored) squares that contain factors of the number you selected. The squares you color can be anywhere on the board.

Example: First player draws 3. The player selects the number 63 and colors square 63 and also squares 1, 3, 7, 9, and 21.

New teacher directions:

To second player: If the first player missed any factors, color in those squares. Then draw a bean and repeat the procedure. If all the squares in your column are colored, draw another bean. The game is over when all of the squares have been colored. The winner is the one with the most squares colored.

Example 5-13
A concept-development activity and a skill-development activity for number theory topics

C/S-3

Materials: Dot paper (Material Sheet 16).

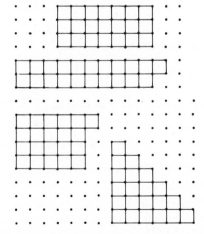

Teacher directions: I've made as many pictures as I can think of to show what I know about 21. Can you think of any others I could make?

21 is not prime.
$21 = 3 \times 7$

21 is an odd number.
$21 = (2 \times 10) + 1$

21 is 1 more than a multiple of 4.
$21 = (4 \times 5) + 1$

21 is a triangular number.
$21 = 1 + 2 + 3 + 4 + 5 + 6$

Now use your paper to represent 36 in as many ways as you can.

New teacher directions: Pick a number less than 100. Tell me as many things as you can about it. Draw pictures to represent your ideas.

The computer corner

In this computer corner, we present GENIE, wonder of the universe. This computer program has captured a surly genie and confined him to the surface of a diskette where he must provide riddles for interested users until the end of time. Run the GENIE program on the MGB Diskette. Read the game directions before you play the game. Then play and see how well you score.

Extend this game to the classroom, with students *playing computer*. Have them make up number rules like those of the genie's, form teams, and try to guess each other's rules, keeping score as the program does.

■ Computer tips

The program GENIE will provide children with endless practice using important abilities—that is, seeing patterns and drawing conclusions. The game format, the personality of the genie, sound, and graphics make the experience a more rewarding one than might be accomplished with pencil and paper. So, when shopping for instructional software, make sure that lessons or activities have more to offer than those on paper. Make sure the software you buy makes best use of your machine's capabilities. Remember, a good computer game can provide a good format for games in the classroom, which encourages many children to exchange ideas.

Ideas for enrichment: Patterns in nature

The Fibonacci (fē'-bō-nä'-chē) sequence is an interesting and challenging topic for many children, an excellent combination of history, mathematics, and art, perfect for independent study. The entire class can appreciate the ideas that have been researched and reported by a few.

Fibonacci, whose real name was Leonardo da Pisa, was an Italian mathematician (1170–1250). His ideas were remarkably advanced: He discussed and wrote about negative numbers 300 years before negative numbers were accepted by most other mathematicians.

The number sequence that bears his name came from patterns observed in rabbit breeding (see *Mathematics, A Human Endeavor*, by Harold R. Jacobs.) The first ten terms of the Fibonacci sequence are:

1   1   2   3   5   8   13   21   34   55

Before you read further, try to determine the next term. Give your children this same opportunity, *the opportunity to discover.*

The mysterious and appealing aspect of the sequence is that these numbers appear *repeatedly* in nature. A daisy head has two spirals of tiny florets. The one going clockwise has 21 florets and the counterclockwise one has 34. Opposing spirals appear in pine-cone scales with 5 going one way and 8 the other. Spirals of 8 and 13 appear in the bumps on pineapples. (From the

third term on, each term is the sum of the two preceding ones.) The Life Science Library volume titled *Mathematics* contains a good discussion of this topic and some beautiful pictures.

In closing . . .

In this chapter, we've presented ideas from number theory that are particularly important for the development of the operations on rational numbers. These ideas have built on the concepts and skills related to multiplication and division of whole numbers discussed in the previous chapters. Applications of the ideas to computation with fractions and decimals will be presented in the next two chapters.

Quite apart from its importance in the se-

quence of elementary school mathematics, number theory is an interesting and challenging area of study for children. As they are encouraged to look for patterns and led to make discoveries, their confidence increases. Teachers who take a relaxed approach to the topic and use motivational activities to reinforce ideas can provide a sound foundation for their students and can also have some fun.

It's think-tank time

1. Determine which of the activities in this chapter involve higher-order thinking skills and could be used to promote problem solving by elementary school children.

2. Make a chart like the one in Example 5-5. Which numbers have an odd number of factors? After 25, the next number with an odd number of factors is 36. Try to predict the next one. See whether you can figure out a way to tell when a number will have an odd number of factors.

3. Determine the prime factorization of each of the following composite numbers:
   a. 54 b. 900 c. 57   d. 128 e. 625
   f. 75 g. 51 h. 10 000 i. 1206 j. 290

4. Explain why the set of composite numbers is infinite.

5. Example 5-6 suggests that there are some interesting patterns in the array of prime and composite numbers. What are some of these patterns? For example:
   a. Where do the multiples of 2 occur?
   b. Where do the multiples of 3 occur?
   c. Where do the multiples of 5 occur?
   d. Where do the multiples of 7 occur?
   e. After the first row, where are all the primes?
   f. How can you express prime numbers greater than 3 in terms of multiples of 6?

6. The divisibility rule for the number 4 can be justified in the following way:
   a. 100 is a multiple of 4.
   b. Any number multiplied by 100 will also be a multiple of 4.

c. A number greater than 100 can be written as the sum of a multiple of 100 and another number less than 100. For example, 1348 = 1300 + 48.

d. If the second number in this sum is a multiple of 4, then the sum is a multiple of 4. For example, in 1348, 1300 is a multiple of 4, since it's 13 × 100; and, since 48 is a multiple of 4, then 1300 + 48, or 1348, is a multiple of 4.

Use this type of reasoning to justify the rule for divisibility by 8.

7. We've noted that in finding the prime factorization of a composite number, the result will be the same regardless of the method used. Make five different factor trees for the number 108 to demonstrate this.

8. Find the greatest common factor of each of the following sets of numbers:
   a. 21 and 24   b. 36 and 45   c. 54 and 12
   d. 104 and 64   e. 4, 6, and 15   f. 30, 42, and 78

9. Find the least common multiple of each of the following sets of numbers.
   a. 18 and 24   b. 54, 12, and 18   c. 22 and 33
   d. 17 and 19   e. 6, 16, and 64

10. Explain two methods you could use to convince a child that 0 is an even number.

11. Examples 5-7, 5-8, 5-11, and 5-12 are games or puzzles that can be fun for children and even nonchildren.

a. Make the materials necessary for these four activities.

b. Try out each activity with at least three people.

c. Write up the results of your tryouts for each activity.

12. Here are three riddles:

A.
> 2 and 4 are factors of me.
> Is 8 a factor of me?

B.
> I am less than 50. I have more factors than any other number less than 50. Who am I?

C.
> I am a multiple of 7. My friend is a multiple of 7. Is our sum a multiple of 7? Is our product a multiple of 7?

a. Analyze each riddle by solving it and listing the concepts that it requires.

b. Make up five more riddles that you could use to reinforce concepts discussed in this chapter.

13. Make a list of the first 20 numbers in the Fibonacci sequence. Every third number in the sequence is a multiple of 2. Use your list to determine which terms are multiples of:
   a. 3   b. 5   c. 8

Suggested readings

Avital, Shumel. "The Plight and Might of Number Seven." *The Arithmetic Teacher*, February 1978, pp. 22–24.

Burton, Grace M., and Knifong, J. Dan. "Definitions for Prime Numbers." *The Arithmetic Teacher*, February 1980, pp. 44–47.

Hoffer, Alan R. "What You Always Wanted to Know about Six but Have Been Afraid to Ask." *The Arithmetic Teacher*, March 1973, pp. 173–180.

Hohlfeld, Joe. "An Inductive Approach to Prime Factors." *The Arithmetic Teacher*, December 1981, pp. 28–29.

Huff, Sara C. "Odds and Evens." *The Arithmetic Teacher*, January 1979, pp. 48–52.

Jeffrey, Neil J. "GCF and LCM on a Geoboard." *The Arithmetic Teacher*, January 1977, pp. 63–64.

Jones, Margaret Hervey, and Litwiller, Bonnie H. "Practice and Discovery: Starting with the Hundred Board." *The Arithmetic Teacher*, May 1973, pp. 360–364.

Karlin, Marvin William. "The Development and Utilization of a Card Game for Teaching Prime Factorization in the Fifth Grade." (Doctoral dissertation, University of Colorado, 1971.) *Dissertation Abstracts International, 33* (1972):80.

Lamb, Charles E., and Hutcherson, Lyndal R. "Greatest Common Factor and Least Common Multiple." *The Arithmetic Teacher*, April 1984, pp. 43–44.

Lichtenberg, Betty Plunkett. "Zero Is an Even Number." *The Arithmetic Teacher*, November 1972, pp. 535–538.

Loomis, Alden H. "The Number 6174: An Arithmetical Curiosity." *The Arithmetic Teacher*, April 1979, pp. 23–24.

Ouellette, Hugh, and Gannon, Gerald. "The Multiple Triangle: Experiences in Discovery." *The Arithmetic Teacher*, January 1979, pp. 34–38.

Padberg, Friedhelm F. "Using Calculators to Discover Simple Theorems—An Example from Number Theory." *The Arithmetic Teacher*, April 1981, pp. 21–23.

Robold, Alice I. "Magic Squares: Concrete to Pictorial to Abstract." *The Arithmetic Teacher*, October 1978, pp. 40–41.

Robold, Alice I. "Patterns in Multiples." *The Arithmetic Teacher*, April 1982, pp. 21–23.

Snover, S. L. "Five-cycle Number Patterns." *The Arithmetic Teacher*, March 1982, pp. 22–26.

Thornton, Carol A. "A Glance at the Power Patterns." *The Arithmetic Teacher*, February 1977, pp. 154–157.

Trigg, Charles W. "Diagonally Magic Square Arrays." *The Arithmetic Teacher*, May 1973, pp. 386–388.

Williams, Gail Atneoson. "Understanding the Check of Nines." *The Arithmetic Teacher*, November 1979, pp. 54–55.

Not All Numbers Are Whole Numbers

Representing, Adding, and Subtracting Rational Numbers

■ This chapter is about representing, adding, and subtracting rational numbers. It begins with a brief, informal account of the development of fractional notation for naming rational numbers. This discussion is important because it emphasizes the relationship between the division expression $a \div b$ and the fraction expression $\frac{a}{b}$.

Specific strategies, models, and activities for intuitively developing key ideas are provided for both fractional and decimal notation. The key ideas are then used to develop systematic methods for comparing numbers, finding equivalent forms, adding and subtracting numbers, and expressing percents. Place-value ideas are stressed for decimal notation, and computation with decimals is compared to algorithms for whole numbers.

We've chosen to combine the discussions of fractions and decimals for two important reasons. First, although the introduction of decimal notation traditionally begins in the later elementary grades, wider acceptance of the metric system of measurement and the use of minicalculators for both business and household purposes require that decimal ideas be developed earlier. Therefore, we've arranged major decimal and fractional concepts in a way that illustrates how they may be presented together. Second, important relationships exist between the fractional and decimal systems for representing rational numbers. These relationships must be made explicit to ensure efficient learning. ■

Objectives for the child

1. Demonstrates the meaning of fractions and decimals by shading regions appropriately.
2. Demonstrates the meaning of fractions and decimals using a set of objects.
3. Identifies equivalent fractions or equivalent decimals.
4. Simplifies fractions.
5. Uses a rule to generate equivalent fractions and equivalent decimals.
6. Rewrites a mixed numeral in fractional form, and vice versa.
7. Finds fractional representations for decimals.
8. Orders rational numbers.
9. Matches sums and differences involving fractions and decimals to appropriate models.
10. Constructs number sentences for addition and subtraction with fractions and decimals.
11. Uses an algorithm to find sums and differences involving fractions and decimals.
12. Interprets story problems that call for addition and subtraction involving fractions and decimals.
13. Estimates the sum or difference of two numbers in decimal form.
14. Identifies a fraction as less than 1, equal to 1, or greater than 1, and uses this idea to order fractions.
15. Associates percents with physical representations.
16. Associates percents with decimals or fractions.

Teacher goals

1. Be able to demonstrate physical or pictorial models for fractions and decimals.
2. Be able to construct models that demonstrate equivalent-fraction ideas.
3. Be able to construct models that demonstrate ideas related to ordering rational numbers.
4. Be able to identify the interpretations for addition and subtraction of rational numbers.
5. Be able to devise problem situations that require children to choose appropriate operations.
6. Be able to construct models that demonstrate the steps in the computational procedures for addition and subtraction with fractions and decimals.
7. Be able to justify the steps in the computational procedures for addition and subtraction with fractions and decimals.
8. Be able to devise interesting practice activities that enable children to compute efficiently.
9. Be able to identify and to sequence student learning objectives that involve addition and subtraction with fractions and decimals.
10. Be able to construct activities that enable children to achieve student learning objectives.

Maggie divided two chocolate pies among three friends. How much did each friend get?

Larry has paid $200 on the stereo system that he bought for $500. What part of the selling price has he paid so far?

Three out of four people in the United States prefer colas to hot tea. What part of the population is that?

Not all numbers are whole numbers, and it's a good thing! Otherwise, solutions to problems like those above would not exist. The answers to these problems belong to a special set of "not-all-whole" numbers. The members of this special set are called *rational numbers*. No doubt you're familiar with many of these numbers, but you probably just refer to them as fractions. Actually, fractions are symbols used for naming numbers. Fortunately, every rational number can be named by a fraction of the form $\frac{a}{b}$ where a and b are integers and b is not 0. The array below actually contains many such names for every nonnegative rational number. In other words, you could determine whether a nonnegative number is rational by checking this list. (This chapter deals only with the nonnegative rational numbers, because negative rational numbers are not usually a part of the elementary school curriculum.)

Fractions of the form $\frac{a}{b}$ where a and b are whole numbers and b is not 0:

$$\frac{0}{1}, \frac{1}{1}, \frac{2}{1}, \frac{3}{1}, \frac{4}{1}, \frac{5}{1} \cdots$$

$$\frac{0}{2}, \frac{1}{2}, \frac{2}{2}, \frac{3}{2}, \frac{4}{2}, \frac{5}{2} \cdots$$

$$\frac{0}{3}, \frac{1}{3}, \frac{2}{3}, \frac{3}{3}, \frac{4}{3}, \frac{5}{3} \cdots$$

$$\cdots \cdots$$
$$\cdots \cdots$$
$$\cdots \cdots$$

If you study the array, you'll see that 1 is a rational number. So are the numbers 1, 2, 3, 4, and so on. Would you believe that $1\frac{1}{4}$ is rational also? How can you tell that these numbers are rational? Consider the number .75. Is this number rational? Of course it is, because .75 can be rewritten as $\frac{75}{100}$, which is in the list.

The need for rational numbers

The need for this very special set of numbers was felt early in the history of civilization, and as interactions and communications became increasingly complex, the need grew. Problems involving the measuring and surveying of land, the assessing of taxes, and the exchanging of goods led to the realization that the whole numbers were not complete enough to supply numerical answers to all of the problems calling for numerical solutions. Concrete situations illustrated that there must be numbers smaller than 1 but greater than 0, numbers between 1 and 2, numbers between 2 and 3, and so on. The problem was to identify these numbers and to find systematic ways of representing them.

After many centuries, the problem was formalized. Mathematicians realized that, although there were whole-number replacements for $\square$ in sentences such as:

$$28 \div 7 = \square \qquad\qquad 12 \div 6 = \square \qquad\qquad 15 \div 3 = \square$$

there were no whole-number replacements for $\square$ in sentences like these:

$$15 \div 4 = \square \qquad\qquad 17 \div 3 = \square \qquad\qquad 27 \div 6 = \square$$

To remedy this situation, mathematicians extended the set of whole numbers by defining some new numbers. These can be used to replace $\square$ in any sentence of the form $a \div b = \square$, where a and b are whole numbers and b is not 0. The replacement for $\square$ is written in the form $\frac{a}{b}$. This form is called a *fraction;* a is called the *numerator* of the fraction, and b is called the *denominator.* The numbers in the set designated by fractions are called *rational numbers.* With this extended set of numbers, it's possible to find replacements for $\square$ in number sentences like the ones given above, specifically:

$$15 \div 4 = \frac{15}{4} \qquad\qquad 17 \div 3 = \frac{17}{3} \qquad\qquad 27 \div 6 = \frac{27}{6}$$

It's easy to memorize that $17 \div 3$ is the same thing as $\frac{17}{3}$, but to be intellectually convinced is another matter. Let's illustrate what we've done graphically. We already know how to associate whole numbers with points on the number line, as shown below.

Our task now is to locate the points of the number line that are associated with the other numbers in the extended set. In Chapter 4, we agreed that one interpretation of the number sentence $6 \div 2 = 3$ is:

If 6 units are separated into 2 parts having the same number of units, then 3 is the number of units in each of the parts.

We can use a similar interpretation for $2 \div 3 = \square$; namely, if 2 units are separated into 3 parts having the same number of units, then $\square$ is the number of units in one of the parts. According to our definition, the replacement for $\square$ is $\frac{2}{3}$. We can relate this result to the number line as follows.

If we "cut" a line segment that is 2 units long into 3 smaller line segments having the same length, then according to our definition, one of the smaller segments will be $\frac{2}{3}$ unit long. We can use one of the smaller segments to find the point associated with $\frac{2}{3}$.

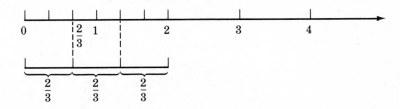

Did you get the point? Then you know we can locate $\frac{3}{4}$ by "cutting" a line segment 3 units long into 4 smaller segments. The smaller segments will each be $\frac{3}{4}$ unit long, since $3 \div 4 = \frac{3}{4}$. Now we can use one of the smaller segments to locate $\frac{3}{4}$ on the number line.

Using this procedure, we can demonstrate that $\frac{1}{2}$ and $\frac{2}{4}$ are associated with the same point.

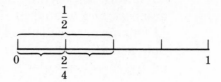

By extending this idea, we can determine that there is an infinite set of fractions associated with a point that represents a rational number.

$$\frac{1}{2} \longrightarrow \left\{\frac{1}{2}, \frac{2}{4}, \frac{3}{6}, \frac{4}{8}, \cdots\right\}$$

Now that we've reviewed the relationship between fractions and division, let's look at how children can use fractions to describe other important mathematical ideas: Fractions can be used for describing parts and for describing ratios.

■ Describing parts

We are emphasizing two kinds of situations in which a rational number is used to describe parts. The first situation involves assigning "1" to some unit that children call "the whole thing." Then, other rational numbers are defined in terms of that 1 unit. Here are some examples using rational numbers in this way.

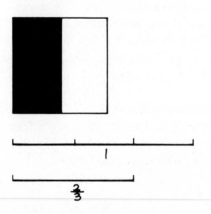

If the area of this square is 1 square unit, then the area of the shaded part is $\frac{1}{2}$ square unit.

If the length of the first line segment is 1 unit, then the length of the second segment is $\frac{2}{3}$ unit.

If a jar of beans is 1 unit, then these jars together indicate $1\frac{1}{2}$, or $\frac{3}{2}$ units.

Of course, the number 1 doesn't have to be assigned to attributes as tangible as length, area, and volume. We could assign the number 1 to the amount of time that it takes to get to Grandma's house—or to anything else that can be treated as a divisible quantity.

Children in the first and second grades can learn to use the fractions $\frac{1}{2}$ and $\frac{1}{3}$ to describe concrete situations, but, for most children, formal instruction begins in the third grade. Associating a fraction—thus a rational number—with a part of an object in order to answer the question "How much?" should be the first idea developed. As with other number ideas, children's appreciation of this concept grows from their experiences in examining, manipulating, and making observations about carefully constructed models. Specifically, the teacher must plan and sequence activities so that children have the following experiences.

1. Children give a sentence to compare an indicated number of parts with the total number of parts into which an object has been separated.

"1 part out of 4 parts having the same size and shape is shaded."

"2 parts out of 3 parts having the same size and shape are shaded."

2. Children associate a fraction with an indicated part and vice versa.

"$\frac{1}{2}$ is shaded." Or "Shade $\frac{1}{2}$."

"$\frac{2}{3}$ is shaded." "Shade $\frac{2}{3}$ of the triangle."

"$\frac{3}{4}$ is shaded." "Shade $\frac{3}{4}$ here."

3. Children select pictures that are appropriate models for a fraction.

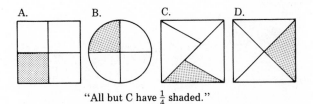

A.   B.   C.   D.

"All but C have $\frac{1}{4}$ shaded."

4. Children use fractions to develop objects that have been separated into parts having the same size but perhaps not the same shape. They should explain why they choose certain fractions for certain objects. This is a much higher level of activity than the preceding ones. The decision to present this kind of activity depends on the conceptual readiness of children; if they can handle it, it's a good experience. It can be supplemented

by paper-folding, tracing, or even cutting activities to allow children to verify their conclusions.

"$\frac{2}{4}$, or $\frac{1}{2}$, is shaded."   "$\frac{1}{2}$ is shaded."   "$\frac{1}{4}$ is shaded."

2 The second type of situation in which rational numbers are used to describe parts involves assigning the number 1 to a set and then using rational numbers to describe parts of the set. Here are some examples.

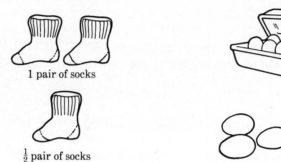

1 pair of socks

1 dozen eggs

$\frac{1}{2}$ pair of socks

$\frac{1}{4}$ dozen eggs

Children should have similar experiences in which they learn to associate a fraction with a part of a set of objects, as in the examples below.

1. Children find the number of smaller sets into which a set of objects has been separated. They determine whether the smaller sets have the same number of objects.

"The cupcakes have been separated into 4 packs. Each pack has the same number of cupcakes."

2. Children give a sentence that compares an indicated number of the smaller sets to the total number of smaller sets.

"3 packs of the 4 packs of cupcakes have chocolate icing."

3. Children associate a fraction with an indicated number of smaller sets.

"$\frac{3}{4}$ of the cupcakes have chocolate icing."

4. Children use sets of objects to construct models for fractions.

"This picture shows $\frac{3}{4}$ of the children holding hands."

"$\frac{17}{28}$ of our class rides the bus."

Whether the teacher is using sets of objects or a single object to illustrate fractional ideas, fractions for numbers smaller than 1 should be emphasized first. After children have had adequate experience finding fractions for numbers smaller than 1, their experiences should be extended so that they can learn to find fractions or mixed numerals to represent numbers greater than 1.

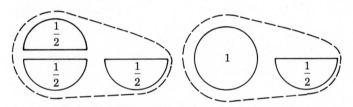

"This picture shows that $\frac{3}{2}$ is the same as $1 + \frac{1}{2}$. I can write $1\frac{1}{2}$ for short."

As you can see, rational numbers are very important when it comes to answering the question "How much?" and it's obvious that whole numbers do not fill the bill.

■ Describing ratios

A ratio is a set of numbers that is given in a certain order to describe the relationship that exists between two or more groups. Here are some examples.

There are 5 toes for every foot.
The ratio (5, 1) describes how toes and feet are related. The ratio (1, 5) describes how feet and toes are related.
For every 3 red marbles in a certain bag, there are 5 blue ones and 8 green ones.
The ratio (3, 5, 8) describes the color distribution of the marbles in the bag.

When a ratio involves only two numbers, we can write the ratio as a fraction. For example, we can write the ratio (1, 5) as $\frac{1}{5}$. So the ratio (1, 5) is associated with a rational number. This association is very important, because we can compute with numbers but not with ratios. To solve a practical problem that involves ratios, we simply find the associated rational numbers and compute. Here is an example to illustrate this idea.

500 toes. How many feet?

Ratio	Rational number	Computation
(1, 5)	$\frac{1}{5}$	$\frac{1}{5} \times 500 = 100$

So, there are 100 feet.

Of course, there are many other ways in which rational numbers are used. The description of parts and the description of ratios, however, are the ones of primary concern in the elementary-school mathematics program. Children should first learn to identify fractions (and thereby, rational numbers) with parts. After success with this concept, they can be introduced to the idea of ratio. By the sixth grade, they should be aware of the associations among a fraction, division, and a point on a number line.

In this section we've made a distinction between fractions and rational numbers: Fractions are symbols that represent rational numbers. (Some fractions may represent numbers that are not rational, such as $\frac{\sqrt{2}}{3}$.) In the next section, we will use "equivalent" to describe fractions that represent the same rational number. Thus, the statement

$$\frac{1}{2} = \frac{2}{4}$$

gives us two important pieces of information: "$\frac{1}{2}$" and "$\frac{2}{4}$" are equivalent fractions, and $\frac{1}{2}$ and $\frac{2}{4}$ are the same rational number. As a professional, you should be aware of the distinction. It is perhaps too subtle for children in elementary school to grasp, so don't worry them with it unless you're convinced it can help. In outlining strategies for teaching important concepts, we'll relax our language and refer to a rational number as a fraction when to do otherwise would only clutter our discussion with unnecessary verbiage.

→ Learning to use rational numbers

■ Exploratory experiences

When children are secure in associating the appropriate fraction or mixed numeral with a model, it's time to begin the development of equivalent-fraction ideas, order ideas, and addition and subtraction ideas. Children don't need to learn new interpretations for addition and subtraction with fractions, since the interpretations are simply an extension of those for addition and subtraction of whole numbers.

Equivalent fractions. Equivalent fractions are fractions that name the same number. The teacher can demonstrate this concept intuitively by folding paper. Have children make a chart to show how many equal-sized pieces the number of folds produces. Similar models can be used to illustrate order ideas and addition and subtraction ideas. Study the examples shown below, and try making up your own models to present these concepts intuitively.

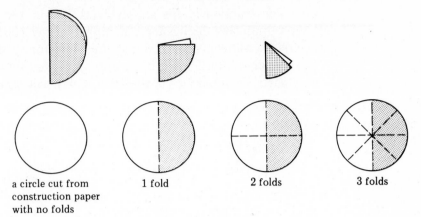

a circle cut from
construction paper
with no folds

1 fold

2 folds

3 folds

"These circles and folds show that $\frac{1}{2} = \frac{2}{4} = \frac{4}{8}$.
What else do they show?"

Order. The circles show that

$$\frac{1}{2} < 1 \qquad \frac{1}{4} < \frac{1}{2} \qquad \frac{1}{8} < \frac{1}{4}$$

So, in order from smallest to largest, we have

$$\frac{1}{8} \qquad \frac{1}{4} \qquad \frac{1}{2} \qquad 1$$

Addition and subtraction.

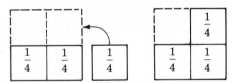

"This picture shows that $\frac{2}{4} + \frac{1}{4}$ is the
same as $\frac{3}{4}$."

"This picture shows
that $\frac{3}{4} - \frac{1}{4}$ is the
same as $\frac{2}{4}$."

■ A more structured approach for developing key concepts

Readiness activities built around the concepts developed so far in this section provide children with the background needed for developing more formal ways of working with fractions. In the beginning, the activities should be exploratory and not too structured. Materials and models should vary from activity to activity so that children see fractions used in many different settings. Eventually, though, you'll want to structure activities in such a way that key ideas are made apparent. You probably should use the same models and materials for these activities, because you'll want children to focus on the key ideas and not be distracted by new materials.

Using the same models and materials will also provide continuity as children move from intuitive ways of working with fractions to more systematic and abstract ways.

The fraction strips on Material Sheets 18A and 18B provide such a manipulative device for teaching the key concepts. By examining each strip, children should determine how many parts the 1 unit has been separated into. Then it must be emphasized that if a strip is separated into 4 parts, each part is $\frac{1}{4}$ of the strip. The whole strip is represented by $\frac{4}{4}$ or 1.

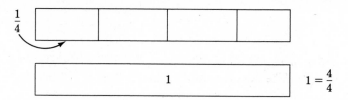

Repeated experiences with each strip reinforce these important concepts and also serve as a basis for related activities.

While children are matching a model with $\frac{1}{4}$ and deciding that $1 = \frac{4}{4}$, they can be led to discover that $\frac{1}{4} + \frac{3}{4} = 1$, or $\frac{2}{4} + \frac{2}{4} = 1$. Similar sentences can be constructed for each strip. Thus, a wealth of information can be drawn from deliberate examination of these models and careful questioning procedures.

The next six activities illustrate this more structured approach. Each activity is planned so that it teaches or reinforces a key idea, and each requires the use of the same model. Children are required to manipulate materials, collect information, and make specific observations. They're encouraged to keep a record of their observations, which will help them to learn new concepts. The activities are presented in the order that seems most effective for presentation to children.

Example 6-1   C-2
An activity in which children find sums when the two fractions have the same denominator

(The teacher should help children make this observation: to find the sum of fractions having the same denominators, add numerators to get the new numerator; keep the same denominator.)

Example: Fraction pair: $\frac{1}{6}, \frac{4}{6}$

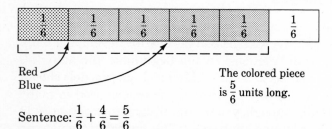

Red
Blue

The colored piece is $\frac{5}{6}$ units long.

Sentence: $\frac{1}{6} + \frac{4}{6} = \frac{5}{6}$

Fraction Pairs: $\frac{2}{4}, \frac{1}{4}$; $\frac{2}{5}, \frac{2}{5}$; $\frac{3}{10}, \frac{4}{10}$; $\frac{2}{8}, \frac{3}{8}$; and so on.

Materials: Charts A and B (see Material Sheets 18A and 18B).

Teacher directions: For each pair of fractions below, do these things:

1. Color a piece of a strip of the chart in red for the first fraction.
2. On the same strip, color a piece in blue for the second fraction. The colored pieces should be end to end.
3. Tell how long the colored strip is.
4. Write a number sentence to describe the colored strip.

Example 6-2

An activity in which children compare their fraction pieces to order fractions

(Children should be guided to make these observations: 1. If two fractions have the same denominators, then the one with the greater numerator has the greater value. 2. If two fractions have the same numerators, then the one with the greater denominator has the lesser value.)

Materials: Fraction pieces (see Material Sheets 18A and 18B).

Teacher directions: Paste 2 of Chart A and 2 of Chart B onto construction paper. Then cut out pieces for each of these fractions:

Example:

$$\frac{1}{2}, \frac{1}{3}, \frac{1}{4}, \frac{1}{5}, \frac{1}{6}, \frac{1}{8}, \frac{1}{10}, \frac{1}{12}, \frac{3}{8}, \frac{5}{8}, \frac{2}{3}, \frac{3}{4}, \frac{2}{5}, \frac{4}{5}, \frac{5}{6}, \frac{7}{8}, \frac{3}{10}, \frac{7}{10},$$

$$\frac{9}{10}, \frac{5}{12}, \frac{7}{12}, \frac{11}{12}$$

Label each of your pieces on the back side with the fraction. Keep your fraction pieces in an envelope labeled "Fraction Pieces."

Find a pair of pieces for each pair of fractions. Tell which of the fraction pieces is longer. Write a number sentence to describe the two pieces.

Fraction pair	*Fraction pieces*	*Sentence*
$\frac{5}{8}, \frac{3}{8}$	$\boxed{\frac{1}{8}}\ \boxed{\frac{1}{8}}\ \boxed{\frac{1}{8}}\ \boxed{\frac{1}{8}}\ \boxed{\frac{1}{8}}$ $\boxed{\frac{1}{8}}\ \boxed{\frac{1}{8}}\ \boxed{\frac{1}{8}}$	$\frac{5}{8} > \frac{3}{8}$
$\frac{1}{2}, \frac{1}{3}$	$\boxed{\qquad\frac{1}{2}\qquad}$ $\boxed{\ \frac{1}{3}\ }$	$\frac{1}{3} < \frac{1}{2}$

☞

Example 6-3

C-2

An activity in which children find sets of equivalent fractions

(Children learn that fractions in a given set have the same value, and they learn to find the simplest name for a set of equivalent fractions.)

Materials: Fraction Charts A and B (see Material Sheets 18A and 18B). Ruler or strip of paper (cut a transparency or file folder into strips).

Teacher directions: The $\frac{1}{2}$ piece is as long as the $\frac{2}{4}$ piece. So $\frac{1}{2}$ can be called $\frac{2}{4}$. Find other names for $\frac{1}{2}$. To do so, lay your ruler alongside the $\frac{1}{2}$ piece. Then check to see what other pieces have the same length. Do this for both of the charts. Then make a list of the fractions that you find.

New teacher directions: Since you see that the pieces for these fractions are the same length, you know that all of these fractions must have the same value (name the same number). So $\frac{2}{4}$ can be called $\frac{1}{2}$ or $\frac{3}{6}$ or $\frac{4}{8}$ or $\frac{5}{10}$, and so on. But $\frac{1}{2}$ is the simplest name for all of these fractions, since its piece is separated into the fewest parts. Find all of the fraction pieces that are as long as the piece for $\frac{2}{6}$. List these fractions, and give the fraction that is the simplest name for the set. (Etc.)

After children have performed several activities like the one in Example 6-3, they should have similar experiences making fraction pieces for fractions such as $\frac{4}{3}$. By comparing these pieces to the unit strip of either chart, they can discover this kind of relationship: $\frac{4}{3} = 1 + \frac{1}{3}$. At this point, it should be explained to children that the standard way for writing $1 + \frac{1}{3}$ is $1\frac{1}{3}$; that is, these expressions mean the same thing.

Whenever children can rename fractions, they are ready to add, subtract, and order fractions that have different denominators. Example 6-4 presents an activity in which children learn to rename fractions that have different denominators and then find their sum. Similar activities could be planned for subtraction and ordering. The development of these activities is left as an exercise for you (see Exercises 6-3 and 6-9).

Example 6-4

C-2

A concept-development activity for addition of fractions

Materials: Fraction pieces, Chart A (see Material Sheets 18A and 18B; also see the activity in Example 6-2).

Teacher directions: You can use your fraction pieces and the chart to help you add. Here's how: Let's find the sum of $\frac{1}{2}$ and $\frac{1}{3}$.

1. Use your fraction pieces for $\frac{1}{2}$ and $\frac{1}{3}$.
2. Start with the top strip on the chart. Check each strip until you find one that can be used to rename both $\frac{1}{2}$ and $\frac{1}{3}$.

3. Lay your fraction pieces end to end under this strip.
4. Write a number sentence to tell what you find out.

Sentence: $\frac{1}{2} + \frac{1}{3} = \frac{3}{6} + \frac{2}{6} = \frac{5}{6}$

Now find these sums:

$\frac{1}{4} + \frac{2}{3} \quad \frac{1}{6} + \frac{1}{3} \quad \frac{3}{8} + \frac{1}{6}$ (Etc.)

Try these sums:

$\frac{3}{4} + \frac{3}{8} \quad \frac{5}{12} + \frac{2}{3} \quad \frac{5}{6} + \frac{1}{2}$

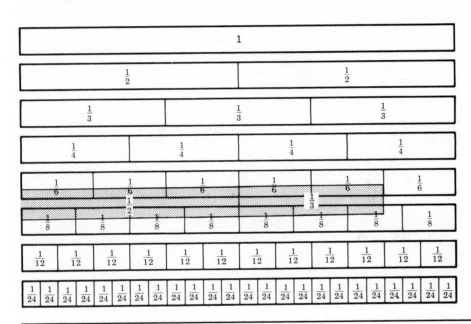

In the next activity, children discover a rule for finding equivalent fractions. We can state the rule in the following way:

$$\frac{a}{b} = \frac{a \times c}{b \times c} \text{ for all } a, b, \text{ and } c, \text{ where } b \text{ and } c \text{ are not } 0.$$

This rule means that multiplying the numerator and the denominator of a fraction by the same nonzero number produces a fraction that represents the same number. Using this rule, we can find families of fractions that represent the same number.

$$\frac{1}{2}, \quad \frac{1 \times 2}{2 \times 2} = \frac{2}{4}, \quad \frac{1 \times 3}{2 \times 3} = \frac{3}{6}, \quad \frac{1 \times 4}{2 \times 4} = \frac{4}{8}, \cdots$$

$$\downarrow \qquad\qquad \downarrow \qquad\qquad \downarrow \qquad\qquad \downarrow$$

$$\frac{1}{2}, \qquad \frac{2}{4}, \qquad \frac{3}{6}, \qquad \frac{4}{8}, \cdots$$

This rule is one of the most important fraction rules that children will learn. An understanding of this rule later will enable children to perform more complicated mathematical tasks.

☞

Example 6-5

C-2

A concept-development activity with equivalent fractions

Materials: Fraction pieces (see the activity in Example 6-2).

Teacher directions: Get out the fraction piece for $\frac{1}{2}$. Now multiply the numerator and the denominator of the fraction $\frac{1}{2}$ by 2. What fraction did you get? Get out the fraction piece for this fraction. How does it compare to the piece for $\frac{1}{2}$? Now multiply the numerator and the denominator of $\frac{1}{2}$ by 3. What fraction did you get? Get out the fraction piece for this fraction. How does it compare to the other two fraction pieces? (The same activity should be carried out with other fractions.)

Question: When you multiply the numerator and the denominator of a fraction by the same number, what happens? (Teacher indicates nonzero whole numbers.)

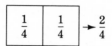

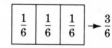

Example 6-6 is an activity that gives children the opportunity to make the associations among a fraction, division, and a point on a number line. Without these associations, the concept of rational numbers would be incomplete.

☞

Example 6-6

C-3

An activity in which children establish the relationship between division of whole numbers and fractional notation

Materials: String. Chart A (Material Sheet 18A).

Teacher directions: Measure and cut pieces of string to find fractions to replace the boxes. Tell what you discover.

$3 \div 4 = \square$ $1 \div 2 = \square$ $1 \div 4 = \square$

$3 \div 2 = \square$ $2 \div 3 = \square$ $1 \div 3 = \square$ Etc.

Example: Replace $\square$ with a fraction: $3 \div 4 = \square$.

1. Using the chart, measure and cut a piece of string 3 units long.
2. Separate the piece of string into 4 pieces that have the same length, by folding and cutting.
3. Use the chart to measure one of the pieces. You should find that it is $\frac{3}{4}$ unit long.

So, $3 \div 4 = \frac{3}{4}$.

Because it has taken us only a few pages to discuss six activities for teaching key concepts about fractions, you may be led to believe that teaching or learning these concepts requires very little time and that once children have successfully completed the activities, they'll have acquired the concepts. Nothing could be further from the truth. It will take most children a great deal of time to permanently fix these concepts and to learn to apply them in various situations. For this reason, the teacher must modify and expand these activities so that they involve a wider range of fractions and can be presented appropriately several times in a given school year and at different grade levels.

The following activity illustrates how the activity in Example 6-3 can be modified. Notice that both activities demonstrate the same concept, but with a different set of fractions. The set of fractions studied in the modified activity is important for developing the background necessary for learning decimal ideas. Notice also that the materials used in the activities are different. The activity in Example 6-7 uses a set of 100 objects, and children associate fractions with parts of the set. The fractions can be written in the ordinary way or, later, as decimals. Also, the concept of percent provides a logical extension of these ideas.

Example 6-7   C-2
A concept-development activity for fractions with denominators of 100

Materials: 100 beans. (Two or three children do this activity together.)

Teacher directions: Spread out the 100 beans. If you separated them into 100 sets having the same number, then each set would be $\frac{1}{100}$ of the beans. How many beans would be in each set? This information is in the chart. Make other sets as shown on the chart. Then complete the chart.

Make this many sets. Each set must have the same number of beans.	How many beans in each set?	What fraction goes with each set?
2 4 5 10 20 . . . 100	50 1	$\frac{50}{100}$ or $\frac{1}{2}$ $\frac{1}{100}$

New teacher directions: I'll give you the fractions; you make sets of beans. Complete the chart. Tell what you find out. The denominators of 10 and 100 are special. These fractions have forms that are much easier to write. Wait until we get to decimals and you'll find out how.

Let's summarize what we've done in this section. First, we presented some relatively unstructured activities that used a variety of models to introduce fractional ideas. These could be classified as "play activities." They're purposely informal, because we want children to enjoy and not be intimidated by their introduction to fractions. Second, we presented activities that are more structured and that employ the use of a single model. These are designed to sharpen children's awareness of key concepts. Third, we suggested expanding and modifying the more structured activities to involve more fractions and different materials. Our purpose here is to broaden children's abilities to apply the ideas that they've learned. At each of these levels of development, children use concrete materials and informal computing techniques. After successfully completing the activities at these three levels, children should give up some of the concrete models and learn to deal with fractional ideas in a more symbolic and systematic way. In the next section, we discuss the continued emphasis on meaning of fractions and the meaning of the operations of addition and subtraction.

Emphasizing meaning

To be retained and to be useful, concepts related to fractions and their addition and subtraction must be tied explicitly to situations that children will meet in real life. We have already discussed the use of fractions in a variety of contexts in this chapter and we discussed situations related to addition and subtraction in Chapter 3. To emphasize the importance of helping children make the connection between mathematics and reality, we will summarize these concepts in the tables below.

Symbol	Interpretation	Problem
$\frac{3}{4}$	Sets of objects	There are 4 cookies in the package; 3 of them are chocolate. What part of the cookies are chocolate?
	Area (square or circular regions, fraction charts)	A playground is separated into 4 parts about the same size; 3 of these parts are grassy. What part of the playground is grassy?
	Length (number lines)	We have 4 meters of red and white yarn for Valentine bows; 3 meters are red. What part of our yarn is red?
	Ratio	There are 3 red cars in the lot and 4 blue cars in the lot. What is the ratio of red to blue?
	Division	We have 3 small pizzas for 4 people. How much pizza will each one get?

Operation	Problem
$\frac{1}{2} + \frac{1}{4}$	I have $\frac{1}{2}$ of a candy bar. My sister gives me $\frac{1}{4}$ of a candy bar that she doesn't want. Now how much do I have?
	By the end of school, you have paid off $\frac{1}{2}$ of your car loan. After working this summer, you paid off another $\frac{1}{4}$ of the loan. How much of your car loan is paid?
$\frac{3}{4} - \frac{1}{2}$	There is $\frac{3}{4}$ of a cake left. Your mom said to leave $\frac{1}{2}$ of a cake for dinner. How much can you eat?
	Latisha has $\frac{3}{4}$ of a milkshake and you only have $\frac{1}{2}$ of a milkshake. How much more does Latisha have than you?

In the next section we discuss the development of procedures for simplifying, for ordering, and for addition and subtraction computation.

■ Developing algorithms

Obviously, children can't walk around for the rest of their lives with fraction pieces and beans in their pockets. Where would they put their combs? At some point, they must give up aids and learn to perform skills using pencil, paper, and patience. That's what this section is all about. We will discuss systematic procedures for simplifying expressions involving fractions, computing sums with fractions, computing differences with fractions, and ordering numbers represented by fractions or mixed numerals. As we mentioned in Chapter 3, such procedures are called *algorithms*. The choice of an algorithm will often depend on the numbers involved.

Among the abilities previously acquired by children, four will be of particular importance for learning these algorithms. These abilities are as follows.

1. The ability to find sums and differences using fractions that have like denominators.
2. The ability to order numbers represented by fractions that have like denominators.
3. The ability to find the common factors, the common multiples, the greatest common factor, and the least common multiple of two numbers.
4. The ability to find equivalent fractions using the *equivalent-fraction rule*. That is: Multiplying the numerator and denominator of a fraction by a nonzero whole number produces an equivalent fraction.

Along with these abilities, it will be helpful if children have experiences that convince them that the same properties that hold for addition and subtraction of whole numbers also hold for addition and subtraction of rational numbers.

Simplifying a fraction. The following algorithm may be used by children to simplify a fraction.

Simplify these two fractions:	$\dfrac{9}{12}$	$\dfrac{15}{12}$
1. Find the **greatest common factor** of the numerator and denominator.	GCF $(9, 12) = 3$	GCF $(15, 12) = 3$
2. Rewrite the numerator and denominator as product expressions involving the GCF.	$\dfrac{3 \times 3}{4 \times 3}$	$\dfrac{5 \times 3}{4 \times 3}$
3. Use the equivalent-fraction rule to rewrite the fraction.	$\dfrac{3}{4}$	$\dfrac{5}{4}$
4. Rewrite as a mixed numeral if possible and if desired.	not possible	$\dfrac{4}{4} + \dfrac{1}{4} =$ $1 + \dfrac{1}{4} = 1\dfrac{1}{4}$

The development in some children's material may replace steps 2 and 3 with division of the numerator and denominator by the greatest common factor.

Of course, sometimes the goal is not to simplify a fraction but rather to find an equivalent fraction for a given fraction or mixed numeral having a specified denominator. Notice that the steps in this algorithm can be performed "backward" to accomplish this goal.

For example, when children must find an equivalent fraction for $\frac{5}{8}$ having a denominator of 24, they can:

Write: $\dfrac{5}{8} = \dfrac{\square}{24}$.

Think: (What whole number, if any, times 8 is 24? $8 \times 3 = 24$.)

So: $\dfrac{5 \times 3}{8 \times 3} = \dfrac{15}{24}$.

Children should practice simplifying fractions and finding equivalent fractions. Most of this should be done in conjunction with other tasks, but it *is* necessary to provide drill exercises specifically for these skills. These don't have to be dull, however, as the next two activities show.

☞
Example 6-8
A skill-development activity for equivalent fractions

Materials: A teacher-made ditto sheet.

Teacher directions: Study each of the regions in this ditto sheet. Color those regions that are labeled with fractions whose simplest name is $\frac{1}{2}$. When you are finished, you'll be able to tell whether you're correct.

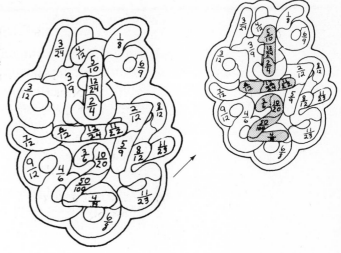

☞
Example 6-9
A skill-development activity with equivalent fractions

Materials: Index cards. Label 13 cards with the following fractions.

$$\frac{1}{2}, \frac{1}{3}, \frac{2}{3}, \frac{1}{4}, \frac{3}{4}, \frac{1}{5}, \frac{3}{5}, \frac{4}{5}, \frac{1}{8}, \frac{5}{8}, \frac{7}{8}, \frac{1}{10}, \frac{1}{12}$$

Then for each card, make three more labeled with equivalent fractions. You should have 52 cards in all.

Example: First make $\boxed{\dfrac{1}{2}}$; then make $\boxed{\dfrac{2}{4}}$, $\boxed{\dfrac{3}{6}}$,

$\boxed{\dfrac{5}{10}}$, and so on.

Directions: Deal five cards to each player. Stack the remainder of the cards face down in the middle of the table, and turn the top card face up. Play five hands of "Quivo." The object of the game is to get a book with three equivalent fractions and a book with two equivalent fractions. At each turn, a player can replace a card in his or her hand by drawing from the deck or from the discard pile. The player must discard the replaced card. The first player to get the required cards calls out "Quivo!" This player is the winner of the hand and gets two points. An extra point is given for each fraction that is in simplest form. The player with the highest score after five hands wins.

Ordering numbers represented by fractions. To decide which of two rational numbers is greater, these are the steps children could use:

Which is greater? $\frac{1}{8}, \frac{3}{10}$

1. Find the least common multiple of the denominators. This number will be called the *least common denominator* (LCD). It is the same as the least common multiple of the denominator.

LCM (8, 10) = 40

2. Find equivalent fractions having the least common denominator.

$$\frac{1 \times \square}{8 \times \square} = \frac{?}{40}$$

$\square$ must be 5, since $8 \times 5 = 40$.

$$\frac{1 \times 5}{8 \times 5} = \frac{5}{40}$$

$$\frac{3 \times \square}{10 \times \square} = \frac{?}{40}$$

$\square$ must be 4, since $10 \times 4 = 40$.

$$\frac{3 \times 4}{10 \times 4} = \frac{12}{40}$$

3. Now, order the numerators.

12 is greater than 5, so $\frac{12}{40}$ is greater than $\frac{5}{40}$. This means that $\frac{3}{10}$ is greater than $\frac{1}{8}$.

Example 6-10   S-3
A high-interest activity for reinforcing concepts related to ordering fractions

Materials: A set of cards labeled with the following fractions (these cards will be called "Fraction Fun Cards").

$$\frac{1}{2}, \frac{2}{4}, \frac{3}{6}, \frac{6}{12}, \frac{1}{3}, \frac{2}{6}, \frac{4}{12}, \frac{4}{12}, \frac{2}{3}, \frac{4}{6}, \frac{8}{12}, \frac{8}{12}, \frac{1}{4}, \frac{3}{12}, \frac{1}{4}, \frac{3}{12}, \frac{3}{4},$$

$$\frac{9}{12}, \frac{3}{4}, \frac{9}{12}, \frac{1}{6}, \frac{2}{12}, \frac{1}{6}, \frac{2}{12}, \frac{5}{6}, \frac{10}{12}, \frac{5}{6}, \frac{10}{12}, \frac{1}{12}, \frac{1}{12}, \frac{1}{12}, \frac{1}{12}, \frac{5}{12},$$

$$\frac{5}{12}, \frac{5}{12}, \frac{5}{12}, \frac{7}{12}, \frac{7}{12}, \frac{7}{12}, \frac{7}{12}, \frac{7}{12}, \frac{11}{12}, \frac{11}{12}, \frac{11}{12}, \frac{11}{12}$$

Example:

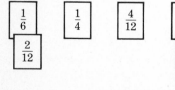 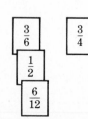

Directions: Play "The Race Is On!" The cards are shuffled, and each player is dealt 10 cards. At the signal, players begin to place their cards in a row so that corresponding numbers are arranged from smallest to largest. Whenever cards are labeled with equivalent fractions, they are placed in a pile.

The first player to complete a row of 10 cards correctly is the winner.

Addition and subtraction with fractions and mixed numerals. Are there algorithms for addition and subtraction when fractions are involved? Of course there are, and the steps for each are very much alike. Just as in the algorithm for ordering, the task in both of these algorithms is to rewrite fractions having unlike denominators as equivalent ones having like denominators. At that point, the computation is almost completed; all that remains is finding the sum or difference of the numerators. It should be pointed out very early to children that some fractions are greater than 1. With circles, we represented $1\frac{1}{2}$ by ① D, or $\frac{3}{2}$. Actually, for a formal procedure to change from a mixed numeral to a fraction, we need to use addition. Thus,

$$1\frac{1}{2} = 1 + \frac{1}{2} = \frac{2}{2} + \frac{1}{2} = \frac{3}{2}$$

Similarly, to express $2\frac{1}{3}$ as a fraction, we perform these steps:

$$\text{Rewrite: } 2\frac{1}{3} = 2 + \frac{1}{3} = \frac{6}{3} + \frac{1}{3} = \frac{7}{3}$$

Addition with fractions. The definition of addition of rational numbers is:

$$\frac{a}{c} + \frac{b}{c} = \frac{a+b}{c} \quad (c \neq 0)$$

In computing the sum of two rational numbers, children can be led through these steps:

Compute the sum.	$\frac{3}{4} + \frac{1}{3}$
1. Find the LCM of the denominators.	LCM $(4, 3) = 12$
2. Find equivalent fractions, using the least common multiple as the least common denominator of the fractions.	$\frac{3 \times 3}{4 \times 3} = \frac{9}{12}$ $\frac{1 \times 4}{3 \times 4} = \frac{4}{12}$
3. Use definition of addition.	$\frac{9}{12} + \frac{4}{12} = \frac{13}{12}$
4. Rewrite as a mixed numeral, if desired.	$\frac{13}{12} = \frac{12}{12} + \frac{1}{12} = 1 + \frac{1}{12} = 1\frac{1}{12}$

Subtraction with fractions. The definition of subtraction of rational numbers is:

$$\frac{a}{c} - \frac{b}{c} = \frac{a-b}{c} \quad (c \neq 0)$$

In computing, the steps for subtraction are the same as those for addition, except that in the third step, the *difference* of the numerators is computed rather than the *sum*.

$$\frac{5}{6} - \frac{3}{10} = \frac{5 \times 5}{6 \times 5} - \frac{3 \times 3}{10 \times 3}$$

$$= \frac{25}{30} - \frac{9}{30}$$

$$= \frac{16}{30} = \frac{8 \times 2}{15 \times 2} = \frac{8}{15}$$

Computation with mixed numerals. The algorithms for addition and subtraction of fractions can be extended to handle computations involving mixed numerals. Children can be convinced of the reasonableness of these extensions by examining simple concrete situations. These are the steps they would follow to compute the sum:

Compute the sum $\qquad\qquad\qquad 2\frac{2}{3} + 1\frac{3}{4}$

1. Find the sum of $\frac{2}{3}$ and $\frac{3}{4}$. Find the $\qquad 2\frac{2}{3} = 2\frac{8}{12}$
   sum of 2 and 1.

$$+ 1\frac{3}{4} = 1\frac{9}{12}$$
$$\overline{\qquad 3\frac{17}{12}}$$

2. Rewrite if the fraction part is $\qquad 3\frac{17}{12} = 3 + 1 + \frac{5}{12} = 4\frac{5}{12}$
   greater than 1.

The "extended" algorithm for subtraction can be difficult for children to learn because of the regrouping procedures that are sometimes required. When we studied division of whole numbers in Chapter 4, we found that there were number sentences that did not have whole-number solutions—for example, $2 \div 3 = \square$. Similar situations occur in the subtraction of positive rational numbers. For example, there is no positive whole number that can be used to replace $\square$ in the sentence $2 - 3 = \square$ and no positive rational number that can be used to replace $\square$ in the sentence $\frac{1}{3} - \frac{4}{5} = \square$. In the elementary curriculum, we usually avoid these situations until children are mathematically mature enough to deal with negative numbers. But a problem like $3 - 1\frac{1}{3}$ cannot be avoided. To compute differences such as $3 - 1\frac{1}{3}$, children must "regroup." In computing the difference, they would

Think: $3 = 2 + 1 = 2 + \frac{3}{3}$

Then $\qquad\qquad\qquad\qquad\qquad\qquad 3 \quad = 2\frac{3}{3}$

$\qquad$ compute $\frac{3}{3} - \frac{1}{3}$. $\qquad\qquad\qquad -1\frac{1}{3} = 1\frac{1}{3}$

$\qquad$ compute $2 - 1$. $\qquad\qquad\qquad\qquad\qquad\overline{\qquad 1\frac{2}{3}}$

In the discussion related to subtraction of whole numbers, the "equal addition" approach was mentioned. To compute

$$\begin{array}{r} 1846 \\ -\ 997 \\ \hline \end{array}$$

We can add 3 to both numbers and compute the much easier one:

$$\begin{array}{r} 1849 \\ -1000 \\ \hline 849 \end{array}$$

This procedure can also be used efficiently with subtraction involving mixed numerals. To compute

$$\begin{array}{r} 5 \\ -3\frac{7}{8} \\ \hline \end{array}$$

Add $\frac{1}{8}$ to both numbers and the computation becomes much simpler:

$$\begin{array}{r} 5\frac{1}{8} \\ -4 \\ \hline 1\frac{1}{8} \end{array}$$

Children should learn to perform computations in which no regrouping is required before being introduced to computations in which it is required. It will also help children if they review subtraction for whole numbers in which regrouping is required before attempting to compute similar differences with mixed numerals. Such a review will help them to realize that the process is not new and that only the units have changed.

The following activity can be used both as a concept-development activity and for reinforcement of ideas related to equivalent fractions, ordering, and addition and subtraction of rational numbers.

Example 6-11   S-3
A skill-development activity for ordering, adding, and subtracting fractions

Materials: Fraction Slide Rule (see Material Sheet 19).

Teacher directions:

1. Find equivalent fractions for the following

$$\frac{1}{2}, \frac{3}{4}, \frac{8}{12}, 1$$

2. Which is greater, $\frac{11}{12}$ or $\frac{5}{6}$?

3. Use your Fraction Slide Rule to find these sums.

$$\frac{1}{12} + \frac{5}{12} \qquad \frac{1}{3} + \frac{1}{4}$$

4. Can you use your slide rule to subtract? Find these differences.

$$\frac{7}{12} - \frac{2}{12} \qquad \frac{3}{4} - \frac{1}{6}$$

A better way to work with rational numbers

We are now in a position to discuss a very important system for naming rational numbers—the decimal system—and to present methods of teaching it. The decimal system is an extension of the place-value system we use for naming whole numbers (see Chapter 2). By "tacking on" to the whole-number system, we obtain a set of numerals for representing rational numbers. These are sometimes called *decimal fractions*, because they're related to the special subset of fractions made up of those fractions that have powers of 10 as denominators. For example, consider the following numeral.

$$5555.555$$

Using place-value ideas, we know that the fourth 5 to the left of the decimal point represents 5 thousands, the third 5 to the left of the decimal point represents 5 hundreds, the second 5 to the left of the decimal point represents 5 tens, and the first 5 to the left of the decimal point represents 5 ones. Notice that each time we move one place to the right, the value of the digit becomes one-tenth as great. So, to be consistent with the system for naming whole numbers, we want the first 5 to the right of the decimal to represent one-tenth of 5, and that's $\frac{5}{10}$. Similarly, we want the second 5 to the right of the decimal point to represent $\frac{5}{100}$, since $\frac{5}{100}$ is one tenth of $\frac{5}{10}$. Accordingly, the third 5 to the right of the decimal point must represent $\frac{5}{1000}$.

Here is a place-value chart for the extended place-value system. Notice that the ones place is in a central position and that the decimal point separates whole-number names from names for numbers less than 1. The decimal point is used to locate the ones place. When this is stressed, it provides the table with a type of symmetry about the ones place.

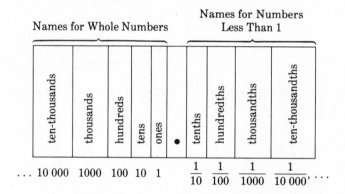

The chart shows that we can use fractions to express any decimal numeral in expanded form. Then, by computing the sum of the fractions, we can associate the decimal numeral with a single fraction or mixed numeral.

$$.25 = \frac{2}{10} + \frac{5}{100} = \frac{20}{100} + \frac{5}{100} = \frac{25}{100}$$

$$.06 = \frac{0}{10} + \frac{6}{100} = \frac{6}{100}$$

$$2.5 = \frac{2}{1} + \frac{5}{10} = 2\frac{5}{10}$$

To read a decimal numeral, we first read the numerals representing the whole number, if any, and then read the part representing a number smaller than 1. The word *and* is used to indicate the position of the decimal point in the numeral. Thus, we read

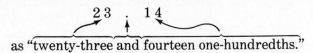

as "twenty-three and fourteen one-hundredths."

■ Exploratory experiences

Children should have experiences in discovering characteristics of our place-value system for naming whole numbers. The teacher can direct these activities by asking appropriate questions, such as the following.

What values do these two digits represent?

Use multiplication to tell how the values compare.

Use a fraction to tell how the values compare.

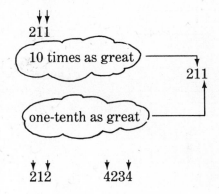

Answer the same questions for these numbers and the indicated digits.

$$\overset{\downarrow}{2}\overset{\downarrow}{1}2 \qquad \overset{\downarrow}{4}2\overset{\downarrow}{3}4$$

Suppose you know that these digits appear in a numeral, but you don't know the places they occupy. What is the size relation between the numbers represented by the 3s?

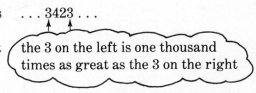

These kinds of experiences enable children to appreciate the way in which our place-value system for naming whole numbers is extended to include rational numbers. Realizing that the place value becomes 10 times greater as we move one digit to the left in a numeral makes it obvious that the only logical value for the first place to the right of the decimal point is tenths. After all, 1 is 10 times as great as $\frac{1}{10}$.

$$\_\ \_\ \_\ \cdot\ \_\ \_$$

Ones   "Tenths" is the only logical designation.

The values assigned to other decimal places are just as logical. Using equivalent fractions, we know that $\frac{1}{10} = \frac{10}{100}$. So we can see that ten $\frac{1}{100}$s make $\frac{1}{10}$. Thus, the second place to the right of the decimal point must represent $\frac{1}{100}$.

Using these ideas, children can develop a place-value chart like the one discussed in this section to define the decimal system of numeration. From the chart they can learn to associate a decimal with a sum involving fractions and, finally, with a single fraction or mixed numeral. These associations are important because children can use them in learning to compute and solve practical problems.

We don't want to give the impression that children's first introduction to decimal notation should immediately involve all of the ideas we've discussed in this section. These are subtle ideas, and children will require much experience before they can understand the entire scheme. Children's first exposure to decimal notation should occur when they understand the meaning of fractions and can associate fractional forms with models. Then certain decimals can be introduced as "other ways to describe parts."

Example: Here is 1 square region. The fraction $\frac{1}{10}$ tells us what part of this region is shaded. We can use the symbol ".1" to tell us the same thing. This symbol is called a *decimal fraction;* the point is called a *decimal point.*

In the second figure, each of the ten rows has been separated into ten small squares. Each square represents $\frac{1}{100}$ of the whole square and the symbol ".01" can be used for one of the small squares.

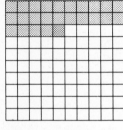

The fraction $\frac{25}{100}$ tells us how much of the square is shaded. We can also use a decimal fraction to tell us the same thing—namely, .25. We can see from the picture that $\frac{25}{100}$ is the same thing as $\frac{2}{10} + \frac{5}{100}$. This means that .25 is the same thing as $\frac{2}{10} + \frac{5}{100}$. The 2 must represent $\frac{2}{10}$ and the 5 must represent $\frac{5}{100}$.

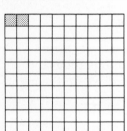

Of this region, $\frac{2}{100}$ is shaded. This fraction is the same as $\frac{0}{10} + \frac{2}{100}$. How can we write a decimal fraction to tell us what part is shaded?

The decimal fraction 2.5 describes the shaded part of these square regions. The 2 represents ones and the 5 represents tenths. It's easy to see that 2.5 is the same as $2\frac{5}{10}$.

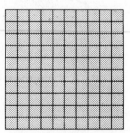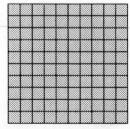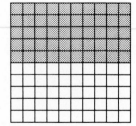

These exploratory experiences can be followed by activities that refine and extend concepts. Here are some examples.

Example 6-12
A concept-development activity for decimals

Materials: A set of cards or a ditto sheet showing square regions that have been partially shaded (see Material Sheet 20).

Teacher directions: Give a decimal fraction to tell how much is shaded. Then write a sentence to tell what you found out using place-value ideas.

Example:

Region	*Decimal*	*Sentence*
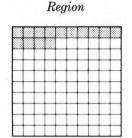	.14	"The decimal fraction .14 is the same as $\frac{1}{10} + \frac{4}{100}$."

Example 6-13
A skill-development activity for decimals

Materials: A ditto sheet showing decimal numerals (see Material Sheet 20).

Teacher directions: Shade a region to show the meaning of each decimal fraction.

.12　.41　.05　.5　.09　.1　.99

Example 6-14
A skill-development activity for decimals

Materials: A set of cards with pictures; each picture occupies a square region separated into 100 smaller square regions (see Material Sheet 20).

Teacher directions: This is a picture of Spiffy the space girl. Use decimals to tell some things about the picture. For example, how much of the picture shows Spiffy's hair? How much shows her lips? How much shows her eyes? How much of the picture shows the background? How much shows the border? Give each answer as a decimal fraction.

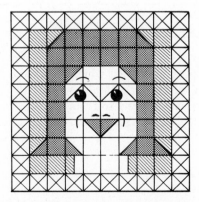

Order. Because children know from experiences with whole numbers that 25 is greater than 8, they often infer that .25 is greater than .8. The use of shaded regions to model each of the numbers represented can help undo this misconception—or, better yet, prevent it.

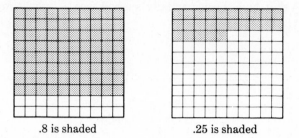

.8 is shaded   .25 is shaded

When children can find the larger of two numbers in fractional form, they should have no difficulty finding the larger of two numbers in decimal form. They can simply rewrite the decimals using fractions and then decide. Of course, it doesn't take long for children to become secure enough to omit the rewriting step and work only with decimals. Even here, most of the work can be done mentally. The essential thing is that children understand the meaning of the decimal numeral involved.

Which is greater? .8, .25

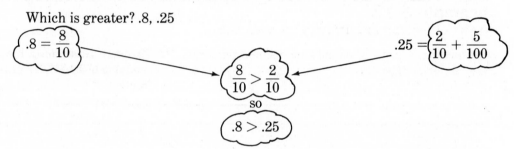

The following activities reinforce order ideas related to decimal notation.

Example 6-15
A skill-development activity for ordering decimals

S-3

Materials: A deck of 30 to 40 cards. The cards should be labeled with decimal numerals, which should be selected in terms of children's capabilities and the number concepts the teacher wishes to reinforce.

Teacher directions: Play "War" (see Chapter 3, Example 3-7). The game can be played by up to four players.

Example 6-16
A skill-development activity for decimals

S-3

Materials: A deck of 30 to 40 cards labeled with decimal numerals.

Directions: Play "In Between." Two, three, or four children can play. Each player draws two cards and announces a number *between* the two numbers indicated by the cards. If the player is correct, he or she keeps the cards; otherwise, the next player can try to name an appropriate number. If the next player names an appropriate number, then he or she can keep the cards as well as taking his or her regular turn. Otherwise, the player loses a turn, and the cards are passed on to the next player. Play continues accordingly. A pair of cards may be passed around once; if no player can name an appropriate number,

the cards are shuffled into the deck. The winner is the player with the most cards when all the cards in the deck are gone.

Example:

"The number .11 is between .1 and .2."

Example 6-17
A skill-development activity for ordering decimals

S-3

Materials: A deck of 30 to 40 cards labeled with decimal numerals.

Teacher directions: Play "The Race Is On!" (See Example 6-10, but ignore the provision for equivalent fractions.)

Example 6-18
A skill-development activity for decimals

S-3

Materials: A set of index cards with place-value positions indicated.

A deck of 9 cards labeled with the numerals from 1 to 9.

Teacher directions: Draw three cards from the deck of numeral cards. Then draw an index card. Use the numeral cards and the index card to name the largest possible number and the smallest possible number.

Example:

Index card   *Numeral cards*   *Answers*
.92 largest
.12 smallest

The calculator, of course, is highly appropriate for use with the topic of decimals. In the next activity, the children themselves will come to the conclusion that decimals are nicer to work with than fractions. They must be prepared for the game, however, by review of the concept that $\frac{1}{2}$ means $1 \div 2$ and by being made aware that this computation on the calculator yields 0.5. Then, with a few hints about ordering decimals, they're ready to go.

Example 6-19

S-3

Materials: Index cards with decimals and fractions. Calculators.

Teacher directions: Play "War." Each person turns up one of his or her cards. If you want to use the calculator to find a decimal fraction for your common fractions, you may. The card with the largest value wins the hand.

Example: Cards

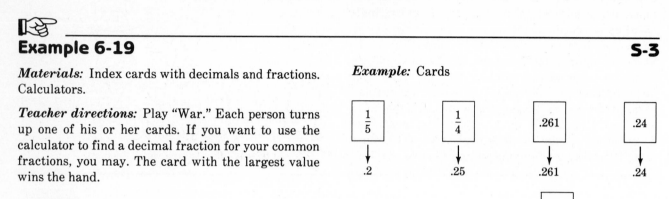

■ Addition and subtraction with decimals

The development of addition and subtraction with decimals should begin with demonstrations using models. Play-money representations are very good for children. As we used a dollar to represent 1 in earlier chapters, we can use a dime for .1 and a penny for .01. These representations are consistent with place-value ideas for whole numbers, and they're familiar. The main objective of the demonstrations must be to convince children that addition and subtraction algorithms for computing with decimals are almost the same as those for computing with whole numbers. At first, the demonstration should be simple and should not involve regrouping. Material Sheet 6 can be helpful in your lessons.

Addition

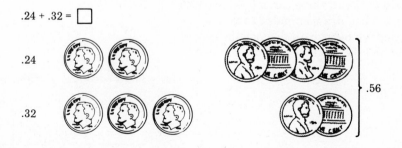

Subtraction

$.53 - .12 = \square$

$.53 - .12 = .41$

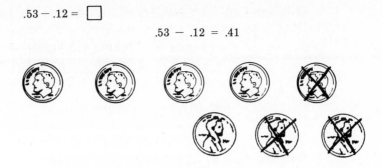

Problems like those above should be followed by ones in which exchanging 10 pennies for 1 dime or 1 dime for 10 pennies is required. Children who have a firm grasp of concepts related to computing sums and differences for whole numbers will have few difficulties. Again, the main idea is to compare the computational processes for decimals with those for whole numbers.

Addition

$.27 + .15 = \square$

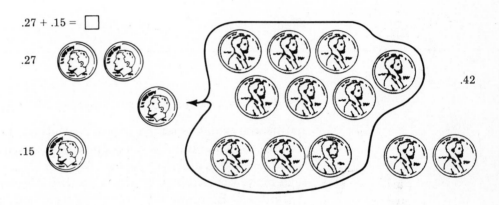

Subtraction

$.32 - .19 = \square$

$.32 - .19 = .13$

■ Addition and subtraction: Expanded form

Eventually, the transition must be made from these experiences to more abstract demonstrations that do not involve using play money. The first step is to find sums and differences using expanded notation. Using the expanded form will help children realize why it is convenient to align decimal points when computing.

Consider $2.32 + .85$. This expression can be written in a vertical computation form in which the numerals are expanded. Places of the same value are aligned, since we will compute sums of like place-value units.

$$\begin{array}{r} 2 \text{ ones} + 3 \text{ tenths} + 2 \text{ hundredths} \\ + 0 \text{ ones} + 8 \text{ tenths} + 5 \text{ hundredths} \\ \hline 2 \text{ ones} + 11 \text{ tenths} + 7 \text{ hundredths} \end{array}$$

or, $(2 \text{ ones} + 1 \text{ one}) + 1 \text{ tenth} + 7 \text{ hundredths}$
or, $3 \text{ ones} + 1 \text{ tenth} + 7 \text{ hundredths} = 3.17$

Now let's look at $7.24 - 5.08$.

$$\begin{array}{rl} 7.24 = & 7 \text{ ones} + 2 \text{ tenths} + 4 \text{ hundredths} \\ = & 7 \text{ ones} + 1 \text{ tenth} + 14 \text{ hundredths} \end{array}$$

Now we can subtract.

$$\begin{array}{r} 7 \text{ ones} + 1 \text{ tenth} + 14 \text{ hundredths} \\ - 5 \text{ ones} + 0 \text{ tenths} + 8 \text{ hundredths} \\ \hline 2 \text{ ones} + 1 \text{ tenth} + 6 \text{ hundredths} = 2.16 \end{array}$$

Finally, we have the shortened forms shown below.

$$\begin{array}{rr} 2.32 & 7.24 \\ + .85 & -5.08 \\ \hline 3.17 & 2.16 \end{array}$$

Children should also be exposed to other models, such as the linear models given on Material Sheet 19. These models, which are in the form of scales, are particularly important because children will eventually have to read many scales that are calibrated with decimals. The following activity is similar to Example 6-11. It can be used both as a concept-development activity and for reinforcement of ideas related to ordering and to addition and subtraction with decimals. If children compare their fraction slide rules with their decimal slide rules, they can discover number sentences such as

$$.5 = \frac{1}{2} \qquad \frac{3}{4} > .7 \quad \text{and} \quad \frac{1}{4} = .25$$

Example 6-20
A concept-development activity for decimals

C-3

Materials: Material Sheet 19. Use the scales to make an addition-and-subtraction slide rule.

Teacher directions:

1. Which is greater, .75 or .8?
2. Use the slide rule to find these sums:
   .24 + .52   .13 + .12   .61 + .32   .2 + .02

Questions: Do you see a pattern? If you wanted to compute .3 + .03, in which of these ways would you set up the computation?

$$
\begin{array}{r} .3 \\ +.03 \\ \hline \end{array}
\qquad
\begin{array}{r} .3 \\ +.03 \\ \hline \end{array}
\qquad
\begin{array}{r} .3 \\ +.03 \\ \hline \end{array}
$$

New teacher directions: Find a way to subtract with the slide rule. Put the bottom of your decimal slide rule against the top of your fraction slide rule.

1. Which is greater, .2 or $\frac{1}{3}$?
2. What is $.25 + \frac{1}{4}$?
3. What is $\frac{3}{4} - .5$?

Continue with questions about *order.*

The relationship between the decimal system of numeration and the U.S. monetary system should also be reinforced through instructional activities. These can be constructed using play money for dollars, dimes, and pennies to reinforce skills for representing, adding, and subtracting rational numbers.

Example 6-21
A concept-development activity for decimals

C-3

Materials: Play money—dollar bills, dimes, and pennies (Material Sheets 5 and 6).

Teacher directions: Use money to help you add and subtract.

$3 + .01 = \square$ $5 - 2.5 = \square$ $.40 - .29 = \square$ (Etc.)

➡ Percent

Our discussion of the teaching of rational numbers in the elementary grades would not be complete if we did not consider special ratios called *percents*. A percent is a ratio that indicates a certain number of "parts" for every "hundred parts." These ratios can be associated with rational numbers expressed either as fractions or as decimals. A percent indicates a fraction with a denominator of 100:

$$12\% = \frac{12}{100} = .12$$

To indicate what part of the figure below is shaded, we could use either percent, decimal, or fractional notation.

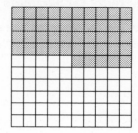

We can say that 45% is shaded, or that .45 is shaded, or that $\frac{45}{100} = \frac{9}{20}$ is shaded.

There are many ways in which percents are applied in our society—for example, in determining discounts, expressing interest rates, finding commissions, and describing statistical information. For this reason, it's important that elementary school children be given readiness experiences to prepare them for dealing with these ideas. The emphasis should be on developing meaning for symbols such as "50%," "25%," and so on, and on associating these symbols with appropriate fractions and decimals. These readiness experiences should be intuitive and involve objects or pictures, leaving the study of related formulas and calculations for the secondary school years. The following are sample activities for the elementary grades.

Example 6-22   C-3
A concept-development activity relating percent, decimals, and fractions

Materials: Chains, each containing 100 colored links (or colored chips, and so on).

Teacher directions: There are 100 links in each chain. Some of the links are red, some blue, some green, and some orange. Count the links in each chain. Use the information to fill in a chart for each chain. Red is already filled in for you.

Color	Number of links	Percent	Decimal	Fraction
Red	17	17%	.17	$\frac{17}{100}$
Blue				
Green				
Orange				

☞

Example 6-23

A skill-development activity for percent

Materials: Material Sheet 20. Colored pencils.

Teacher directions: Use your imagination and the information in this chart to make creative designs.

	Color		
Picture	*Red*	*Blue*	*Green*
1	42%	28%	30%
2	16%	47%	37%
3	25%	40%	35%
4	10%	20%	70%

➡ Estimation

At every level of development, children should be convinced of the reasonableness of the correct responses. Teachers who emphasize estimation can help children establish that arithmetic "makes sense."

To estimate solutions to computations involving addition and subtraction of rational numbers, children must first know something about the relative sizes of the numbers. Ordering the fractions $\frac{17}{18}$, $\frac{19}{19}$, and $\frac{21}{20}$ shouldn't require any computation. We can tell immediately that $\frac{17}{18} < 1$, $\frac{19}{19} = 1$, and $\frac{21}{20} > 1$. In general, we can decide without computation whether any nonnegative rational number is less than, equal to, or greater than 1 by comparing the numerator and denominator of the fraction.

Decisions about the relative size of rational numbers also can be made on the basis of rather informal methods. For example, we can decide that $\frac{3}{8} < \frac{15}{28}$ on the grounds that $\frac{3}{8}$ is a little less than $\frac{1}{2}$ and $\frac{15}{28}$ is a little more than $\frac{1}{2}$. Children should be encouraged to use reasoning procedures such as these, which not only are useful in arithmetic class but also are practical in the real world. The following chart contains a few of the estimation procedures that could be used by children to see whether their answers are reasonable.

Computation	*Estimation procedures*
$3\frac{1}{3} + 4\frac{1}{4}$	The answer must be greater than 7, since $3 + 4 = 7$. It must be less than 9, since $3\frac{1}{3} < 4$, $4\frac{1}{4} < 5$, and $4 + 5 = 9$. It must be less than 8, since $\frac{1}{3} + \frac{1}{4} < 1$.
$5\frac{1}{6} - 4\frac{5}{8}$	The answer must be less than 1, since $5\frac{1}{6} - 4\frac{1}{6} = 1$ and $\frac{5}{8} > \frac{1}{6}$.

———————————————————————

The computer corner

The next computer program we present will provide practice for very important skills that many school students fail to acquire. This program, called PARTS, uses a game format and requires students to build visual models for percents, fractions, and decimal fractions.

Here's how the program works. Students are given numbers in the form of decimal fractions, fractions, or percents. They are told to use each number to fill a "part" of a box that will hold one hundred □'s. They can only use tenths and hundredths to do the job, so the going gets sticky. The goal and the player's status in reaching the goal are constantly displayed on the screen. The program will not let the player exceed the goal. When the goal is achieved the player is congratulated and rewarded with another exercise. The fun goes on and on. In the meantime, students really learn relationships that exist among different number forms!

 ———————————————————————————————

Example 6-24   C-2/3
Building models for percents, fractions, and decimal fractions

Materials: The PARTS program on the MGB Diskette.

Teacher directions: In this computer game you are to fill boxes. Each box will hold 100 □'s. The program will give you a number in the form of a percent, decimal fraction, or fraction. Your job is to use tenths and hundredths to fill the amount indicated by the

given number. Be sure to watch the screen for information on how you are doing. Write number sentences to show relationships that you learn. Compare with your classmates.

Example: The computer gives the fraction $\frac{3}{20}$. You press "H" for hundredths and "T" for tenths to fill $\frac{3}{20}$ of the box.

Just right!

$$\frac{\frac{3}{20}}{\text{GOAL}} \quad \frac{.15}{\text{STATUS}}$$

You write $\frac{3}{20} = .15$

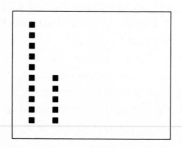

———

■ Computer tips

Computer programs can save the teacher a great deal of time when it comes to providing children with opportunities to practice concepts. To make this practice most effective, children should be encouraged to write number sentences for some of the relationships discovered while running the program.

Ideas for enrichment: Scientific notation

The topic of scientific notation can be challenging to many of the children in your classes. A combination of ideas about decimals and powers of ten are required to appreciate the topic, which provides insight into the mathematics of very large or very small numbers.

We already saw how exponents can be used efficiently as a kind of shorthand notation for large numbers in our base-ten numeration system.

For example, $10^6 = 10 \times 10 \times 10 \times 10 \times 10 \times 10$.

Remember the googol in Chapter 2? It is 10^{100}, or 1 followed by 100 0's. We can always express large numbers as a product of a small number and a power of ten.

$$287 = 2.87 \times 10 \times 10 \qquad\qquad = 2.87 \times 10^2$$
$$2876 = 2.876 \times 10 \times 10 \times 10 \quad = 2.876 \times 10^3$$
$$2876945 \qquad\qquad\qquad\qquad = 2.876945 \times 10^6$$

In scientific notation, the "small" number we use is always between 1 and 10; everyone using scientific notation will follow that same procedure. It is a powerful idea that can be introduced to some of your students early in their mathematical study.

Large numbers seem to be interesting to large numbers of children, not just academically talented ones. While some study independently, everyone can discuss the question, "How big is big?" or more specifically, "How big is a million?" (The last number is close to 3 million.) "Can you find a million of something? How about a million Cheerios?" Questions such as these can spark the interest of most children. Of course, they need to decide on their own verification procedures. An interesting reference book to have on hand is *The Lore of Large Numbers*, by Philip Davis.

In closing . . .

In this chapter, we have described methods for teaching children to represent, add, and subtract rational numbers in both fractional and decimal form, and we have associated these forms with percents. Children should learn the computational algorithms for fractions, but applications of these algorithms should be confined to relatively simple, realistic situations. Computing $\frac{2}{3} + \frac{3}{5}$ requires the same procedures as computing $\frac{17}{19} + \frac{18}{31}$, but the chances are remote of confronting a situation in which the latter sum is needed. Because of contemporary technology and because of the relationship between computation with decimals and computation with whole numbers, computation with decimals should be stressed.

Interpretations of problem situations did not receive much attention in this chapter because of the extensive coverage of this topic in Chapter 3. But whether computing with whole numbers or with rational numbers, whether using fractions or decimals, children should be able to interpret problem situations. Therefore, writing number sentences, choosing appropriate operations, and solving story problems should be integral parts of the development of rational-number ideas.

It's think-tank time

1. Determine which of the activities in this chapter involve higher-order thinking skills and could be used to promote problem solving by elementary school children.

2. Sketch figures (circles or rectangles) that would help a child to see that the following are true.

   a. $\frac{3}{5} = \frac{6}{10}$ b. $\frac{2}{3} = \frac{4}{6}$ c. $\frac{1}{2} = \frac{4}{8}$

   d. $.1 = .10$ e. $.5 = .50$

3. Show how you can use fraction pieces and Charts A and B (Material Sheets 18A and 18B) to order the following numbers from smallest to largest.

   $$\frac{5}{6} \quad \frac{1}{2} \quad \frac{1}{3} \quad \frac{2}{3} \quad \frac{3}{8} \quad \frac{5}{12}$$

4. Show how you can use the 10×10 squares (Material Sheet 20) to order the following numbers from smallest to largest.

   $$.1 \quad .01 \quad .17 \quad .7 \quad 1 \quad .71$$

5. You can order the rational numbers $\frac{17}{30}$ and $\frac{15}{32}$ by comparing each number to $\frac{1}{2}$; $\frac{17}{30} > \frac{1}{2}$ and $\frac{15}{32} < \frac{1}{2}$, so $\frac{17}{30} > \frac{15}{32}$. Use this type of reasoning to decide which number is greater in each of the following pairs.

   a. $\frac{7}{10}, \frac{5}{12}$ b. $\frac{13}{22}, \frac{11}{24}$

6. Which is greater, $\frac{15}{31}$ or $\frac{14}{29}$? You should have to compute only two products. Which two? Why?

7. Explain how you can convince children in a meaningful way that $3\frac{3}{8}$ is the same as $\frac{27}{8}$.

8. Give four different meanings for $\frac{2}{3}$. Sketch figures to illustrate.

9. Show how you can use fraction pieces and Charts A and B (Material Sheets 18A and 18B) to find solutions to the following.

   a. $\frac{2}{3} + \frac{1}{4}$ b. $\frac{3}{4} - \frac{1}{4}$ c. $\frac{3}{8} + \frac{1}{2}$

   d. $\frac{1}{2} - \frac{1}{3}$ e. $\frac{2}{5} + \frac{7}{10}$ f. $\frac{2}{3} - \frac{1}{6}$

   g. $\frac{1}{6} + \frac{3}{4}$ h. $\frac{5}{6} - \frac{3}{4}$

10. Use the fractions $\frac{3}{4}, \frac{5}{6}$, and $\frac{5}{8}$ to illustrate the associative property of addition.

11. Using fractions, give a specific example to illustrate the commutative property of addition.

12. Using fractions, give a specific example to show that subtraction is not commutative.

13. a. Complete the following computations.

    a. $2\frac{1}{3} + 3\frac{2}{5}$ b. $5\frac{5}{6} + 1\frac{7}{12}$ c. $5.46 + 1.5$

    d. $5\frac{1}{6} - 3\frac{5}{12}$ e. $.038 + .29$ f. $4.2 - 3.18$

    g. $4\frac{3}{4} - 2\frac{1}{6}$ h. $.037 - .02$

    b. Now, for each computation, list all of the concepts that a child must have attained before being able to perform the computation.

14. Tell whether the following sentences are true or false. If a sentence is false, tell why it is false.

    a. If a and b are nonzero whole numbers, $\frac{a}{a} + \frac{b}{b} = \frac{a+b}{a+b}$.

    b. $2\frac{3}{4} = \frac{2+3}{4}$ c. $\frac{5}{16} > \frac{5}{17}$ d. $\frac{7}{20} = \frac{6}{19}$

    e. $\frac{7}{16} > \frac{5}{16}$ f. $\frac{20}{41} < \frac{21}{43}$

    g. $\frac{1}{1\,000\,000} = .000001$

    h. $.472 = (4 \times 10) + (7 \times 100) + (2 \times 1000)$

    i. A major emphasis in the teaching of concepts about decimals should be place value.

    j. Decimal fractions and regular fractions have very little in common and therefore should be taught separately.

15. How could you use play money to demonstrate the following?

    a. $.1 = .10$ b. $.27 + .32 = .59$

    c. $1.58 + 2.43 = 4.01$ d. $.5 - .03 = .47$

    e. $5 - 4.21 = .79$ f. $.27 = .2 + .07$

16. Using a meter tape or a meterstick that is graduated in millimeters, verify each of

the following relationships. Explain the procedure you use for each.

a. $\frac{1}{4}$ is .25 or 25%.

b. $\frac{1}{3}$ is about .333 or 33.3%.

c. $\frac{1}{1000}$ is .001 or .1%.

17. Make a chart like the one for fractions (see pages 184-185) to illustrate the meaning of decimals and of addition and subtraction with decimals. (Be sure to include story problems.)

18. Express the following numbers in scientific notation.
    a. 5346 b. 678902 c. 100034 d. 23456789

19. Design a classroom activity that extends the teaching strategy used in the program PARTS on the MGB Diskette. Prepare cards that show relationships between fractions, decimal fractions, and percents. Have children shade regions in *hundredths* and *tenths* to illustrate the relationships.

➡ ━━━━━━━━━━━━━━━━━━━━━━━━━━━

Suggested readings

Ashlock, Robert B. "Introducing Decimal Fractions with the Meterstick." *The Arithmetic Teacher*, March 1976, pp. 201-206.

Behr, Merlyn J., Wachsmuth, Ipka, Post, Thomas R., and Lesh, Richard. "Order and Equivalence of Rational Numbers: A Clinical Teaching Experiment." *Journal for Research in Mathematics Education*, November 1984, pp. 323-341.

Brown, Christopher N. "Fractions on Grid Paper." *The Arithmetic Teacher*, January 1979, pp. 8-10.

Carpenter, Thomas P., Corbitt, Mary Kay, Kepner, Henry S., Jr., Lindquist, Mary Montgomery, and Reys, Robert E. "Decimals: Results and Implications from National Assessment." *The Arithmetic Teacher*, April 1981, pp. 34-37.

Coburn, Terrence Gordon. "The Effect of a Ratio Approach and a Region Approach on Equivalent Fractions and Addition/Subtraction for Pupils in Grade Four." (Doctoral dissertation, University of Michigan, 1973.) *Dissertation Abstracts International*, *34A* (1974):4688.

Cook, Nancy. "Fraction Bingo." *The Arithmetic Teacher*, March 1970, pp. 237-239.

Crouse, Richard. "Concentration." *The Arithmetic Teacher*, December 1975, p. 636.

Dana, Marcia, and Lindquist, Mary Montgomery. "From Halves to Hundredths." *The Arithmetic Teacher*, November 1979, pp. 4-8.

Faires, Dano Miller. "Computation with Decimal Fractions in the Sequence of Number Development." (Doctoral dissertation, Wayne State University, 1962.) *Dissertation Abstracts International*, *23* (1963):4183.

Glatzer, David J. "Teaching Percentages: Ideas and Suggestions." *The Arithmetic Teacher*, February 1984, pp. 24-26.

Hall, Robert Tremaine, Jr. "A Case Study of the Teaching of Fractions." (Doctoral dissertation, University of Georgia, 1983.) *Dissertation Abstracts International*, *44A* (1984):3311A.

Hauck, Eldon. "Concrete Materials for Teaching Percentage." *The Arithmetic Teacher*, December 1954, pp. 9-12.

Hollis, L. Y. "Teaching Rational Numbers—Primary Grades." *The Arithmetic Teacher*, February 1984, pp. 36-39.

Jacobson, Marilyn Hall. "Teaching Rational Numbers—Intermediate Grades." *The Arithmetic Teacher*, February 1984, pp. 40-42.

Kieren, Thomas. "Knowing Rational Numbers: Ideas and Symbols." In *Selected Issues in Mathematics Education*, National Society for the Study of Education and National Council of Teachers of Mathematics, 1980, pp. 69-82.

Lester, Frank K., Jr. "Teacher Education, Preparing Teachers to Teach Rational Numbers." *The Arithmetic Teacher*, February 1984, pp. 54-56.

Lichtenberg, Betty K., and Lichtenberg, Donovan R. "Decimals Deserve Distinction." In *Mathematics for the Middle Grades*, Forty-seventh Yearbook of the National Council of Teachers of Mathematics. Reston, Va.: The Council, 1982.

Payne, Joseph N. "Curricula Issues: Teaching Rational Numbers." *The Arithmetic Teacher*, February 1984, pp. 14-17.

Phillips, E. Ray, and Kane, Robert B. "Validating Learning Hierarchies for Sequencing Mathematical Tasks in Elementary School Mathematics." *Journal for Research in Mathematics Education*, May 1973, pp. 141-151.

Phillips, E. Ray, and Uprichard, A. Edward. "An Analysis of Rational Number Addition: A Validation Study." *Journal for Research in Mathematics Education*, January 1977, pp. 7-17.

Post, Thomas. "Fractions: Results and Implications from National Assessment." *The Arithmetic Teacher*, May 1981, pp. 26-31.

Reys, Robert, Carpenter, Thomas, Corbitt, Mary Kay, Kepner, Henry S., Jr., and Lindquist, Mary Montgomery. "Decimals: Results and Implications from National Assessment." *The Arithmetic Teacher*, April 1981, pp. 34-37.

Scott, Wayne R. "Fractions Taught by Folding Paper Strips." *The Arithmetic Teacher*, January 1981, pp. 18–21.

Skypek, Dora Helen. "Special Characteristics of Rational Numbers." *The Arithmetic Teacher*, February 1984, pp. 10–12.

Rode, Joann. "Make a Whole—A Game Using Simple Fractions." *The Arithmetic Teacher*, February 1972, pp. 116–118.

Usiskin, Zalman P. "The Future of Fractions." *The Arithmetic Teacher*, January 1979, pp. 18–20.

Van de Walle, John, and Thompson, Charles S. "Let's Do It, Fractions with Fraction Strips." *The Arithmetic Teacher*, December 1984, pp. 4–9.

Zawojewski, Judith. "Initial Decimal Concepts: Are They Really So Easy?" *The Arithmetic Teacher*, March 1983, pp. 52–56.

Security Is Knowing Why

Multiplying and Dividing Rational Numbers

■ *In this chapter, we discuss multiplication and division with rational numbers in both fractional and decimal forms, placing a great deal of emphasis on the interpretations for these operations. Initially, our discussion of the interpretations treats fractional forms. Our strategy is to explore situations that are as simple as possible, discover patterns, and draw generalizations that can be used to develop corresponding ideas about decimals.*

Because of the diversity of real-life problems whose solutions depend on multiplication and division of rational numbers, characterizing general interpretations is difficult. Nevertheless, general interpretations must be given since the understanding of such interpretations provides the link between solving problems and selecting appropriate algorithms. The development and justifica-

tion of algorithms given in this chapter are based on these interpretations. Formal mathematical justification is an integral part of the development of algorithms, but it is not an overriding one, since such justification is not sufficient to establish the connection between applications and algorithms.

Elementary school children are not expected to acquire all of the concepts developed in this chapter. More extensive coverage of these concepts and opportunities for developing computational speed and accuracy should be provided at the secondary level. But elementary teachers should be knowledgeable about the full range of ideas discussed here so that they can provide children with appropriate developmental experiences. ■

Objectives for the child

1. Matches products involving fractions and decimals with appropriate models.
2. Uses an algorithm to find products involving fractions and decimals.
3. Interprets story problems that call for multiplication involving fractions and decimals.
4. Matches quotients involving fractions and decimals to appropriate models.

5. Finds equivalent division expressions using the equivalent-fraction rule.
6. Uses an algorithm to find quotients involving fractions and decimals.
7. Interprets story problems that call for division involving fractions and decimals.
8. Estimates products and quotients involving decimals.

Teacher goals

1. Be able to identify the interpretations for multiplication and division of rational numbers.
2. Be able to devise problem situations for children that require them to choose appropriate operations.
3. Be able to construct models to demonstrate various interpretations of multiplication and division of rational numbers.
4. Be able to construct models to demonstrate steps in the computational procedures for multiplication and division with fractions and decimals.
5. Be able to justify the steps in computational procedures for multiplication and division with fractions and decimals.

6. Be able to devise interesting concept-development activities to reinforce the meaning of multiplication and division.
7. Be able to devise interesting practice activities that will lead to computational efficiency.
8. Be able to diagnose common computational errors and to provide appropriate remedial strategies.
9. Be able to identify and sequence student learning objectives for multiplication and division with fractions and decimals.
10. Be able to construct activities enabling children to achieve student learning objectives.

So now we've covered the use of fractions and decimal numerals to name rational numbers, and we've found fairly simple methods for finding sums and differences with either fractions or decimals. We've also discovered strategies for teaching these concepts, and you've probably come up with some good ideas of your own. Now we turn to multiplication and division with fractions and decimal numerals. To help children understand these ideas, we must pay particular attention to physical representations that are convincing. Explaining the computational procedures in a sensible way is a harder task for teachers.

Multiplication of rational numbers

■ The additive interpretation for multiplication with fractions

We can see that in expressions such as $3 \times \frac{1}{2}$ or $2 \times 3\frac{1}{2}$, the additive interpretation fits the ideas developed with whole-number multiplication. Thus, if we can think of 3×2 as $2 + 2 + 2$, then we can extend this to thinking of $3 \times \frac{1}{2}$ as $\frac{1}{2} + \frac{1}{2} + \frac{1}{2}$. Similarly, if 2×3 can be interpreted as $3 + 3$, then we can think of $2 \times 3\frac{1}{2}$ as $3\frac{1}{2} + 3\frac{1}{2}$. This explanation begins to present difficulties as soon as the first factor is not a whole number. How can we add 2 together $3\frac{1}{2}$ times in the expression $3\frac{1}{2} \times 2$, or what on earth would $1\frac{1}{2} \times 3\frac{1}{4}$ mean in terms of repeated addition?

Many of the ideas for multiplication with combinations of whole numbers, fractions, and mixed numerals can be illustrated by the use of circles and circle pieces. In each case, the first step is to associate the number 1 with the corresponding unit.

This determines the size of other fractions you'll use. Children should always be able to see the circle that represents 1.

Let's consider an illustration first for $\frac{1}{2} \times 3$.

You make $3 an hour. You only work $\frac{1}{2}$ hour. How much do you make?

Replace $\square$ in this sentence: $\frac{1}{2} \times 3 = \square$.

1. Find the pieces associated with 3.

2. Now find $\frac{1}{2}$ of this.

3. The total number of units associated with our result is $\frac{3}{2}$. So $\frac{1}{2} \times 3 = \frac{3}{2}$

and $\frac{3}{2} = 1\frac{1}{2}$.

Now, let's look at $\frac{1}{2} \times \frac{1}{2}$.

Replace □ in this sentence: $\frac{1}{2} \times \frac{1}{2} = \square$.

1. If this represents 1, ⟶

> You get paid $\frac{1}{2}$ dollar an hour. You only work for $\frac{1}{2}$ hour. How much do you make?

then the piece associated with $\frac{1}{2}$ is

2. Find $\frac{1}{2}$ of this. If we separate the $\frac{1}{2}$ piece into 2 parts having the same size, then 1 of the parts looks like this.

3. The number of units for this is $\frac{1}{4}$.

So, $\frac{1}{2} \times \frac{1}{2} = \frac{1}{4}$.

Another example we will illustrate is $3\frac{1}{2} \times 2$.

Replace □ in this sentence: $3\frac{1}{2} \times 2 = \square$.

1. Find the pieces associated with 2.

> You get paid 2 dollars an hour. You work $3\frac{1}{2}$ hours. How much money do you get?

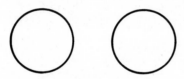

2. Now find $3\frac{1}{2}$ of these.

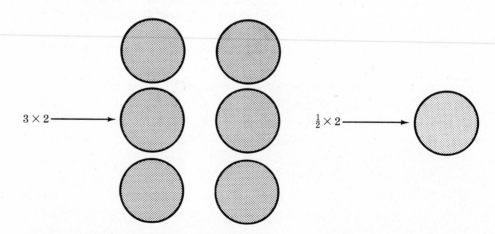

$3 \times 2 \longrightarrow$ $\frac{1}{2} \times 2 \longrightarrow$

3. There are 7 in all. So, $3\frac{1}{2} \times 2 = 7$.

Finally, this type of illustration can be used with even the messiest example, $1\frac{1}{2} \times 3\frac{1}{2}$.

Replace □ in this sentence: $1\frac{1}{2} \times 3\frac{1}{2} = \square$.

1. Find the pieces associated with $3\frac{1}{2}$.

You get paid $3\frac{1}{2}$ dollars an hour. You work $1\frac{1}{2}$ hours. How much do you get paid?

2. Find $1\frac{1}{2}$ of these.

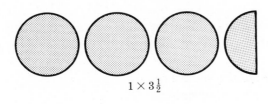

$1 \times 3\frac{1}{2}$

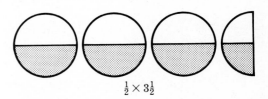

$\frac{1}{2} \times 3\frac{1}{2}$

3. Find the total number of units. Rearrange the pieces.

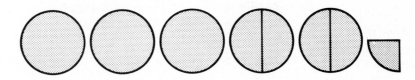

4. There are $5\frac{1}{4}$ units in all. So, $1\frac{1}{2} \times 3\frac{1}{2} = 5\frac{1}{4}$.

This example is a good place to refer to the distributive property of multiplication over addition. With whole numbers, there are four products involved in 23×45: 20×40, 20×5, 3×40, and 3×5. There are also four products involved in $1\frac{1}{2} \times 3\frac{1}{2}$: 1×3, $1 \times \frac{1}{2}$, $\frac{1}{2} \times 3$, and $\frac{1}{2} \times \frac{1}{2}$.

A situation that children can appreciate and that fits any of the possible combinations of numbers is the earning of money. In this sentence,

$$\bigcirc \times \triangle = \square$$

$\bigcirc$ tells how many hours you worked, $\triangle$ tells how much money you received per hour, and □ tells how much you were paid. Since we can work parts of an hour $\left(\frac{1}{2} \text{ or } 3\frac{1}{4}\right)$ and receive parts of a dollar $\left(\frac{1}{2} \text{ or } 3\frac{1}{4}\right)$ per hour, we have a sensible problem situation.

■ The area interpretation for multiplication with fractions

In Chapter 4, we used regions to illustrate multiplication of whole numbers; that is, we verified that $3 \times 5 = 15$ by constructing a rectangle 3 units by 5 units and observing that the area of the rectangle was 15 square units. We can also use regions to illustrate multiplication of rational numbers. This approach to multiplication is very valuable, because it clearly illustrates the reasonableness of the algorithm that children will learn.

Consider this square region. Suppose we say that the length of a side is 1 unit. Then the area is 1 square unit.

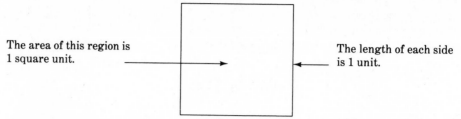

The area of this region is 1 square unit.

The length of each side is 1 unit.

To illustrate $\frac{1}{3} \times \frac{1}{2}$, we must find a rectangular region with a height of $\frac{1}{3}$ unit and a base of $\frac{1}{2}$ unit.

1. Separate the height of the square region into 3 parts having the same length. Each part is $\frac{1}{3}$ unit.

$\frac{1}{3}$ unit

2. Separate the base of the square region into 2 parts having the same length. Each part is $\frac{1}{2}$ unit.

$\frac{1}{3}$ unit

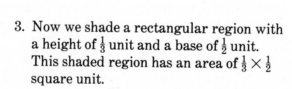

$\frac{1}{2}$ unit

3. Now we shade a rectangular region with a height of $\frac{1}{3}$ unit and a base of $\frac{1}{2}$ unit. This shaded region has an area of $\frac{1}{3} \times \frac{1}{2}$ square unit.

The area of this region is $\frac{1}{3} \times \frac{1}{2}$ square unit.

4. The dashed lines that we have drawn separate the square region into 6 regions, each having the same size and shape. Therefore, each region is $\frac{1}{6}$ of the square unit. Thus, $\frac{1}{3} \times \frac{1}{2} = \frac{1}{6}$.

The area of each of these is $\frac{1}{6}$ square unit.

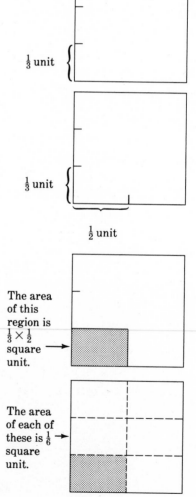

Now let's illustrate this expression: $\frac{2}{3} \times \frac{3}{5}$.

1. Separate the height of the square unit into 3 parts having the same length. Each part is $\frac{1}{3}$ unit.

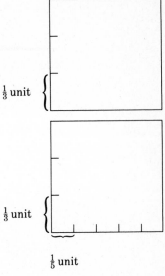

$\frac{1}{3}$ unit

2. Separate the base of the square unit into 5 parts having the same length. Each part is $\frac{1}{5}$ unit.

$\frac{1}{3}$ unit

$\frac{1}{5}$ unit

3. Now shade a rectangular region with a height of $\frac{2}{3}$ unit and a base of $\frac{3}{5}$ unit. This shaded region has an area of $\frac{2}{3} \times \frac{3}{5}$ square unit.

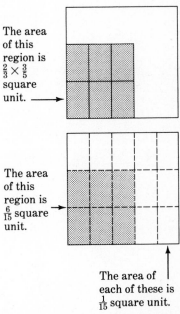

The area of this region is $\frac{2}{3} \times \frac{3}{5}$ square unit. →

4. The dashed lines separate the square region into 15 regions, each having the same size and shape. Therefore, each region is $\frac{1}{15}$ of the square unit, and 6 of these regions are shaded. Thus, $\frac{2}{3} \times \frac{3}{5} = \frac{6}{15}$.

The area of this region is $\frac{6}{15}$ square unit. →

The area of each of these is $\frac{1}{15}$ square unit.

Paper folding can be used to illustrate some of the simpler multiplication ideas. For example, to illustrate $\frac{1}{2} \times \frac{1}{3}$, we can fold the paper into thirds, then take $\frac{1}{2}$ of that by folding again. We've separated the paper into six pieces that are the same size, and we may be able to convince children that $\frac{1}{2} \times \frac{1}{3} = \frac{1}{6}$. Try these techniques with other multiplication expressions, such as $\frac{1}{2} \times \frac{1}{4}$ or $\frac{1}{3} \times \frac{1}{2}$. For such expressions as $1\frac{1}{2} \times 3\frac{1}{4}$, however, paper folding is not very effective at all!

■ Algorithms for multiplication with fractions

Children won't have to examine very many multiplication expressions and corresponding regions to discover the rule for multiplication with fractions. The definition of multiplication of rational numbers is:

$$\frac{a}{b} \times \frac{c}{d} = \frac{a \times c}{b \times d} \qquad \text{for all whole numbers } a, b, c, d, \text{ where } b \text{ and } d \text{ are not } 0.$$

So how do we deal with computations such as $3 \times \frac{1}{2}$ or $2\frac{1}{2} \times 3\frac{1}{4}$? One way is to express all of the numbers as fractions and then use the definition of multiplication.

$$\frac{3}{1} \times \frac{1}{2} = \frac{3 \times 1}{1 \times 2} = \frac{3}{2} = 1\frac{1}{2} \qquad 2\frac{1}{2} \times 3\frac{1}{4} = \frac{5}{2} \times \frac{13}{4} = \frac{5 \times 13}{2 \times 4} = \frac{65}{8} = 8\frac{1}{8}$$

We can also demonstrate these products with regions, as we did with multiplication of whole numbers. In doing so, we discover another method of computing—one that relies heavily on the distributive property. As an example, let's illustrate $2\frac{1}{3} \times 3\frac{1}{2}$.

A. Construct a rectangular region that is $2\frac{1}{3} \times 3\frac{1}{2}$. Separate it into parts. Associate the parts with product expressions.

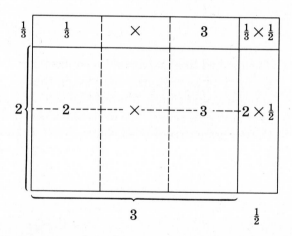

B. Express $2\frac{1}{3} \times 3\frac{1}{2}$ as a sum of the product expressions. Then compute the sum.

$$2\frac{1}{3} \times 3\frac{1}{2} = (2 \times 3) + \left(2 \times \frac{1}{2}\right) + \left(\frac{1}{3} \times 3\right) + \left(\frac{1}{3} \times \frac{1}{2}\right)$$
$$= 6 + \frac{2}{2} + \frac{3}{3} + \frac{1}{6}$$
$$= 6 + 1 + 1 + \frac{1}{6}$$
$$= 8 + \frac{1}{6}$$
$$= 8\frac{1}{6}$$

Both of these methods for finding products with mixed numerals give correct results. The first is probably the more frequently used. But the decision to rewrite the numerals in fractional form or to use the distributive property depends on the fractions involved and on which method is easier to use in a particular case. You wouldn't want to convert $60\frac{1}{2} \times 40\frac{1}{3}$ to fractional form to compute.

Obviously, children must be given many experiences that enable them to acquire these ideas about multiplication with fractions. Here are some examples of concept-development activities.

Example 7-1   C-3
A concept-development activity for multiplication with fractions

Materials: Construction paper circles and parts of circles (halves, fourths, and eighths).

Teacher directions: Use your circle pieces to show:

$$\frac{1}{2} \times \frac{1}{2} = \square \qquad \frac{1}{2} \times 2 = \square \qquad 2 \times 1\frac{1}{2} = \square \qquad \text{Etc.}$$

Example: $\frac{1}{2} \times \frac{1}{2} = \square$

Start with a piece for $\frac{1}{2}$.

Take $\frac{1}{2}$ of it.

How much?

So, $\frac{1}{2} \times \frac{1}{2} = \frac{1}{4}$

Example 7-2   C-3
A concept-development activity for multiplication with fractions

Materials: Teacher-made ditto showing multiplication exercises and square regions. A table for recording information.

Example:

Teacher directions: Use the square regions to help you find fractions to replace the boxes. Record the results in the table. See whether you can discover a way to multiply without using square regions.

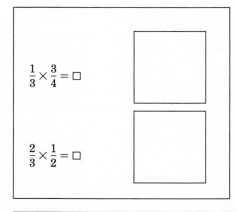

$$\frac{1}{3} \times \frac{3}{4} = \square$$

$$\frac{2}{3} \times \frac{1}{2} = \square$$

Factors	Product
$\frac{1}{3}, \frac{3}{4}$	
$\frac{2}{3}, \frac{1}{2}$	

☞
Example 7-3
C-3
A concept-development activity for problem solving with fractions

Materials: Index cards with multiplication sentences. Index cards with story problems. There should be two or more appropriate story problems for each number sentence and two inappropriate ones for each.

Teacher directions: Match the number sentences with the right story problem.

Example:

$$\frac{1}{3} \times \frac{3}{4} = \frac{3}{12} = \frac{1}{4}$$

Appropriate   *Inappropriate*

Sandy got $\frac{3}{4}$ of the pie that she and Alan baked. She ate $\frac{1}{3}$ of her piece. How much of the whole pie did she eat?

Sandy got $\frac{3}{4}$ of a pie. She gave her brother $\frac{1}{3}$ of a pie. How much did she have herself?

■ Multiplication with decimals

The area approach to multiplication with fractions can be used with decimals as well. To illustrate .3 × .7, we can find a height that is .3 unit and a width that is .7 unit. Then, a corresponding rectangle has an area of .3 × .7 square units.

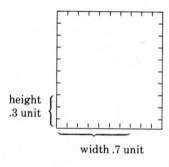

height
.3 unit

width .7 unit

The area is .3 × .7 square unit.

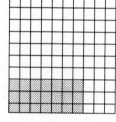

The area is .21 square unit.

When analyzing the area of the rectangle, we see that there are 21 of the small squares shaded and 100 small squares in the whole unit. So, the area is $\frac{21}{100}$ or .21 square units.

A good way to demonstrate multiplication with decimals is to use what children have already learned about fractions. Teachers can demonstrate the computation of the products below, for example, by rewriting the numerals as fractions and then computing accordingly.

$$.3 \times .7 = \underbrace{\frac{3}{10}} \times \underbrace{\frac{7}{10}} = \underbrace{\frac{3 \times 7}{10 \times 10}} = \frac{21}{100} = .21$$

$$(1, 1) \longrightarrow 2 \longrightarrow \text{(2 decimal places)}$$

$$.43 \times .2 = \frac{43}{100} \times \frac{2}{10} = \frac{43 \times 2}{100 \times 10} = \frac{86}{1000} = .086$$

$$\overbrace{10 \times 10} \quad \overbrace{10} \quad \overbrace{10 \times 10 \times 10}$$

$$(2, \quad 1) \longrightarrow 3 \longrightarrow \text{(3 decimal places)}$$

$$.07 \times .003 = \frac{7}{100} \times \frac{3}{1000} = \frac{7 \times 3}{100 \times 1000} = \frac{21}{100\,000} = .00021$$

$$\overbrace{10 \times 10} \quad \overbrace{10 \times 10 \times 10} \quad \overbrace{10 \times 10 \times 10 \times 10 \times 10}$$

$$(2, \quad 3) \longrightarrow 5 \longrightarrow \text{(5 decimal places)}$$

After many demonstrations or exercises such as these, children can discover the rule below. But the rule should never be introduced prematurely or mechanically. To find the product of two numbers represented by decimal numerals:

1. Multiply the numbers as though they were whole numbers.
2. Count the numbers of decimal places in the factors. The sum of these numbers will be the number of decimal places in the product.

The key word is *discover*. Children should not be allowed to memorize and apply the rule without prior concept development. The following activity is designed to provide practice for children once they have been convinced that the rule is reasonable.

☞

Example 7-4
A skill-development activity for multiplication with decimals

S-3

Materials: A teacher-made ditto sheet showing a "Math Mod" and a set of exercises involving multiplication with decimals.

Teacher directions: Compute the products. Color the regions that have the correct answers.

Example:

a. .3 × .4
b. .04 × .03
c. 1.2 × 6
d. .12 × .2
e. .2 × .15

f. .16 × 10
g. 8 × 1.4
h. .3 × .009
i. .9 × .01
j. 2.4 × 1.2

→ ▬▬▬▬▬▬▬▬▬▬▬▬▬▬

More about percent

In Chapter 6, we looked at the meaning of percent. Notice that the operations of addition and subtraction discussed in that chapter do not have application when it comes to percent. But now that we're discussing multiplication, percent is easily included.

Children in elementary school should not be rushed into computation when dealing with concepts that are easily misunderstood. Children do need almost all of their experiences to deal with meaning. The common fractions that you will first use with them are:

$$\frac{1}{2} \quad \frac{1}{4} \quad \frac{1}{5} \quad \frac{1}{10} \quad \frac{3}{4}$$

It is relatively easy to find equivalent fractions with denominators of 100 for these five fractions. Then it is certainly simple to write them as percents.

With an understanding of the place-value ideas involved, decimals are even simpler. So children need much experience with the *concepts* at this level.

Applications that deal with percent usually involve multiplication. The skills associated with multiplication of fractions and decimals are all that are needed to compute the results of many problems. Children will have experience with percent in their daily lives while in elementary school.

The topic is an excellent one for combining genuine applications and problem solving. The following activity includes all of these aspects and is a motivational example.

Example 7-5
A concept-development activity for percents

Materials: Newspaper advertisements and other advertising pamphlets that are mailed or delivered, some prepared by clipping and pasting on heavier paper. For example:

> TAPE SALE
>
> REGULAR PRICE
> $12
>
> NOW 50% OFF

Teacher directions: Lead them through a careful series of questions such as the following:

What was the price before the sale?
What does 50% off mean?
What fraction is this?
What is $\frac{1}{2}$ of 12?
So, what do we have to pay?

Now, look through these materials and find something that has a certain percent marked off. Write a story problem for your ad that has several questions to answer. Answer the questions on another page.

Let's discuss our problems.

Reciprocals

If the product of two numbers is 1, the numbers are said to be *reciprocals* of each other. For instance, since $\frac{2}{3} \times \frac{3}{2} = \frac{2 \times 3}{3 \times 2}$, or 1, $\frac{2}{3}$ is the reciprocal of $\frac{3}{2}$, and $\frac{3}{2}$ is the reciprocal of $\frac{2}{3}$. To find the reciprocal of $\frac{1}{2}$, we must find $\square$ in the sentence $\frac{1}{2} \times \square = 1$. The reciprocal here is 2, since $\frac{1}{2} \times \frac{2}{1} = \frac{1 \times 2}{2 \times 1}$, or 1. Notice that 0 has no reciprocal, because there is no number that we can multiply by 0 to get 1.

Children will quickly see that it's an easy task to find a reciprocal for a number, but the teacher must be sure that they can demonstrate *why* two numbers are reciprocals.

Division of rational numbers

■ Interpretations for division with fractions

In our discussion of division with whole numbers (Chapter 4), we noted that either of two interpretations could be assigned to a statement such as $12 \div 4 = \square$. The *subtractive* interpretation involves asking, for example, "How many sets with 4 objects can be made from a set having 12 objects?" The *distributive* interpretation involves asking "If we separate 12 objects into 4 sets with the same number of objects, then how many objects will be in each set?" Now we must consider interpretations for sentences such as $\frac{1}{2} \div \frac{1}{4} = \square$. The two interpretations we will consider are very much like those for whole numbers but are not nearly so intuitive.

One interpretation is similar to the subtractive interpretation for whole numbers. It involves asking, for example, "How many $\frac{1}{4}$s are in $\frac{1}{2}$?" The answer is 2. Children must think of how big the whole unit is before they can know what $\frac{1}{2}$ is.

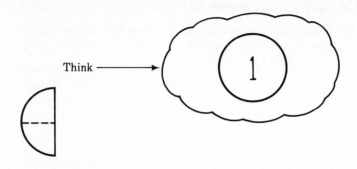

This shows that there are two $\frac{1}{4}$s in $\frac{1}{2}$. So $\frac{1}{2} \div \frac{1}{4} = 2$.

Notice that the sentence $\frac{1}{2} \div \frac{1}{4} = 2$ can be rewritten in terms of multiplication: $\frac{1}{2} = 2 \times \frac{1}{4}$.

The second interpretation, which is somewhat like the distributive interpretation for whole numbers, is more difficult to see intuitively. Consequently, you may want to postpone the use of models until you're sure children understand the first one.

How many quarters are in one-half dollar?

Here are some illustrations for each of the interpretations. We will use

$\left(\; 1 \;\right)$ as our unit.

How many one-half dollars can you get for $1\frac{1}{2}$ dollars?

$1\frac{1}{2} \div \frac{1}{2} = \dfrac{\square}{\square}$

$\left\{ \begin{array}{l} \text{How many } \frac{1}{2}\text{'s} \\[4pt] \text{are in } 1\frac{1}{2}? \\[4pt] \text{Answer:} \quad 3 \end{array} \right.$

How many groups of 3 quarters could you get for $3?

$3 \div \dfrac{3}{4} =$

$\left\{ \begin{array}{l} \text{How many } \frac{3}{4}\text{'s} \\[4pt] \text{are in } 3? \\[4pt] \text{Answer:} \quad 4 \end{array} \right.$

The following example will allow children to verify these interpretations with materials they can manipulate.

☞

Example 7-6

C-3

Concept-development activity for division with fractions

Materials: Construction paper circles and parts of circles (halves, fourths, and eighths).

Teacher directions: Use circle pieces to show:

$$1 \div \frac{1}{2} \quad \frac{3}{4} \div \frac{1}{4} \quad \frac{1}{2} \div \frac{1}{4} \quad 2 \div \frac{1}{2} \quad \frac{1}{2} \div 2$$

Draw a sketch of your work.

We can push intuition only so far, and some children have keener intuitions than others. At some stage, concrete or pictorial models become mind-boggling. So eventually we must provide symbolic means for systematically finding quotients. The primary reason we want children to examine the interpretations for simple cases is that these reflect problem types that occur in everyday life. Once children have learned systematic procedures for computing quotients of rational numbers, we want to be sure that they can apply them appropriately.

■ An algorithm for division with fractions

Let's look more carefully at the first interpretation of division with fractions with the object of developing an algorithm for dividing with fractions. So far, we have found that the following statements are true.

$$\frac{2}{4} \div \frac{1}{4} = \frac{2}{1} \qquad \frac{3}{2} \div \frac{1}{2} = \frac{3}{1} \qquad \frac{12}{4} \div \frac{3}{4} = \frac{4}{1}$$

Do you see a pattern? Notice that if the denominators of the two fractions are the same, the quotient of the fractions is the quotient of their numerators; that is,

$$\frac{a}{c} \div \frac{b}{c} = a \div b = \frac{a}{b} \qquad \text{for whole numbers } a, b, \text{ and } c \\ \text{where } b \text{ and } c \text{ are not } 0.$$

For example,

$$\frac{5}{9} \div \frac{3}{9} = 5 \div 3 = \frac{5}{3}$$

If the denominators are not the same, then we need to find equivalent fractions with common denominators. We're not necessarily interested in the *least* common denominator, however.

Suppose we have this quotient.
$$\frac{2}{3} \div \frac{4}{5}$$

We can find equivalent fractions with common denominators by multiplying the numerator and denominator of each fraction by the denominator of the other fraction.
$$\frac{2 \times 5}{3 \times 5} \div \frac{4 \times 3}{5 \times 3}$$

Then we can divide numerators. (There's no need to compute standard names for the denominators, since the product expressions show that they are the same.)

$$\frac{2 \times 5}{4 \times 3} = \frac{2 \times 5}{3 \times 4}$$

We can also rewrite this in another form.

$$\frac{2}{3} \times \frac{5}{4}$$

Now, if you compare this expression with the one we started with, you'll see that we've demonstrated a rule—namely, to compute the quotient of two fractions, invert the divisor and multiply. Of course, it would have been easy just to state the rule at the start. But it's important for you to see that an interpretation for division with fractions can motivate an algorithm. Then you'll be able to properly explain the relationship between the algorithm and the interpretation to children in the upper elementary grades. We can write the rule in general form.

The definition of division of rational numbers is:

$$\frac{a}{b} \div \frac{c}{d} = \frac{a}{b} \times \frac{d}{c} \qquad \text{for all whole numbers } a, b, c, \text{ and } d \\ \text{where } b, c, \text{ and } d \text{ are not } 0.$$

■ Extending the equivalent-fraction rule

The value of the equivalent-fraction rule has been demonstrated several times in this chapter, primarily in connection with developing reasonable ways to compute with fractions. So far, however, the rule has been verified only for whole numbers; that is,

$$\frac{a}{b} = \frac{a \times c}{b \times c} \qquad \text{where } a, b, \text{ and } c \text{ are whole numbers} \\ \text{and } b \text{ and } c \text{ are not } 0.$$

One of the nice things about this rule, however, is that it's also valid for rational numbers. It leads us, once more, to a justification of the rule for division of rational numbers. Mathematically, we say:

For all rational numbers p, q, and r, where q and r are not 0 $\qquad (p \div q) = (p \times r) \div (q \times r)$

Before we verify this extension of the equivalent-fraction rule, let's review the meaning of $\frac{a}{b}$.

In Chapter 6, we noted that the two expressions $a \div b$ and $\frac{a}{b}$ have the same meaning. So,

if $\frac{a}{b} = \frac{a \times c}{b \times c}$, then $a \div b = (a \times c) \div (b \times c)$.

The following is a specific instance of this statement.

If $\frac{6}{3} = \frac{6 \times 2}{3 \times 2} = \frac{12}{6}$, then $6 \div 3 = (6 \times 2) \div (3 \times 2) = 12 \div 6$.

Of course, $\frac{12}{6}$ *is* the same as $12 \div 6$. (Can you arrive at other expressions for $12 \div 6$?) Now let's look at an instance that involves rational numbers.

Suppose that $p = \frac{2}{3}$ and $q = \frac{3}{4}$. Consider:

$\dfrac{2}{3} \div \dfrac{3}{4} =$	We can multiply p and q by $\dfrac{4}{3}$, the reciprocal of q.
$\left(\dfrac{2}{3} \times \dfrac{4}{3}\right) \div \left(\dfrac{3}{4} \times \dfrac{4}{3}\right) =$	We can multiply p and q by $\dfrac{4}{3}$, the reciprocal of q.
$\left(\dfrac{2}{3} \times \dfrac{4}{3}\right) \div 1 =$	Now we're dividing by 1 (the product of q and its reciprocal, r).
$\dfrac{2}{3} \times \dfrac{4}{3}$	Dividing by 1 gives that number.

So, we can conclude that dividing by a nonzero number is the same as multiplying by its reciprocal.

Children should be presented with situations that require them to decide whether to multiply or divide. This kind of experience is just as important as learning of the computational procedures themselves. The following activity is designed to provide such experience.

Example 7-7
C-3

A concept-development activity for problem solving with fractions

Materials: Cards with story problems written on them. Problems should require either multiplication or division with fractions.

Teacher directions: Deal the cards so that every player has the same number of cards. When I say "Go," separate your cards into two stacks, one for multiplication and the other for division. The first one who finishes is the winner. You must be able to tell why you stacked the cards the way you did.

Example:

> $\frac{1}{2}$ of a cake.
> Joe ate $\frac{1}{4}$ of this piece.
> What part of the whole cake did Joe eat?

> $\frac{1}{3}$ meter of ribbon.
> We need $\frac{1}{6}$ meter to make a bow.
> How many bows can we make?

■ Division with decimals

The interpretations for sentences such as $.5 \div 3.5 = \square$, $.35 \div .5 = \square$, and so on are the same as those for sentences such as $\frac{5}{10} \div 3\frac{5}{10} = \square$ and $\frac{35}{100} \div \frac{5}{10} = \square$. That the interpretations are the same isn't surprising, since the sentences in the first group are equivalent to the sentences in the second group. The only difference between the two kinds of sentences is notation. This means that if children understand the interpretations for division of rational numbers using fractions, they can extend their understanding to decimal notation. The reverse is probably true also. The important thing to remember is that before children use either system of notation, they should know what the symbols represent and how they are related.

With the interpretations out of the way, the next step is to find meaningful ways of computing quotients with decimals. This task presupposes these important abilities:

1. the ability to compute products with decimals,
2. the ability to apply the equivalent-fraction rule in situations involving rational numbers, and
3. the ability to compute quotients of whole numbers.

■ Developing an algorithm

Let's examine how these abilities are related to computing quotients with decimals. To do so, we must consider several types of quotients that occur in everyday applications. First, we'll consider computations with a remainder of 0. (No zeros need to be annexed in the dividend.)

Type One. $42 \overline{).714}$ The divisor is a whole number.

Type Two. $.42 \overline{).714}$ The divisor is not a whole number.

Type One. The divisor is a whole number. Here are two computations.

$$2 \overline{\smash{).46}} = .23 \qquad 42 \overline{\smash{).714}} = .017$$

These computations are each related to multiplication sentences, so we can use what we know about multiplication to help us gather information about these quotients.

$$2 \times .23 = \frac{2}{1} \times \frac{23}{100} = \frac{2 \times 23}{1 \times 100} = \frac{46}{100} = .46$$

$$42 \times .017 = \frac{42}{1} \times \frac{17}{1000} = \frac{42 \times 17}{1 \times 1000} = \frac{714}{1000} = .714$$

In the first place, we know that the actual computation for multiplication with decimals is done with whole numbers. The decision about where to place the decimal point in the product is based on the number of decimal places in the factors, and this decision can be made after the computation is performed. Doesn't it seem reasonable, then, because of the relationship between division and multiplication, that actual computations for quotients like the ones above can also be carried out with whole numbers? If so, then we must determine how to locate the correct place for the decimal point once the computation has been completed. Again, we can look at the corresponding multiplication statements. Notice that since the first factor is a whole number, the product must have the same number of decimal places as the second factor. But in each case, the second factor is the same as the quotient in the original problem, and the product is the same as the dividend. Thus, we can make these two observations about computations of this type:

1. The actual computation can be performed with whole numbers.
2. The number of decimal places in the quotient will be the same as the number of places in the dividend. In other words, we can just move the decimal point straight up in the computation.

Type Two. The divisor is not a whole number. Consider the following examples:

$$4.2\,\overline{)71.4} \qquad .42\,\overline{).714} \qquad .042\,\overline{)7.14}$$

Here we must apply the equivalent-fraction rule to rewrite the indicated divisions as equivalent ones in which the divisors are whole numbers. This rewriting isn't difficult; in each instance, selecting the appropriate power of 10 is all that's required.

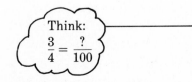

Multiply by 10.

Multiply by 100.

Multiply by 1000.

$71.4 \div 4.2 = (71.4 \times 10) \div (4.2 \times 10) = 714 \div 42$
If 4.2 is multiplied by 10, the result is a whole number.

$.714 \div .42 = (.714 \times 100) \div (.42 \times 100) = 71.4 \div 42$
If .42 is multiplied by 100, the result is a whole number.

$7.14 \div .042 = (7.14 \times 1000) \div (.042 \times 1000) = 7140 \div 42$
If .042 is multiplied by 1000, the result is a whole number.

Of course, it isn't necessary to write out these statements when computing, as long as the process is understood. Actually, this step can be performed mentally and indicated in the computation with arrows.

$$4.2.\,\overline{)71.4.} \qquad .42.\,\overline{).71.4} \qquad .042.\,\overline{)7.140.}$$

Once this step has been completed, the computations are the same as those of the first type and are performed accordingly. So, we can make these observations about computations for quotients of this type:

1. Multiply both numbers by a power of ten to make the divisor a whole number.
2. The actual computation can be performed with whole numbers.
3. The number of decimal places in the quotient will be the same as the number of decimal places in the new dividend. Move the decimal point straight up in the computation.

In some cases, zeros must be annexed to the dividend. Below are examples of this type.

Think:
$\frac{3}{4} = \frac{?}{100}$

$$4\,\overline{)3} \qquad .8\,\overline{).7} \qquad .042\,\overline{)7.14}$$

In the first case, the divisor is a whole number. In the other cases, the divisor is not a whole number, but we could easily rewrite these examples as equivalent ones having a whole-number divisor. But if we compute with the numbers given above, the results will not be very satisfying unless we want the quotients expressed only to the nearest whole number.

$$4 \overline{)\begin{array}{l} 0 \\ 3 \\ \underline{0} \\ 3 \end{array}}$$ The quotient, to the nearest whole number, is 1.

$$.8. \overline{)\begin{array}{l} 0 \\ .7. \\ \underline{0} \\ 7 \end{array}}$$ The quotient, to the nearest whole number, is 1.

To obtain more precise results, we must annex 0s to the right of the decimal point in the dividend.

Think:
Sure. $\dfrac{3}{4} = \dfrac{75}{100} = .75$

$$4 \overline{)\begin{array}{l} 0.75 \\ 3.00 \end{array}}$$ $$.8. \overline{)\begin{array}{l} 0.875 \\ .7.000 \end{array}}$$

This procedure gives us an equivalent decimal numeral for the quotient. Ideas related to rounding off whole numbers certainly apply with decimals. Thus, we can obtain quotients expressed to the nearest tenth, nearest hundredth, and so on. Now we're ready to state the algorithm systematically.

1. Multiply both numbers by a power of ten to make the divisor a whole number.
2. Annex 0s to the right of the decimal in the dividend so that the desired precision can be obtained.
3. Compute as if the numbers were whole numbers; round off to an appropriate place.
4. Place the decimal in the quotient so that it has the same number of places as the dividend.

Are you convinced that learning to compute quotients is a long and delicate process? Well, don't despair! Children aren't expected to complete the process in a single year; in fact, many children will not have completed it when they enter high school. For most, the junior high school years are the ones in which achievement in learning to compute quotients is greatest. But by the end of the sixth grade, most children should be able to deal with different types of quotients as long as the computation involved is simple. The teacher's goals should be to strengthen the prerequisite abilities and to develop sound computational procedures as the logical outcome of these prerequisites.

To accomplish these goals, instruction must begin early. As soon as children have learned to deal with models for decimal numerals, they should begin to consider such questions as, "How many 2s in 6?" "How many .2s in .6?" "How many .02s in .06?" "How many .02s in .3?" Here are some sample concept-development activities that illustrate these ideas.

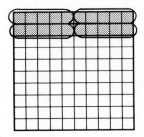

Example 7-8
A concept-development activity for division with decimals

Materials: Index cards showing models illustrating division with decimals. Cards with division sentences.

Teacher directions: Match each of the pictures with the right sentence.

Example:

$$.2 \div .05 = 4$$

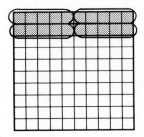

Example 7-9
A concept-development activity for multiplication and division with decimals

Materials: Index cards showing models that illustrate division with decimals. Cards with multiplication and division sentences.

Teacher directions: Match each picture with two sentences.

Example:

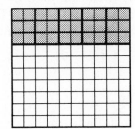

$$.3 \div .02 = 15$$

$$15 \times .02 = .3$$

Children also can use play money to find simple quotients. In these cases, the divisors should be whole numbers, and there should be exact decimal representations for quotients.

👉

Example 7-10
A concept-development activity for division with decimals

C-3

Materials: Play money—pennies, dimes, and dollars. Envelopes labeled with numerals. Some play money is placed in each envelope.

Teacher directions: Separate the money in each envelope into piles. Use the number on the envelope to tell you how many piles to make. Then write a division sentence to show the results. (You may have to exchange dollars for dimes or dimes for pennies.)

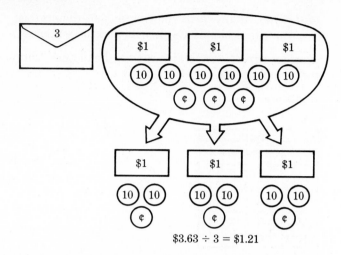

$$\$3.63 \div 3 = \$1.21$$

It's also a good strategy to have children verify that the distributive method that they have learned to use in computing quotients for whole numbers also works with decimals when the divisor is a whole number. Again, computations should be simple. The preceding activity demonstrated this with play money. Another model may involve a place-value chart and counters. To illustrate $.42 \div 3$, for example, we'd have children represent .42 as 4 tenths and 2 hundredths.

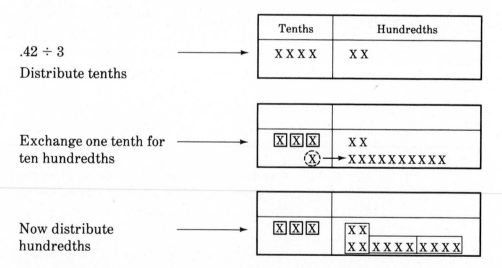

.42 ÷ 3
Distribute tenths

Exchange one tenth for ten hundredths

Now distribute hundredths

Our conclusion is that if we distribute .42 to 3 people, each would receive .14. Our model fits the computational procedure:

$$.42 \div 3 = .14$$

$$
\begin{array}{r}
.14 \\
3\overline{)\,.42} \\
\underline{3} \\
12 \\
\underline{12} \\
0
\end{array}
$$

While they are using the distributive method for computing simple quotients, children should be confronted with situations in which the remainder is not zero and should often be required to consider alternatives for dealing with the remainder.

$$
\begin{array}{r}
2.1 \\
5\,\overline{)10.7} \\
10\ 0 \\
\hline
7 \\
5 \\
\hline
2
\end{array}
$$

Notice that we have 2 tenths left over. What shall we do with them? Shall we represent them as hundredths and distribute again or shall we round off?

Other activities should require children to use multiplication and fraction ideas to write and solve division sentences. Children should be encouraged to find patterns and to make generalizations.

☞

Example 7-11   C-3
A concept-development activity for division with decimals and fractions

Materials: Models for decimals 1, .1, and .01. Cards labeled with simple division sentences involving fractions.

Example:

.01

.1

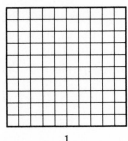

1

$$\frac{1}{2} \div \frac{1}{4} = 2$$

$$\frac{4}{5} \div \frac{1}{5} = 4$$

$$\frac{1}{5} \div \frac{1}{20} = 4$$

Teacher directions: How good are you at translating? Try your luck! Translate the sentences on the cards into sentences using decimals. Use the models to help you.

Example:

$$\frac{1}{2} \div \frac{1}{4} = 2 \longrightarrow .5 \div .25 = 2$$

$$\frac{4}{5} \div \frac{1}{5} = 4 \longrightarrow .8 \div .2 = 4$$

$$\frac{1}{5} \div \frac{1}{20} = 4 \longrightarrow .2 \div .05 = 4$$

After children have successfully completed many activities like the ones presented in the last few pages, they are ready to learn formal methods for computing and estimating quotients. A good deal of time will be required for developing these methods, so, as we said earlier, the emphasis should not be on difficult computation but rather on the logical development of ideas.

Estimating

■ Products

When estimating products involving fractions and decimals, we use techniques similar to those used for whole numbers. We can approximate numbers so that basic facts (or very simple products) can be used to find estimates. Estimating techniques involving decimals can be very helpful in determining where to place the decimal point.

Examples:

$$\begin{array}{r} 3.592468 \\ \times\ .93342568 \end{array} \longrightarrow \begin{array}{r} 3.6 \\ \times\ 1 \\ \hline 3.6 \end{array} \qquad \begin{array}{r} 24.653 \\ \times\ .49621 \end{array} \longrightarrow 24 \times \frac{1}{2} = 12$$

$$12\frac{1}{2} \times \frac{7}{16} \longrightarrow 12\frac{1}{2} \times \frac{1}{2} = 6\frac{1}{4} \qquad (\dots \text{a little more than } 6)$$

$$\begin{array}{r} 14.109 \\ \times\ 3.62 \end{array} \longrightarrow \begin{array}{r} 14 \\ \times\ 4 \\ \hline 56 \end{array} \qquad \begin{array}{r} .2167 \\ \times .312 \end{array} \longrightarrow \begin{array}{r} .2 \\ \times .3 \\ \hline .06 \end{array}$$

$$2\frac{7}{8} \times 4\frac{1}{3} \quad (\dots \text{between } 2 \times 4 \text{ and } 3 \times 5 \dots \text{close to } 3 \times 4)$$

For two *whole* numbers a and b, let's look at the product $a \times b$.

1. If a or b is 0, then $a \times b = 0$.
2. If a or b is 1, then the product is the other factor.
3. Otherwise, $a \times b > a$ and $a \times b > b$ (the product is greater than either one of them).

But with *rational* numbers, children must be carefully shown that the third statement does not hold. Sometimes the product is smaller than one factor and sometimes it's smaller than both of them. Generalizations must be carefully examined and the size of the two factors in comparison to the number 1 and to each other must be considered.

■ Quotients

There are several ways to estimate quotients; the decision to use one method rather than another depends on the numbers.

1. Use the equivalent-fraction rule. Multiply the divisor and the dividend by a power of 10 so that the equivalent quotient involves only whole numbers.

To estimate $.4 \div .052$, multiply by 1000. We get $400 \div 52 \approx 8$.

2. Compare the quotient to a familiar quotient that is close in value.

To estimate $5 \div 49$, compare it with $5 \div 50$. We get $5 \div 49 \approx 5 \div 50 = .1$.

$$1 \div \frac{11}{12} \rightarrow 1 \div 1 = 1$$

To estimate $1 \div 21$, compare it with $1 \div 20$. We get $1 \div 21 \approx 1 \div 20 = .05$.

$$\frac{1}{2} \div \frac{3}{16} \rightarrow \frac{1}{2} \div \frac{1}{4} = 2$$

For whole numbers we know that, if $b \neq 0$

1. and if $a = b$, then $a \div b = 1$.
2. and if $a > b$, then $a \div b > 1$.
3. and if $a < b$, then $a \div b < 1$.

These same statements are true for rational numbers, as well. If $\frac{a}{b}$ and $\frac{c}{d}$ are positive rational numbers and b and d are not 0, then

1. if $\frac{a}{b} = \frac{c}{d}$, then $\frac{a}{b} \div \frac{c}{d} = 1$.
2. if $\frac{a}{b} > \frac{c}{d}$, then $\frac{a}{b} \div \frac{c}{d} > 1$.
3. if $\frac{a}{b} < \frac{c}{d}$, then $\frac{a}{b} \div \frac{c}{d} < 1$.

For example,

1. $\frac{2}{3} \div \frac{4}{6} = 1$
2. $\frac{5}{8} \div \frac{9}{16} > 1$ (barely)
3. $\frac{7}{9} \div \frac{9}{10} < 1$ (again, just by a little bit)

The following strategies for estimating some products and quotients may be helpful:

Computation	*Estimation procedures*
$\frac{4}{5} \times \frac{1}{3}$	The product must be less than $\frac{1}{3}$, since $1 \times \frac{1}{3} = \frac{1}{3}$ and $\frac{4}{5} < 1$.
$1\frac{1}{4} \times \frac{1}{3}$	The product must be greater than $\frac{1}{3}$, since $1 \times \frac{1}{3} = \frac{1}{3}$ and $1\frac{1}{4} > 1$. The product must be less than $1\frac{1}{4}$, since $\frac{1}{3} < 1$.
$\frac{1}{3} \div \frac{1}{3}$	The quotient must be 1, since any number except 0, divided by itself, is 1.
$\frac{1}{3} \div \frac{1}{6}$	The quotient must be greater than 1, since $\frac{1}{6} < \frac{1}{3}$. I know that $\frac{1}{3} \div \frac{1}{3} = 1$, and now I'm dividing by a smaller number, so the result will be bigger.
$\frac{1}{3} \div \frac{5}{8}$	The quotient must be less than 1, since $\frac{5}{8} > \frac{1}{3}$. I know that $\frac{1}{3} \div \frac{1}{3} = 1$, but I'm dividing by a bigger number, so the result will be smaller.

Children should have many experiences using the methods described in this section to estimate products and quotients. In fact, they should be encouraged to avoid errors by making estimates before performing computations. Useful activities can be planned to help them acquire these skills; for example, a minicalculator can be used to determine the accuracy of estimates. Here are some examples of different kinds of activities.

Example 7-12   S-3
A skill-development activity for decimals

Materials: Index cards labeled with multiplication and division sentences involving decimals. Decimal points are not placed in products or quotients. Calculators.

Teacher directions: We're going to play a game called "Don't Miss the Point." First draw a card. Then use estimation to place a decimal point in the product. You aren't allowed to compute.

Sample cards:

$120.965 \times .498$ $= 6024057$

$30.3596289 \div 12.123$ $= 25043$

New teacher directions: Now use the calculators to check your decimal-point placement. Did you miss the point?

Play "Beat the Calc." with these cards. You get to use your heads; I have to use the calculator. You should beat me every time.

Example 7-13   S-3
A skill-development activity for decimals and fractions

Materials: Cards showing division expressions that involve fractions or decimals.

Example:

$\frac{1}{2} \div \frac{3}{4}$

$\frac{7}{8} \div \frac{7}{8}$

$.5 \div .2$

$.4 \div .79$

Teacher directions: Deal the cards so that everyone has an equal number. Sort your cards into three stacks, one having quotients less than 1, one stack having quotients equal to 1, and one stack having quotients greater than 1. You won't have time to compute!

When to teach what

The content in this chapter, for the most part, should be taught to children in the later grades. All of the activities are labeled "3" to tell you that this placement is most appropriate. Careful questioning procedures in second and third grades can reinforce the meaning of these operations. For example, you can ask "How can you fit three $\frac{1}{2}$s together?" or "How many $\frac{1}{4}$s in $\frac{1}{2}$?" without introducing the expressions $3 \times \frac{1}{2}$ or $\frac{1}{2} \div \frac{1}{4}$.

Ideas related to multiplication and division with fractions and decimals will be retaught, reviewed, and extended in middle school and in junior high school. Mastery of all of these concepts is not expected at the elementary level. So, for elementary school children, experiences that are sensible, reasonable, and convincing should be provided. These experiences should be preparatory to formal development of the system of rational numbers.

☆ ━━━━━━━━━━━━━━━━━━━━━━━━━━━━ ☆

The computer corner

How good are you at estimating? The next program we present will shape you up in no time! ESTIMATE generates exercises that require users to estimate the value of number expressions involving several arithmetic operations. Of course, the numbers are not all whole numbers—that would be too easy. After you enter your estimate for an exercise, you will be told whether or not you are within 10% of the actual value. Your job is to increase your *accuracy!*

To use this program effectively, students should be challenged to beat their records. A systematic procedure should be adopted in the classroom for identifying and rewarding students who show progress.

■ Computer tips

A few mathematical concepts and skills promise to be with us for some time. Estimation concepts and skills are among these. Learning to estimate well takes lots of practice. For a teacher to provide this practice would require hours that could be used for other teaching activities. Actually, it is more cost efficient for the computer to provide the practice. When practice activities are carried out by the computer, however, there should be challenge, record-keeping, and, most important, concepts worth practicing.

Ideas for enrichment: Terminating and repeating decimals

The topic of repeating decimals can be challenging for the more talented students in your classes. Results of their explorations should be extremely useful and valuable. Study of this topic will provide a great deal of insight into the system of rational numbers.

Any rational number can be written as a decimal. We saw earlier that for many of them, this is a pretty easy process. We can use the equivalent fraction rule and find an equivalent fraction with a denominator of 10 or 100 or any other power of ten. For example:

$$\frac{1}{2} = \frac{5}{10} = .5 \qquad \frac{3}{4} = \frac{75}{100} = .75 \qquad \frac{5}{8} = \frac{625}{1000} = .625$$

You can help your students with the last one by asking "What do you multiply 8 by to get 1000? (125.) So, multiply 5 by 125 also to get an equivalent fraction."

The procedure for fractions that aren't so "easy" involves interpreting a fraction as meaning division. Five-eighths means 5 divided by 8 and this results in (as you expected) .625. Those three were nice.

When you consider another kind of fraction, it is apparent that you will never find a power of ten that is a multiple of its denominator. When you use division with this kind of fraction, you will never get a remainder of 0. One-third is like this.

Decimals such as .5 and .625 are called "terminating decimals." Decimals such as .333 . . . are called "repeating decimals." All rational numbers can be represented by either terminating decimals or repeating decimals.

Now, for an independent study project. Have students find decimals for fractions, systematically starting with 2, 3, 4, 5, 6, and so on in the denominator. The students should find out which fractions are represented by terminating decimals and which ones are represented by repeating decimals. Try to lead the students to discover that terminating decimals represent fractions with denominators that are powers of 2 or powers of 5 or some combination of 2s and 5s. Other fractions will be represented by repeating decimals.

A preceding chapter on prime numbers can be useful in their exploration. It turns out that all we have to do to determine whether a fraction in lowest terms will be represented by a terminating decimal or a repeating decimal is to find the prime factorization of the denominator. If there are other prime factors besides 2s and 5s in the denominator, then it will be a repeating decimal. Otherwise, it is a terminating decimal.

The topic can lead to some interesting discoveries that will, in turn, lead to other questions and suggest many possible explorations. Many classmates can benefit from the reports given by other students.

$$\frac{1}{3} = \overline{?} \qquad 3\overline{\big)\,1.000}^{\,.333...}$$
$$\underline{9}$$
$$10$$
$$\underline{9}$$
$$10$$
$$\underline{9}$$
$$1...$$

In closing . . .

In this chapter, we have considered the operations of multiplication and division for rational numbers in both fractional and decimal forms. The various interpretations of these operations have been emphasized and have even been used to develop the algorithms. Teachers should remember that an understanding of multiplication and division with decimals will depend on how explicitly the connection is made between decimals and fractions.

An extensive amount of material has been covered; choices concerning how much can be expected of a particular group of children must be made by the teacher. The activities in this chapter, like those in other chapters, can be modified to fit many instructional objectives.

Chapters 3, 4, 6, and 7 have dealt with the four operations on *nonnegative* rational numbers and the associated computational procedures; the following chapters deal with applications related to these numbers. The development of negative rational numbers and irrational numbers has been omitted for several reasons. First, realistic applications for elementary school children are limited or, at best, contrived. Second, the cognitive development of children is such that these ideas are more appropriate at the secondary level. Finally, we believe that there are so many other appropriate and fruitful ideas to be developed in the elementary school that time spent on these numbers would be counterproductive.

It's think-tank time

1. Determine which of the activities in this chapter involve higher-order thinking skills and could be used to promote problem solving by elementary school children.

2. Use the fractions $\frac{1}{2}$, $\frac{3}{5}$, and $\frac{7}{8}$ to illustrate the associative property of multiplication.

3. Give a specific example, using fractions, to show that division is not associative.

4. Give a specific example, using fractions, to illustrate the commutative property of multiplication.

5. Sketch circle pieces or fraction strips to demonstrate the following multiplication sentences.

   a. $4 \times \frac{1}{3}$ b. $2 \times 1\frac{1}{2}$ c. $\frac{1}{2} \times 4$

   d. $2\frac{1}{2} \times 2$ e. $\frac{1}{2} \times \frac{2}{3}$ f. $1\frac{1}{2} \times 2\frac{1}{2}$

6. We used the distributive property of multiplication over addition of whole numbers to compute 23×45 by finding $(20 \times 40) + (20 \times 5) + (3 \times 40) + (3 \times 5)$. Use the distributive property to compute $1\frac{1}{2} \times 2\frac{2}{5}$.

7. Use the area approach to demonstrate $\frac{2}{5} \times \frac{2}{3}$.

8. To find $6 \div 2 = \Box$, children may think "What must I multiply 2 by to get 6?" or "$6 = 2 \times$ what number?" To find $\frac{4}{15} \div \frac{2}{3} = \Box$, we can think

   $$\text{``} \frac{4}{15} = \frac{2}{3} \times \boxed{\frac{2}{5}}$$
   ← since 2×2 is 4 and
   ← since 3×5 is 15."

   Use this approach to find $\frac{9}{10} \div \frac{3}{5} = \Box$, and explain your procedure.

9. You could use a number line to find $1\frac{1}{2} \div \frac{3}{4}$, as follows.

   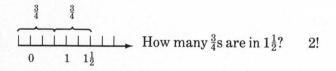
   How many $\frac{3}{4}$s are in $1\frac{1}{2}$?   2!

   Draw a number line to illustrate $2 \div \frac{2}{3}$.

10. The quotient $\frac{1}{2} \div 4$ can be demonstrated with circle pieces in this way.

       Separate into 4 parts of equal size.

    Since there are 4 of these parts in $\frac{1}{2}$, there are 8 in 1 whole unit. Thus, each part is $\frac{1}{8}$. So $\frac{1}{2} \div 4 = \frac{1}{8}$. Use circle pieces to demonstrate these quotients.

    a. $\frac{2}{3} \div 2$ b. $\frac{1}{4} \div 2$ c. $\frac{3}{4} \div 3$ d. $\frac{3}{4} \div 2$

11. Explain how you can correctly place the decimal point in the following examples by estimating (without any computation or rules).
    a. $36.8082 \div 2.34 = 015730$
    b. $140.2 \times 2.010 = 281802$

12. Explain why $\frac{5}{7} \div \frac{3}{4}$ is the same as $\frac{5}{7} \times \frac{4}{3}$, using methods discussed in this chapter.

13. How could you convince a child that $.3 \times .6 = .18$ and that $.2 \times .004 = .0008$?

14. Write two story problems for each of the following:
    a. multiplication of fractions
    b. multiplication of decimals
    c. division of fractions
    d. division of decimals
    Share with others in your class to collect an item bank.

15. Which of these fractions can be represented by terminating decimals and which can be represented by repeating decimals?

    a. $\frac{1}{15}$ b. $\frac{5}{12}$ c. $\frac{7}{64}$ d. $\frac{37}{15625}$

16. Design a record-keeping system for the program ESTIMATE, so that the students are aware of their daily and weekly improvements. Also create a reinforcement method for the students who increase their accuracy, and a *keep-trying* message for those who don't.

→ Suggested readings

Bergen, Patricia M. "Action Research On Division of Fractions." *The Arithmetic Teacher*, April 1966, pp. 293–295.

Bidwell, James King. "A Comparative Study of the Learning Structures of Three Algorithms for the Division of Fractional Numbers." (Doctoral dissertation, University of Michigan, 1968.) *Dissertation Abstracts International*, *229A*(1968):830.

Bruni, James V., and Silverman, Helene J. "Using Rectangles and Squares to Develop Fractions." *The Arithmetic Teacher*, February 1977, pp. 96–102.

Dilly, Clyde A., and Rucker, Walter E. "Division with Common and Decimal Fractional Numbers." *The Arithmetic Teacher*, May 1970, pp. 438–441.

Ettline, J. Fred. "A Uniform Approach to Fractions." *The Arithmetic Teacher*, March 1985, pp. 42–44.

Firl, Donald H. "Fractions, Decimals, and Their Futures." *The Arithmetic Teacher*, March 1977, pp. 238–240.

Flournoy, Frances. "A Consideration of Pupils' Success with Two Methods of Placing the Decimal Point in the Quotient." *School Science and Mathematics*, June 1959, pp. 445–455.

Green, George F., Jr. "A Model for Teaching Multiplication of Fractional Numbers." *The Arithmetic Teacher*, January 1973, pp. 5–9.

Hales, Barbara Budzynski, and Nelson, Marvin N. "Dividing Fractions with Fraction Wheels." *The Arithmetic Teacher*, November 1970, pp. 5–9.

Heddens, James W., and Hynes, Michael. "Division of Fractional Numbers." *The Arithmetic Teacher*, February 1969, pp. 99–103.

Karau, Earl A. "Arithmetic Football." *The Arithmetic Teacher*, November 1956, pp. 212–213.

Kolesnik, Theodore S. "Illustrating the Multiplication and Division of Common Fractions." *The Arithmetic Teacher*, May 1963, pp. 268–271.

Lazerick, Beth Ellen. "The Conversion Game." *The Arithmetic Teacher*, January 1961, pp. 54–55.

Lyvers, Donald B. "A Fraction Circle." *The Arithmetic Teacher*, April 1956, pp. 119–121.

Nelson, Diane, and Nelson, Marvin N. "Pegboard Multiplication of a Fraction by a Fraction." *The Arithmetic Teacher*, February 1969, pp. 142–144.

Olberg, Robert. "Visual Aid for Multiplication and Division of Fractions." *The Arithmetic Teacher*, January 1967, pp. 44–46.

Price, Jack. "Why Teach Division of Common Fractions?" *The Arithmetic Teacher*, February 1969, pp. 111–112.

Shokoohi, Gholam-Hossein. "Readiness of Eight-Year-Old Children to Understand the Division of Fractions." *The Arithmetic Teacher*, March 1980, pp. 40–43.

Silvia, Evelyn M. "A Look at Division with Fractions." *The Arithmetic Teacher*, January 1983, pp. 38–41.

Sweetland, Robert D. "Understanding Multiplication of Fractions." *The Arithmetic Teacher*, September 1984, pp. 48–52.

Thompson, Charles. "Teaching Division of Fractions with Understanding." *The Arithmetic Teacher*, January 1979, pp. 24–27.

Trafton, Paul R., and Zawojewski, Judith S. "Teaching Rational Number Division: A Special Problem." *The Arithmetic Teacher*, February 1984, pp. 20–22.

Zytkowski, Richard Thomas. "A Game with Fraction Numbers." *The Arithmetic Teacher*, January 1970, pp. 822–83.

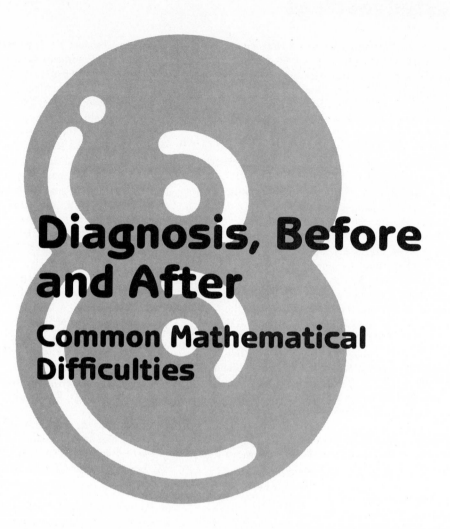

Diagnosis, Before and After

Common Mathematical Difficulties

■ In this chapter, we describe methods for diagnosing children's mathematical difficulties. A distinction is made between the difficulties children encounter and the specific errors they make. We believe that errors are usually related to conceptual difficulties and that any realistic approach to remediation must be done in terms of the conceptual difficulty. We believe further that if teachers are to effectively diagnose in the ordinary classroom, then they must know, prior to diagnosing, the difficulties children might have and the errors that reveal these difficulties. In the first part of the chapter, we define diagnosis of errors related to whole numbers and describe methods for performing related tasks. Then we list common difficulties that children experience, provide associated errors, and give suggestions for remediation. In the second part of the chapter, we will do the same for rational number concepts. We will also discuss some reading and writing difficulties that influence the learning of mathematics.

Our motto is "Prevention over remediation." We encourage teachers to become aware of the difficulties that children frequently experience and plan learning activities that will help children avoid these difficulties. Since the next best thing to prevention is remediation, we also provide materials that help teachers develop skills for quickly spotting and remediating difficulties. ■

Teacher goals

1. Be able to identify and sequence learning objectives for arithmetic of rational numbers.
2. Be able to identify difficulties children encounter and related errors they make with these learning objectives.
3. Be able to diagnose difficulties children are having and design appropriate remediation activities.
4. Be able to implement remediation activities.

In recent years, there has been much discussion among educators about diagnosing mathematical difficulties encountered by learners. The results have been both positive and confusing. Currently, most mathematics educators agree that a mathematics instructional program should include systematic procedures for achieving these goals:

Identifying learner competencies
Diagnosing learner difficulties
Prescribing efficient learning experiences

Researchers have undertaken studies concerning how best to implement these goals, while practitioners have adopted programs to achieve them. These results are certainly positive when we consider the way most instructional programs operated prior to the 1970s. In general, learners at a given level were provided a standard course of study, and they were expected to proceed through it at approximately the same rate. Attention was not focused on the specific difficulties individual learners experienced, the specific competencies they had achieved, or their individual learning styles. The extent of learner differentiation was to designate a student as slow, average, or accelerated. Usually, these designations were based on achievement-test scores, rather than on data indicating specific strengths and weaknesses.

On the other hand, attempts to achieve these goals, no matter how noble, have led to confusion. Research efforts have led to the development of diagnostic teaching models as well as lists of common errors that children make. But questions concerning how to remediate difficulties or how to implement diagnostic teaching models in the ordinary classroom have not been adequately dealt with.

Publishing companies have also made contributions by developing many types of instructional systems labeled as diagnostic. But their efforts also have not been adequate to meet the needs of students and teachers in ordinary classrooms. The commercial systems range in quality from good to poor. Some deliver sound programs for diagnosing, whereas others consist of complicated surveys related to limited learning objectives. Most of the programs focus on computational skills and ignore other, more important topics, such as solving problems or estimating, and most provide insufficient guidance or materials for remediation. Thus, many school districts have established programs that are expensive and difficult to manage and that rest on shaky theoretical ground. Teachers complain that so much time is spent on testing and record-keeping that no time is left for the important task of teaching. They further claim that good instructional materials are either too expensive to provide in classrooms with a large student-to-teacher load or are too hard to find. Teachers also complain that most materials provided for remediation are primarily traditional text materials that have merely been "repackaged" to look unique.

The degree to which these criticisms are accurate is not as important as the fact that many teachers are experiencing frustration. Many do believe that the goals for individualized instruction are worthy ones. Nonetheless, they experience frustration because they do not have enough answers to implement workable systems in classrooms with 35 or so students.

Certainly, one chapter cannot solve all the problems related to diagnosing mathematical difficulties. But many practical questions can be answered and pulled together to provide a realistic framework within which the classroom teacher can function. The purpose of this chapter is to do just that—give practical answers to some important questions, then pull these answers together to provide the framework for diagnosing and remediating in the ordinary classroom. What are these questions? Here goes!

Getting ready to diagnose

■ What is diagnosis?

When we diagnose mathematical difficulties, we determine areas of weakness a child has, we study specific errors the child is frequently making, and we attempt to explain why these errors are being made. The explanation we give must be specific enough to allow us to plan specific remediation activities. It is important here to distinguish between *surveying* and *diagnosing*. When we survey, we simply determine whether a child has acquired a particular set of mathematical competencies; for example, we may give a rather exhaustive test on addition and subtraction only. Suppose we conclude that a child is weak in both topics. Our conclusion may be based on the fact that the number of correct responses for each topic fell below some accepted standard.

At this point, we have only *surveyed* two arithmetic topics, we have not *diagnosed*. To diagnose, we carefully study the errors the child is making and decide why the errors are being made. If we merely identify errors that a child is making, we are not diagnosing. Errors are simply symptoms, and we must determine the cause of these symptoms. Suppose we study the wrong answers that the child has made with respect to addition and find that every mistake has to do with a basic fact. The child seems to understand the mechanics of computation—whether or not regrouping is involved—but does not recall basic facts correctly. Have we diagnosed? Maybe, maybe not. We may again study these incorrect responses, more closely, to find that every error involves zero. More specifically, the child consistently uses the incorrect relation $n + 0 = 0$. Now we have diagnosed. We have collected and studied the set of errors that the child has made, and we have classified these errors in terms of one *difficulty*. Now, efficient remediation is more easily implemented than had we simply decided that the child "can't add" or that the child "doesn't know the basic facts." To summarize, we carry out these steps when we diagnose:

we survey broad areas of knowledge to identify weaknesses,
we study specific errors in the area of weakness, and
we classify these errors in terms of specific difficulties.

■ What learner characteristics can be diagnosed?

Many different types of abilities can be diagnosed. Abilities that are important to the teacher include memory abilities, attending abilities, visual perceptual abilities, auditory perceptual abilities, general motor abilities, and academic abilities. Although mathematics teachers are primarily concerned with the development of mathematical abilities, they cannot ignore the fact that difficulties related to other types of abilities affect the learning of mathematics. For example, when assigning a page of word problems, the teacher cannot ignore the fact that some children are poor readers. When assigning a timed test, the teacher cannot ignore the fact that children with poorly developed motor skills may work more slowly than other children. A teacher also cannot expect children with poorly developed attending abilities to sit passively and listen to lengthy explanations. Suggestions both for diagnosing and for remediating difficulties will take into account other difficulties that children may be experiencing.

■ What skills must teachers have?

Busy classroom teachers do not have hours to spend mulling over the incorrect responses of children to discover the difficulties they're having. Busy teachers must have trained ears and eyes that enable them to determine on the spot what kinds of errors are being made and why. They must be able to make these determinations no matter what kind of classroom activity is underway—a test, a discussion session, a routine practice activity, or even a game. To do this efficiently, teachers must be able to do the following:

Identify specific learning objectives,
develop learning activities for these objectives, and
predict difficulties and errors for learning objectives.

In plain terms, a good teacher knows the objectives of the curriculum and the most common pitfalls that are associated with a particular learning objective. Right now, you may be feeling anxious. You may fear that for each learning objective, there will be a huge list of difficulties. Not so; we human beings are more alike than we are different, and most of us make the same kinds of mistakes whether we are learning a mathematical concept or learning to ride a bicycle. For most learning objectives, there are just a few difficulties to be identified that will account for most of the errors being made. For example, one difficulty that could be associated with the learning objective "Computes sums of numbers between 9 and 100," is the ability to apply regrouping procedures rationally. Many different errors made for the same exercise may all point to the same difficulty—the inability to make regrouping decisions.

Sally Ann had $27 and got $35
for her birthday. Now how much
does she have?

$$\begin{array}{r} 27 \\ + \ 35 \\ \hline 512 \end{array}$$

Maria paid $32 for a dress and
$25 for a pair of shoes. How
much did she spend?

$$\begin{array}{r} 32 \\ + \ 25 \\ \hline 67 \end{array}$$

Another fear you may have is that each of the errors, although associated with the same difficulty, must be remediated separately. Nothing could be further from the truth. The errors in the preceding example point to the fact that children attempted to learn regrouping procedures by rote. Patching up each error with some mechanical routine is superficial and will not have lasting results. If several children suffer from the same difficulty, even though the symptoms are slightly different, then it is probable that they all need the same cure. In this case, the teacher should get out play money and review the following place-value concepts:

1. You can associate numerals with place-value models.
2. When you have 10 ones, you should trade them in for 1 ten. Be sure, *if you trade*, to write it down.

The teacher should demonstrate each of these steps with the place-value model. Such a plan for remediation might involve a small group or a large one and could easily be carried out in the ordinary classroom.

So far we have said that many errors can be associated with a single difficulty. To remediate them, you remediate the difficulty. To make things even simpler, a difficulty that is associated with one learning objective may also be associated with other learning objectives. For example, the inability to make regrouping decisions is associated with learning objectives for representing numbers, as well as for adding, subtracting, multiplying, dividing, and rounding numbers.

So, the advantages of correcting a difficulty are that correction transfers to other learning objectives rather than just patching up an error that resulted from using a mechanical procedure. Let's summarize:

1. To diagnose mathematical difficulties efficiently, the teacher must be able to predict the most common difficulties that are associated with a given learning objective.
2. For most learning objectives, the identification of a few major difficulties will account for almost all errors that children make.
3. There may be a number of errors associated with a given difficulty. But the remediation for each error will usually be the same, since remediation is done in terms of the difficulty.
4. The same difficulty may be related to several learning objectives. So, correcting it will affect many learning objectives other than the one under consideration.

■ What methods are used for diagnosing?

Before you begin to diagnose specific difficulties, it is convenient to identify areas of content in which individual children are weak. Sometimes classroom teachers are responsible for developing and giving survey tests that evaluate student competencies at a given level. More often than not, the school, the local district, or the state board administers survey tests and supplies classroom teachers with checklists that indicate broad competencies that students have or have not acquired. In any event, classroom teachers should collect and organize information concerning student competencies in general areas before they begin to look at specific content to detect difficulties.

One of the most efficient and popular methods to survey mathematics content is to use criterion-referenced tests. A criterion-referenced test is one that is based upon a set of specifically stated learning objectives. Usually, the set of objectives has been accepted by a school or school district as being appropriate for a particular grade or school level. The objectives for the test relate mathematical concepts or skills to "acceptable student behaviors." Then, items are written that allow children to demonstrate the acceptable behaviors.

Suppose, for example, that a criterion-referenced test was developed for a set of objectives including this one:

Identifies a pictorial model for a number sentence involving a basic fact of multiplication.

Items would have to be included that require children to demonstrate the ability to identify appropriate pictorial models. Here is a sample item:

Which picture goes with this number sentence?
$3 \times 5 = 15$

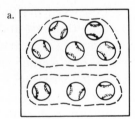 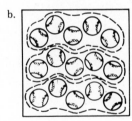 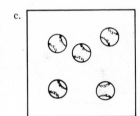

Don't let the term "criterion-referenced test" frighten you. If you had to prepare such a test, you would follow these steps:

1. Decide on the learning objectives that are to be surveyed.
2. Decide which student behaviors are necessary for demonstrating each learning objective.
3. Decide how many items are necessary for each objective and how many should be completed correctly to demonstrate mastery.

Then:

4. Obtain items from various sources: textbooks, worksheets, old tests, and so forth.
5. Establish a plan for recording data obtained for individual students.

After you have identified the weaknesses of individual students, you will have to study the errors to pinpoint difficulties. There are many techniques to accomplish this goal. The following discussion describes the most useful ones for the elementary teacher.

Observation. One of the best ways for a busy teacher to collect helpful information is to watch children as they carry out daily activities. Whether children are playing a game, manipulating materials, doing boardwork, or completing practice exercises, the discerning eye can quickly identify children who have acquired certain objectives and those who have shown specific difficulties.

Suppose that a teacher, when checking papers, finds these errors on a child's practice exercises:

$$\begin{array}{r} 25 \\ + \ 27 \\ \hline 412 \end{array} \qquad \begin{array}{r} 27 \\ + \ 18 \\ \hline 315 \end{array}$$

The teacher may conclude that the child cannot regroup, which is correct, but perhaps not correct enough. The child may also be computing from left to right. After recording a sum in the tens place, the child is at a loss as to how to handle the 1 ten that results from computing with the digits in the ones position. The only efficient way for a teacher to catch this difficulty is to, first, *suspect* it, and, second, *observe* it. So, to make worthwhile observations, teachers must be tuned in; that is, they must be aware of the difficulties children might experience and must be ready to identify them whenever and wherever the difficulties occur.

Question-and-answer sessions. Asking the right questions at the right time and listening to children's answers can be as revealing as studying errors made on tests. In most instances, asking questions and listening carefully to children's responses will be less time-consuming than preparing, giving, and grading a test. For example, most young children who have trouble solving subtraction word problems have more trouble with comparison and missing-addend situations than with take-away situations. When they confront either a comparison or missing-addend problem, they may add instead of subtract. A few well-chosen problems given orally, a few good questions, and careful listening may reveal these difficulties immediately. The key, of course, to using question-and-answer sessions is to know what questions to ask, and teachers who are aware of difficulties that children frequently encounter have a good chance of asking the right questions.

Diagnostic inventories. Another way to pinpoint children's difficulties is to use diagnostic inventories. Unlike a survey test, a diagnostic inventory involves only a few learning objectives. The items are selected to represent the range of possible items that could be specified for the learning objectives. Usually, the time required for administering an inventory is short. Testing only a few objectives allows a careful look at specific errors that are made so that difficulties can quickly be identified. If the inventory can be completed quickly, children will be less likely to experience intimidation and test fatigue.

A diagnostic inventory may only require pencil and paper, or it may involve activities where children manipulate concrete materials. It may be a test that can only be administered to one child at a time, or it may be one that is suitable for use with a small or large group. A diagnostic inventory may involve printed materials, or it may involve audio-visual materials such as tape recorders and filmstrips. However the inventory is constructed, children should not perceive it as different from a routine classroom activity. Neither should classroom teachers. The goal is an instructional one—attempting to systematically determine difficulties children have so that appropriate learning experiences can be planned. The goal is not one of collecting "scores"! To construct a diagnostic inventory, you must follow these steps:

1. Select objectives. Select specific learning objectives to be tested and decide what behaviors must be exhibited to demonstrate mastery of the objectives.

2. Choose items. Choose items for each objective. The number of items will depend on the objectives, but at least three items should be chosen for each objective. This will ensure that a persistent error will be caught. Whatever the number of items, you will want to make sure that your inventory represents all the items that could be specified for the objectives under consideration. Suppose, for example, that you are to choose items for this objective: "Student computes sums for numbers between 10 and 100, no regrouping." You must choose items that involve 0, items that don't involve 0, and items that involve sums of more than two addends.

$$
\begin{array}{cccccccc}
 & & & & & & 14 & \\
20 & 40 & 63 & 55 & 62 & 74 & 21 & \\
+30 & +25 & +10 & +23 & +17 & +24 & +53 & \text{etc.}
\end{array}
$$

If you find that too many items are required to represent an objective, then the objective may be too broad. If you can only come up with a few items for an objective, then the objective may be too narrow. You want to select items that allow the greatest amount of information to be obtained with the least amount of testing.

3. Develop a format. Organize the items so that the format of the inventory is easy to follow. Make sure that directions are clear and that all necessary art work is uncluttered.

4. Plan for error analysis. Devise a system so that errors can be easily spotted and analyzed.

5. Prepare a key.

6. Prepare student folder and classroom checklist. A student folder contains specific information regarding an individual student's acquisition of learning objectives; a classroom checklist is a list of students and learning objectives with columns for recording student acquisition.

In this section, we have discussed three methods for diagnosing student difficulties: observation, question-and-answer sessions, and diagnostic inventories. The key to making these methods work for the busy teacher is to use each one appropriately. Difficulties for some learning objectives can be spotted easily during routine instructional tasks. On the other hand, identifying some difficulties will require the use of an inventory. Practice will enable you to choose the most expedient method.

Common mathematical difficulties involving whole numbers and suggestions for remediation

Now that you know what procedures to use when diagnosing mathematical difficulties, you need to know specifically what difficulties to look for. You should also have some plan for remediating these difficulties. We will first describe some general characteristics of mathematical difficulties, and then list some common difficulties and associated errors that occur when learning place value, addition, subtraction, multiplication, and division concepts for whole numbers. We then make some suggestions for remediation. This book is filled with activities for use in remediation; thus, your task is to locate specific activities to meet the remediation suggestions.

Because of space limitations, we have not been able to list specific difficulties for all the topics discussed in the book. Choices had to be made. We have chosen to list difficulties for topics in arithmetic, not because these are more important than other topics, but because other topics such as mathematical concepts for early childhood, geometry, and measurement have been developed in terms of the child's perception of mathematical ideas. We have also included extensive discussion on related difficulties.

■ What kind of difficulties are there?

Most difficulties that children have with arithmetic stem from a lack of understanding of important place-value concepts. Children may have difficulties adding, subtracting, multiplying, or dividing, and their difficulties may have little to do with the operations themselves. They simply do not sufficiently understand the decimal (base ten) system of numeration to use its numerals to compute. Difficulties in computing also are often related to the particular operation being used. For example, a child who makes errors such as those below cannot use the distributive property of multiplication over addition to compute products:

$$\begin{array}{r} 48 \\ \times\ 21 \\ \hline 88 \end{array} \qquad \begin{array}{r} 37 \\ \times\ 32 \\ \hline 104 \end{array}$$

Some operations depend on prerequisite skills involving other operations. For example, completing a long-division computation requires subtraction. A child who cannot subtract efficiently will not be able to divide efficiently.

It is important to make these distinctions so that proper remediation can be planned. If a child has place-value difficulties, he or she must be retaught the appropriate place-value concepts, even though the difficulties are discovered while the child is attempting to add. Likewise, you should not try to patch up division errors by reteaching division if the child's problem is subtraction. You must put division aside and reteach subtraction.

Some difficulties are clerical. Suppose, for example, that you find the following on a child's practice sheet:

$$
\begin{array}{cccc}
& \overset{\text{\large|}}{} & & \\
39 & 48 & 25 & 38 \\
-18 & +32 & +76 & +47 \\
\hline
21 & 80 & 91 \;\text{X} & 85
\end{array}
$$

The error is probably clerical. Can you tell why?

In the following sections, lists of common difficulties are given. As you study these difficulties, try to decide which type each is. Don't be surprised to find several difficulties that appear in several lists for different topics.

Place value

Naming, comparing, and rounding numbers using the decimal system of numeration is based on place-value concepts. Furthermore, all computational procedures involving decimal representations—whole number or rational number—depend on place-value concepts. Thus, the place-value difficulties that children experience will affect most of the arithmetic concepts they learn. Regardless of the grade level, it is wise to check for place-value difficulties immediately when a child demonstrates weaknesses in arithmetic.

As you read this section, you may occasionally want to refer to Chapter 2 for detailed discussion of place-value concepts and related activities.

1. Difficulty associating place-value models with numerals
   Errors:
   a. Children may give an incorrect numeral for a given place-value model.

   How many? (place-value model) ⟶ **32**

   b. Children who have memorized counting sounds but have not based counting techniques on place-value concepts may make counting errors. For example, a child may say the number following 143 is 153 or that the next number after 4968 is 5968.
   c. Children may be unable to make digit comparisons of two numbers to find the larger. For example, they do not reason that 87 is greater than 68 because 8 tens is more than 6 tens. Instead, they would say "87 is greater than 68 because 87 comes after 68 when I count." When this strategy is used, errors are more likely as numbers become larger.

   Discussion and suggestions for remediation. Children should learn to associate place-value models that show objects arranged in groups of 1s, 10s, and 100s for numbers less than 1000. For numbers greater than 1000, they should learn to make associations involving models such as play money. Children should learn to count using these place-value models.

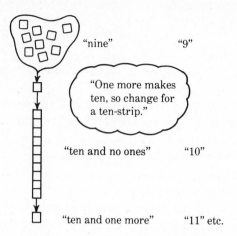

"nine"   "9"

"One more makes ten, so change for a ten-strip."

"ten and no ones"   "10"

"ten and one more"   "11" etc.

Children should learn to order numbers by making digit comparisons. For example, deciding which of the numbers below is larger by thinking in terms of counting is difficult. It is easy to determine, however, when one starts on the left and compares digits in like place-value positions.

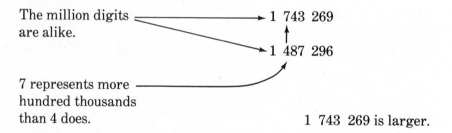

The million digits are alike.   1 743 269

                                                        1 487 296

7 represents more hundred thousands than 4 does.

                                        1 743 269 is larger.

2. Difficulty using zero when writing numerals
   Errors:
   a. Children may not see the need for recording a 0 digit when writing the numeral for a given mode. Example: Children may write 4 for the model on the left but 43 for the one on the right.

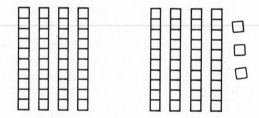

   b. Children may record too many zeros by writing what they hear. For example, a child may write "1002" for the oral expression "one hundred two."

Discussion and suggestions for remediation. Each of the digits 0 through 9 can be used as a place holder in a numeral. In the numeral 40, the 4 is a place holder in the tens place and indicates 4 tens. Zero is also a place holder, indicating the number of ones. The 0 digit is unique, not because it can be used as a place holder, but because it is sometimes indicated and sometimes understood. In the numeral 4, it is understood there are 0 tens. In the numeral 40, the zero must be written; otherwise, we would confuse 4 with 40.

This subtlety often confuses children as they learn to represent numbers greater than 9. Perhaps they reason that if they don't *always* have to record zeros to indicate tens, hundreds, thousands, and so on, then they are at liberty to pick the occasions when they will record zeros.

Some children want to record zeros when they are not necessary. This usually occurs when children must represent numbers greater than 99 and must rely on memorization rather than place-value concepts to come up with numerals. Again, for example, the child writes "1001" for "one hundred one."

To remediate these difficulties, the teacher must plan activities with place-value models and place-value charts. These activities should require children to distinguish between situations in which zeros are necessary and are to be recorded in a numeral and those in which zeros are understood. They should be spread throughout the year as "quickie" place-value reviews.

3. Difficulty using regrouping concepts to represent numerals
   Errors:
   a. Children may be unable to find the next numeral in a counting sequence. For example, "210" may follow "28, 29," or "910" may follow "98, 99."
   b. Children may be unable to round numbers appropriately. When asked to round 496 to the nearest ten, the child may fail to regroup, giving responses such as the following: 4100, 410, 4106, or 400.

   Discussion and suggestions for remediation. Children should learn to use regrouping procedures before learning computational procedures that require regrouping. Specifically, they should learn that regrouping is necessary to generate the numbers that follow each of these numerals: 19, 29, 99, 239, 999, and so forth. Place-value models such as Happy Face pieces or play money should be used to explicitly demonstrate this concept.

4. Difficulty naming place-value positions in a numeral
   Error:
   a. Children may not be able to write numerals for the numbers indicated by their word names, especially in cases involving zeros. When asked to write numerals for these word names, children may respond as indicated:

twenty thousand, twenty-one	2021
four hundred one thousand	4 001 000
thirty thousand, eighty-two	3082

   Discussion and suggestions for remediation. To remediate, teachers should make sure children can associate numerals with place-value models. Using these models and place-value charts, they should provide activities that stress the following idea: *Each three digits, starting at the right of a numeral, contain a group that contains hundreds, tens, and ones.* This association enables children to locate positions indicated by word names and allows them to decide where certain digits, especially zeros, must be recorded.

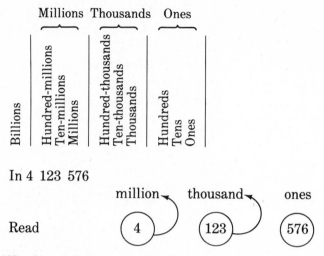

In 4 123 576

Read

5. Difficulty giving nonstandard place-value representations for a numeral (for example, 4753 = 475 tens and 3 ones)
   Error:
   a. Children may not round numbers appropriately. They may round 367 to 360 or to 368 or to 400 when asked to round 367 to the nearest ten.

   Discussion and suggestions for remediation. Children who depend on memorized routines for rounding numbers make numerous errors. Children who can make nonstandard interpretations for numerals can use a flexible strategy when rounding a number such as 367 to the nearest ten. These children can think: "367 is the same as 36 tens and 7 ones. Seven ones is nearer to 10 than zero. So 367 is closer to 37 tens than 36 tens. So 367 rounded to the nearest ten is 370." Thus, children should be taught to make nonstandard representations for numerals and to use these representations when rounding whole numbers. A mini-number line can be very helpful. In this case, it would contain whole numbers from 360 to 370.

Addition and subtraction

We have identified seven difficulties that children have when learning to compute sums and differences and when solving related word problems. These difficulties occur whether children work with two-digit, three-digit, or *n*-digit numerals. It is usually the lack of understanding of important addition and subtraction concepts or place-value concepts that causes difficulties. Remediation at the second- and fifth-grade levels may involve the same techniques, but the frame of reference may be different. Aids for fifth-graders may appear to be more sophisticated, and numbers used in demonstrations and exercises may be larger.

As you read this section, you may want to refer to Chapter 3 for detailed discussions of addition and subtraction concepts and activities.

1. Difficulty identifying addition or subtraction situations
   Errors:
   a. Children may select the wrong operation for a story problem. This occurs frequently for subtraction. Often children add when they encounter subtraction situations that involve comparison or missing-addend interpretations.
   b. Children may not be able to write appropriate number sentences for story problems. When writing a number sentence for a subtraction story problem, children may write $5 - 7 = \square$, instead of $7 - 5 = \square$.
   c. Children may choose the wrong information when extra information is included in a story problem.
   d. Children may not be able to suggest reasonable information when information is missing in a story problem.

   Discussion and suggestions for remediation. Children who have these difficulties are ones whose introduction to addition and subtraction probably began with learning basic facts. Readiness experiences that involved learning the interpretations for the operations were not sufficient. To remediate this difficulty, children must learn the interpretations using concrete materials and must learn to associate these situations with appropriate number sentences.

2. Difficulty using counting to find basic addition facts
   Errors:
   a. Children may equate $n + 1$ with n (for example, $n + 1$ is confused with $n \times 1$).
   b. Children may make counting errors when finding facts with sums less than 10.
   c. Children may make counting errors when finding facts with sums greater than 10.

   Discussion and suggestions for remediation. Learning the basic facts is no easy task. As we said in Chapter 3, it takes practice, practice, practice. It also takes motivation, motivation, motivation! So, to remediate this difficulty, teachers must decide which error the child is making. Then the teacher must design motivational activities that allow children to learn and practice the following concepts:
   1) If n is a whole number, then $n + 1$ is the whole number that follows n.
   2) Numbers 10 or less should be thought of in terms of families of facts. For example, the family of facts for 6 is:

   $$6 + 0 = 6 \quad 5 + 1 = 6 \quad 4 + 2 = 6$$
   $$0 + 6 = 6 \quad 1 + 5 = 6 \quad 2 + 4 = 6 \quad 3 + 3 = 6$$

3) To find facts with sums greater than 10, we should use place-value concepts and family-of-facts ideas; for example,

$$8 + 6 = 8 + (2 + 4) = (8 + 2) + 4 = 10 + 4$$

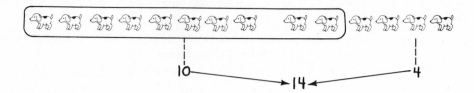

3. Difficulty with zeros in computations
   Errors:
   a. Children may fail to record a zero in a sum or difference.

$$
\begin{array}{r} 20 \\ + 30 \\ \hline 5 \end{array}
\qquad
\begin{array}{r} 204 \\ + 502 \\ \hline 76 \end{array}
\qquad
\begin{array}{r} 60 \\ - 40 \\ \hline 2 \end{array}
\qquad
\begin{array}{r} 604 \\ - 102 \\ \hline 52 \end{array}
$$

   b. Children may compute $n + 0$ or $n - 0$ as 0.

$$
\begin{array}{r} 28 \\ + 40 \\ \hline 60 \end{array}
\qquad
\begin{array}{r} 39 \\ - 20 \\ \hline 10 \end{array}
$$

Discussion and suggestions for remediation. Sometimes children reason that "0 represents 'nothing,' so why bother recording it in a computation?" These children need to review addition and subtraction computation using place-value models. They should be encouraged to record a digit for each place-value position. When they have completed their computations, they can decide whether some zeros have been recorded that are not necessary.

Children who make errors like those given in 3b probably have $n \times 0$ and $n + 0$ confused. They need experiences that convince them that $n + 0$ or $n - 0$ are the same as n. Concrete materials can be used to demonstrate $4 + 3 = 7$, then $4 + 2 = 6$, and $4 + 1 = 5$. Each time the sum is "one less," so $4 + 0$ must be 4. Likewise, $4 - 3 = 1$, $4 - 2 = 2$, and $4 - 1 = 3$. Each time the difference is one more, so $4 - 0$ must be 4.

4. Difficulty using counting to find differences that are related to addition facts
   Errors:
   a. Children may make counting errors when finding differences that involve related addition facts having a sum of 10 or less.
   b. Children may make counting errors when finding differences that involve related addition facts having a sum greater than 10.

Discussion and suggestions for remediation. Most children make these errors because they do not relate addition and subtraction. These children depend on counting, which is inefficient. To remediate, basic addition facts must be taught systematically. As they are, activities involving concrete models should be used to demonstrate the relationship between addition and subtraction. Finally, children should learn that every addition sentence or computation is related to two subtraction sentences or computations.

$$3 + 5 = 8 \longrightarrow 8 - 5 = 3 \text{ and } 8 - 3 = 5$$

Then addition and related subtraction facts should be practiced together.

5. Difficulty regrouping when computing sums and differences
   Errors:
   a. Regroups when not necessary.

$$\begin{array}{r} \overset{1}{34} \\ + 62 \\ \hline 106 \end{array} \qquad \begin{array}{r} \overset{3}{\cancel{4}5} \\ - 23 \\ \hline 12 \end{array}$$

   b. Does not account for regrouping step.

$$\begin{array}{r} 37 \\ + 48 \\ \hline 75 \end{array} \qquad \begin{array}{r} 92 \\ - 68 \\ \hline 34 \end{array}$$

   c. Subtracts the smaller digit from the larger.

$$\begin{array}{r} 312 \\ - 149 \\ \hline 237 \end{array} \qquad \begin{array}{r} 600 \\ - 463 \\ \hline 263 \end{array}$$

   d. Does not regroup. (May also compute from left to right.)

$$\begin{array}{r} 48 \\ + 56 \\ \hline 914 \end{array}$$

Discussion and suggestions for remediation. After children achieve competence with basic addition facts and related subtraction sentences, the next big hurdle is to apply regrouping procedures in order to compute sums and differences. The errors given above are typical ones that children make when they have memorized routines for regrouping and do not really understand why steps are performed as they are. Children who make errors such as those in 5c have probably heard comments such as "always subtract the smaller number from the larger." The problem is, however, that these children do not know a number from a digit, let alone how to regroup.

There's just no shortcut. Remediation for these errors involves doing what should have been done to prevent the errors:

1) Teach children to associate numerals with place-value models.
2) Teach children to use regrouping concepts to find numerals that come after 19, 29, 39, and so forth.
3) Extend regrouping ideas to computations involving sums, then differences. Use place-value models to demonstrate each step in a computation. Have children *practice, practice, practice* with place-value models.

6. Difficulty when the two numerals in an exercise have a different number of digits

Errors:

a. Children add or subtract the ones digit from the tens digit, hundreds digit, and so on.

$$
\begin{array}{r}
43 \\
+\ \ 6 \\
\hline
109
\end{array}
\qquad
\begin{array}{r}
89 \\
-\ 5 \\
\hline
34
\end{array}
$$

b. Children may not complete a computation.

$$
\begin{array}{r}
43 \\
+\ 6 \\
\hline
9
\end{array}
\qquad
\begin{array}{r}
89 \\
-\ 5 \\
\hline
4
\end{array}
$$

c. Children may add all the digits.

$$
\begin{array}{r}
43 \\
+\ 6 \\
\hline
13
\end{array}
$$

Discussion and suggestions for remediation. Again, it is apparent that children have memorized ways to compute and do not understand what steps are to be taken and why. In these cases, it is probable that children are not really sure what the addition and subtraction symbols mean. To remediate, the teacher should review the meaning of addition or subtraction—whichever applies—to be sure that children understand the meaning of the operation. Then teachers should get out the good old place-value pieces and demonstrate each step that occurs in a computation. Children should practice computing using the place-value pieces. They should be required to justify each step they take and each number they record, using the place-value models.

Notice that these examples would be easier in horizontal form. This is a perfect spot for mental arithmetic activities.

$43 + 6 = \boxed{49}$ $\circ$ $\circ$ $^\circ$ (How many ones? 9 / What else? 40 / $40 + 9 = 49$)

$89 - 5 = \boxed{84}$ $\circ$ $\circ$ $_\circ$ ($9 - 5 = 4$ / What else? 80 / $80 + 4 = 84$)

Children don't need a pencil to do these computations. They need to be encouraged to *not* use a pencil. These would have been preceded with questions such as "How much is $43 + 1$? ... $43 + 2$?" or "How much is $89 - 1$? $89 - 2$?"

7. Difficulty when a sum involves several addends or when a sum or difference involves large numbers

Error:

$$
\begin{array}{cc}
\text{a.} & 48 \\
& 53 \\
& \underline{69} \\
& |7|
\end{array}
\qquad
\begin{array}{cc}
\text{b.} & 5000 \\
& \underline{-\ 1843} \\
& 4|57
\end{array}
$$

Discussion and suggestions for remediation. Looking at these errors, we can see that it is more of the same. But the more addends in a sum, or the larger the numbers in a sum or difference, the greater the chance that children will make clerical errors. So, as well as remediating basic difficulties, the teacher should provide aids or suggestions that help a child minimize clerical errors. Here are two examples:

1) Encourage making examples easier and using mental arithmetic.

$$48 + \underbrace{53 + 69} =$$

$$48 + \underbrace{52 + 70} =$$

$$48 + 122 =$$

$$40 + 10 + 120 = 170$$

Write things down when you *need* to.

2) Have children follow this line of reasoning to compute differences like the one shown at the right.

$$
\begin{array}{r}
5000 \\
-\ 1843
\end{array}
$$

I can think of 5000 as 5000 ones,

or 500 tens, or

50 hundreds, or

5 thousands. ⎯⎯⎯⎯⎯⎯⎯→

$$
\left\{
\begin{array}{l}
5000 \text{ ones} \\
5000 \text{ tens} \\
5000 \text{ hundreds} \\
5000 \text{ thousands}
\end{array}
\right.
$$

In this case, it is easiest to think of it as 500 tens, since I need to regroup 1 ten as 10 ones. ⎯⎯⎯→

$$
\begin{array}{r}
5000 \text{ tens} \\
-\ 1843
\end{array}
$$

If I regroup one of the tens as 10 ones, there will be 499 tens left and a set of 10 ones. Now I can subtract easily. ⎯⎯⎯⎯⎯→

$$
\begin{array}{r}
\overset{499\,1}{\cancel{5000}} \\
-\ 1843
\end{array}
$$

Multiplication and division

We have identified eight basic difficulties that children have when learning to compute products and quotients and to solve related word problems. Whether children are working with two-digit, three-digit, or *n*-digit numerals, these difficulties can occur. It is usually the lack of understanding of important multiplication, division, or place-value concepts, rather than the size of numbers, that causes the difficulties. Remediation at the fourth-grade and sixth-grade levels may involve the same techniques, but the frame of reference may be different. Aids for sixth-graders may appear to be more sophisticated, and numbers used in demonstrations and exercises may be larger.

1. Difficulty identifying multiplication and division situations
   Errors:
   a. Children may select the wrong operation for a story problem. This occurs frequently for division when problems are stated in a missing-factor form; for example, "Craig delivers 3 times as many papers as Jack. If Craig delivers 369 papers, how many does Jack deliver?" In cases such as these, children may multiply instead of divide.
   b. Children may not be able to write appropriate number sentences for story problems. When writing a number sentence for a division problem, children may write $3 \div 369 = \square$ when they mean $369 \div 3 = \square$.
   c. Children may choose the wrong information when extra information is included in a story problem.
   d. Children may not be able to suggest reasonable information when it is missing in a story problem.

   Discussion and suggestions for remediation. Children who have these difficulties are ones whose introduction to multiplication and division probably began with learning basic facts. These children probably did not have sufficient readiness experiences in which to learn the interpretations for the operations. To remediate this difficulty, you must teach the children the interpretations using concrete materials and teach them to associate these situations with appropriate number sentences or computations (see Chapter 4).

2. Difficulty determining the basic facts
   Errors:
   a. Children may compute $n \times 0$ as n, or $0 \div n$ as n.
   b. Children who have not memorized the facts and who mentally add to find products may use too many or too few addends; for example, 4×6 is given as "$6 + 6 + 6 = 18$."
   c. Children may confuse facts that have close products; for example, 7×8 and 9×6 may be confused because they are close.

   Discussion and suggestions for remediation. There are no shortcuts for learning the basic multiplication facts. As we said in Chapter 4, it takes teacher effort and lots of student practice. The same steps required for learning the facts are required for remediation—so back to Chapter 4!

3. Difficulty using the basic multiplication facts to find related quotients
   Error:
   a. Children may make estimation errors when finding quotients; for example, "49 divided by 6 is about 7."

   Discussion and suggestions for remediation. Most children make these errors because they are unsure of their multiplication facts or because they do not relate multiplication facts to division sentences. These children depend on guessing; for example, "How many 6s are in 49?" To remediate, basic multiplication facts must be taught systematically. As these facts are taught, activities involving concrete models should be used to demonstrate the relationship between multiplication and division. Finally, children should learn that every multiplication sentence or computation is related to two division sentences, as long as zero is not one of the factors.

$$3 \times 5 = 15 \longrightarrow 15 \div 3 = 5 \text{ and } 15 \div 5 = 3$$

4. Difficulty in applying place-value concepts and basic facts to obtain products and quotients of multiples of ten
   Errors:
   a. Children who use rules such as "Multiply the nonzero digits and annex the number of zeros in the factors" make errors such as the following:

   $$50 \times 40 = 200$$

   5×4=20, I NEED 2 ZEROS, I HAVE ONE, SO ALL I NEED IS ONE MORE!

   b. Because children are not sure what to do when multiplying, they make related errors in division:

   $$60 \overline{\smash{)}3600} = 600 \quad \text{or} \quad 60 \overline{\smash{)}3600} = 6$$

   Discussion and suggestions for remediation. Children should develop sound concepts for finding products of multiples: "30 × 500 means 3 tens × 5 hundreds; 3 × 5 is 15, and 10 × 100 is 1000, so the product is 15 000."
      Then these concepts must be applied to division: "420 ÷ 60 can be thought of as 60 × □ = 420, or 6 tens × □ = 42 tens. Since 6 × 7 = 42, and 10 × 1 is 10, the answer has to be 7 ones. So, 420 ÷ 60 = 7."

5. Difficulty using zeros in a product or quotient
   Errors:
   a. Children may compute $n \times 0$ as n in a product, or $0 \div n$ as n in a quotient.

   $$\begin{array}{r} 302 \\ \times \ 3 \\ \hline 936 \end{array} \qquad 3 \overline{\smash{)}603} = 231$$

b. Children may omit zeros in a quotient or product.

$$\begin{array}{r} 402 \\ \times \quad 3 \\ \hline 126 \end{array} \qquad \begin{array}{r} 8\,4 \\ 3\,\overline{\smash{\big)}\,2412} \end{array}$$

c. Children may ignore zeros in a divisor.

$$\begin{array}{r} 4\,1 \\ 60\,\overline{\smash{\big)}\,246} \end{array}$$

Discussion and suggestions for remediation. In each of these situations, children's concepts are based on memorization without understanding. In 5a, children confused $n \times 0$ and $0 \div n$ with $n \times 1$ and $n \div 1$. A sound explanation of why $n \times 0$ and $0 \div n$ are 0 is necessary. These explanations are provided in Chapter 4. In 5b children do not know that they must use zero to tell how many tens. Children who have this difficulty sometimes reason that since zero represents "nothing," there is no reason to record it. These children should review computational procedures and learn to justify the recording of each digit in a product or quotient. For example, "3×402 means 3×2 ones $+ 3 \times 0$ tens $+ 3 \times 4$ hundreds. We record 0 in the tens place for the product to show the results of 3×0 tens." When computing $2412 \div 3$, children who have learned rote routines for dividing may say, "3 won't go into 2, so go to the next number." With a division concept such as this, what prevents them from saying "3 goes into 24 eight times, but 3 won't go into 1, so I'll go to the next number"? Nothing prevents them! And their result is:

$$\begin{array}{r} 8\,4 \\ 3\,\overline{\smash{\big)}\,2412} \end{array}$$

When dividing, children should be encouraged to make a record of each place-value position indicated in the dividend. They should be required to justify each step. When they have completed their computation, they can decide whether any of the zeros are unnecessary.

$$\begin{array}{r} \cancel{0}804 \\ 3\,\overline{\smash{\big)}\,2412} \end{array}$$

After children have used this method a number of times to complete division computations, they will be able to decide beforehand which zeros need to be recorded.

Children who make errors like those in 5c need to perform sets of computations involving divisors that are multiples of 1, 10, 100, and so on. They should review and use the distributive method of division to justify their procedures. Then they should compare the results of their computations.

Again, the strategy of mental arithmetic is appropriate.

$$\begin{array}{l} 2412 \div 3 = (2400 + 12) \div 3 \\ \qquad\qquad\quad\; 800 + 4 \qquad = 804 \end{array}$$

This would have been preceded by questions such as, "How much is 600 ÷ 3?" No pencils allowed!

6. Difficulty using the distributive property of multiplication over addition when computing products
Error:
a. Children may not complete a multiplication exercise.

$$\begin{array}{r} 32 \\ \times\ \ 4 \\ \hline 8 \end{array} \qquad \begin{array}{r} 41 \\ \times\ 37 \\ \hline 287 \end{array}$$

Discussion and suggestions for remediation. Arithmetic experience prior to multiplication involves addition and subtraction. In these experiences, children are constantly reminded to "keep their columns straight" when computing. They are told to "add ones to ones, tens to tens," and so on. They are told to "subtract ones from ones, tens from tens." Unless development or remediation of multiplication involves models illustrating why it makes sense to distribute when computing, children will continue to use rules they learned for addition and subtraction.

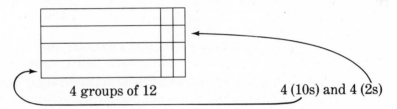

4 groups of 12   4 (10s) and 4 (2s)

7. Difficulty regrouping when computing products and quotients
Errors:
a. Does not regroup.

$$\begin{array}{r} 25 \\ \times\ \ 3 \\ \hline 615 \end{array} \qquad \begin{array}{r} 1221 \\ 3\overline{\smash{)}4683} \end{array}$$

b. Regroups incorrectly.

$$\begin{array}{r} ^1 39 \\ \times\ \ 9 \\ \hline 288 \end{array} \qquad \begin{array}{r} ^{3}{}_2 37 \\ \times\ 54 \\ \hline 158 \\ 205 \\ \hline 2208 \end{array}$$

c. Does not account for regrouping step.

$$\begin{array}{r} 42 \\ \times\ \ 6 \\ \hline 242 \end{array}$$

Discussion and suggestions for remediation. We're sure that by now you've gotten the message. You must explain and demonstrate regrouping steps anytime you teach computation of whole numbers. First, you must teach regrouping as it applies to naming numbers, then you must extend the regrouping ideas to the operation involved. Furthermore, demonstrations must be done with concrete models. Have children estimate whenever possible and *think.*

1) Three quarters equals 75 cents.
   Distribute $4683 (in play money) to you and two friends.
2) 39×9 $40 \times 10 = 400$ .... 288 isn't close.
   37×54 .... Use a calculator.
3) 42×6 .... 6 (40s) and 6 (2s) = 240 + 12 = 252

8. Difficulty aligning partial products
   Error:
   a. Aligns incorrectly.

$$
\begin{array}{r}
56 \\
\times\ 28 \\
\hline
448 \\
112 \\
\hline
560
\end{array}
$$

Discussion and suggestions for remediation. Children make this mistake when they do not understand place-value concepts and how these concepts apply to multiplication. Sometimes children are taught to say "two times six" when referring to computing the indicated step in this computation:

$$
\begin{array}{r}
56 \\
\times\ 28 \\
\hline
448 \\
1120 \\
\hline
1568
\end{array}
$$

What they should learn to think is:

This is 8×56 and then 20×56.
Estimate: $30 \times 60 = 1800$.
560 isn't even close.
That's only 10×56.

There are other difficulties that children may encounter, but these have been covered in one way or another in our previous discussions. For example, a child may make errors in division because he or she cannot multiply or subtract; so it is back to multiplication and subtraction.

■ Problem solving

The crucial aspect of whole number arithmetic is not *how* we add, subtract, multiply, and divide but *when*. This is what we mean when we say that children have trouble with interpretations of the operations. They cannot match situations with operations effectively. In other words, they don't know when to do what.

Diagnosis is easy. They have selected the wrong operation. Remediation is obvious. They need frequent experiences that require them to do this. The 0–9 cards and ASMD cards described in Chapter 4 can be used with the following story line, which can be changed to fit any level.

Latisha has four big tulips and three rows of roses with two in each row.

How do you find out?	How many roses?	M
	How many flowers?	M A
	Are there more roses or tulips?	
Sample story problems:	How many more?	M S
	If you shared them all with Mom and Grandma, how many would each one get?	M A D

Common difficulties involving fractions and decimals and suggestions for remediation

So far you have learned which procedures to use when diagnosing mathematical difficulties, and you have learned to identify difficulties for whole-number concepts. To complete this picture, you must learn to identify the difficulties that children encounter when learning concepts involving fractions and decimals. First, we start with fractions. Again, we will list some major difficulties, identify associated errors, and suggest remediation techniques. Here goes!

■ Difficulties associated with fractions

Children make a lot of errors when learning to compute and solve problems with fractions. Curiously enough, though, a great many of these errors can be classified in terms of only a few difficulties. The first of these difficulties is so subtle that it is often missed by even an experienced teacher.

Difficulty associating meaning with a fraction
Errors:
a. Children may make the following incorrect associations.

3 problem

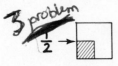

These errors are made when children think of a fraction such as $\frac{2}{5}$ as representing 2 out of 5 parts. These children do not require the 5 parts to have the same size or the same number.

b. Children may make the following incorrect associations.

These errors are made when children think of the denominator and the numerator as representing regions or sets that are distinct.

c. Children may not be able to correctly associate a fraction with division. For example, they may not think that "$\frac{3}{4}$ is the same as $3 \div 4$," or they may incorrectly write:

$$\frac{3}{4} = 4 \div 3 \text{ or } 3\overline{)4}$$

Discussion and suggestions for remediation. In many instructional programs, not enough emphasis is given to learning the meanings that any fraction has. Often children are quickly introduced to the idea that fractions represent numbers and then are rushed into computing before they appreciate *how* fractions represent numbers. With such a limited understanding of fractions, children make many errors. These occur when computing, renaming, and ordering fractions. Here are some examples:

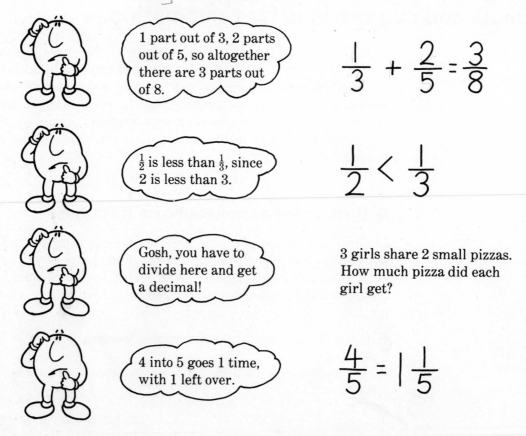

These errors and many similar ones could be corrected or, better yet, prevented by spending more time on systematically teaching three important ideas before introducing complicated computational procedures.

1) If a and b are whole numbers and $b \neq 0$, then the fraction $\frac{a}{b}$ represents a parts out of b parts where each part has the same size.

3 parts out of 4 parts. Each part has the same size.

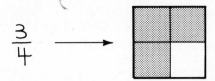

2) The fraction $\frac{a}{b}$ represents a sets out of b sets where each set has the same number.

3 sets with 2, out of 4 sets with 2.

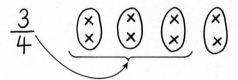

3) The fraction $\frac{a}{b}$ represents the quotient $a \div b$.

Mmm, 3 divided by 4.

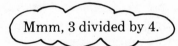

Only when children understand these ideas should they be introduced to more complicated ideas that involve computing, renaming, and ordering.

Another major difficulty that children have is related to understanding and using the equivalent-fraction rule. This rule is given in Chapter 6, but we will repeat it here:

If a, b, and c are whole numbers where b and c are not zero, then

$$\frac{a}{b} = \frac{a \times c}{b \times c} \qquad \frac{c}{c} = 1$$

You might say that this rule is as important to fractional concepts as place value is to whole-number concepts. This is because to order, add, subtract, multiply, divide, or simplify, you may in some way have to use the equivalent-fraction rule. So, the second difficulty for fractions is the following.

2. Difficulty using the equivalent-fraction rule

Errors:

a. Children may not be able to use appropriate methods to find equivalent fractions.

$$\frac{3}{4} = \frac{\square}{8} \longrightarrow ?$$

Student may choose 3 for $\square$, since 3 is the numerator of $\frac{3}{4}$. Student may choose 2 for $\square$, since $8 \div 4 = 2$. Student may choose 4 for $\square$, since the denominator of $\frac{3}{4}$ is 4.

b. Children may not simplify appropriately.

$$\frac{4}{8} = \frac{\square}{\square} \longrightarrow ?$$

Children may say 2, since $8 \div 4$ is 2.

c. $\frac{11}{8} = 3\frac{1}{8} \longrightarrow ?$

Children may say "8 into 11 goes 1 time with 3 left over, so it's $3\frac{1}{8}$."

$1\frac{1}{8} = \frac{8}{9}$

Children may say "$1 \times 8 = 8$ and 1 more makes 9, so $1\frac{1}{8} = \frac{8}{9}$."

Discussion and suggestions for remediation. The inability to correctly use the equivalent-fraction rule makes it impossible for children to order fractions and to compute with fractions correctly.

Remediation can make use of a sheet of plain paper folded into eighths.

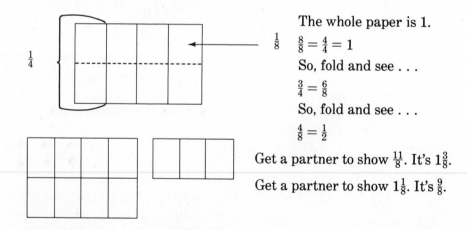

The whole paper is 1.

$\frac{1}{8}$ $\frac{8}{8} = \frac{4}{4} = 1$

So, fold and see . . .

$\frac{3}{4} = \frac{6}{8}$

So, fold and see . . .

$\frac{4}{8} = \frac{1}{2}$

Get a partner to show $\frac{11}{8}$. It's $1\frac{3}{8}$.

Get a partner to show $1\frac{1}{8}$. It's $\frac{9}{8}$.

We cannot overemphasize the importance of learning the equivalent-fraction rule. By "learning," however, we do *not* mean that children should simply memorize a statement of the rule. By "learning," we mean that children should be convinced that the rule works and should be able to demonstrate specific instances of the rule using concrete or pictorial models like the ones just given. There are many more activities in Chapter 6 that help children develop this awareness. Children should also learn to recognize instances when this rule is being applied; for example, "When I add fractions that have unlike denominators, I must use the equivalent-fraction rule to find equivalent fractions with the same denominator." Mechanical procedures for performing a skill involving the rule also should be avoided unless that procedure is backed up with thorough understanding.

A third difficulty that children have when learning fractional ideas is with the meaning of the term *common denominator*. Oh, they learn to say the term all right, and they learn mechanical routines that involve finding common denominators. But unfortunately, they often do so without purpose or understanding.

3. Difficulty applying appropriate uses of a common denominator
   Errors:
   a. Children may not be able to order fractions correctly.

   $\mathbf{3}$ $\quad \dfrac{3}{4} < \dfrac{5}{8}$ since $4 < 8$ and $3 < 5$

   b. Children may not add fractions appropriately.

   $\mathbf{3}$ $\quad \dfrac{2}{3} + \dfrac{1}{4} = \dfrac{2}{12} + \dfrac{1}{12} = \dfrac{3}{12} = \dfrac{1}{4}$ $\qquad \dfrac{2}{3} + \dfrac{1}{4} = \dfrac{3}{7}$

   c. Children may not subtract fractions appropriately.

   $\mathbf{3}$ $\quad \dfrac{2}{5} - \dfrac{1}{2} = \dfrac{1}{3}$ $\qquad \dfrac{4}{5} - \dfrac{2}{3} = \dfrac{4}{15} - \dfrac{2}{15} = \dfrac{2}{15}$

   Discussion and suggestions for remediation. In most practical instances, a fraction represents a measure. Thus, when we compare, add, or subtract fractions, we are in a sense comparing, adding, or subtracting measures. To do this accurately, we must be concerned with the unit of measure. Obviously, we can't compare, add, or subtract measures that involve different units without first translating those measures to equivalent ones involving the same unit. When we find a common denominator for two or more fractions, we are doing just that; we are translating two measures to equivalent ones that involve the same unit. Telling children to "find a common denominator" will not enable them to internalize this concept. Activities using models, such as the fraction strips (see Material Sheets 18A and B), that involve finding a common denominator in order to compare, add, or subtract fractions are necessary. Further, children should be able to explain that when they attempt to find fractions with a common denominator, they are looking for equivalent fractions where the denominators indicate "parts" that have the same size.

4. Difficulty making appropriate interpretations for forms such as

   $2\dfrac{2}{3}$ or $5\dfrac{3}{8}$

Error:

a. Children may incorrectly associate mixed numerals with fractions or other mixed numerals.

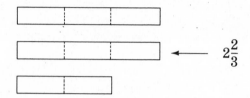

$$5\frac{3}{8} = \frac{23}{3} \qquad 4\frac{7}{8} = 1\frac{3}{8}$$

Discussion and suggestions for remediation. Children often don't understand that forms such as $2\frac{2}{3}$ mean the same as $2 + \frac{2}{3}$. So, when changing such forms to fractions, they use mechanical routines that are easily confused or forgotten. These children need experience with models such as fraction strips, where they compare models for expressions such as $2 + \frac{2}{3}$, $2\frac{2}{3}$, and $\frac{8}{3}$ and draw conclusions. (See Material Sheets 18A and B).

$$\longleftarrow \quad 2\frac{2}{3}$$

5. Difficulty with the meaning of the operations
   Error:
   a. The student may compute as follows:

$$20 \div \frac{1}{2} = 10$$
$$\frac{1}{3} \times \frac{3}{4} = \frac{4}{12}$$
$$2\frac{1}{3} \times 3\frac{3}{4} = 6\frac{3}{12}$$
$$\frac{3}{4} \div \frac{1}{4} = \frac{4}{3} \times \frac{1}{4} \text{ or } \frac{3}{4} \times \frac{1}{4} \text{ or } \frac{3}{4} \div \frac{4}{1} \text{ or } \frac{4}{3} \times \frac{4}{1}$$

Discussion and suggestions for remediation. Any of the computations just indicated could be done without the use of a formal algorithm if children knew the meaning of the operations. Here's how:

Correct computation

$$20 \div \frac{1}{2} = 40$$

Appropriate meaning

This does not mean $\frac{1}{2}$ of 20. It means "how many" $\frac{1}{2}$s are in 20. Since there are two $\frac{1}{2}$s in 1, there must be forty $\frac{1}{2}$s in 20. Gee, that's the same as multiplying by 2!

$$\frac{1}{3} \times \frac{3}{4} = \frac{1}{4}$$

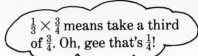

$\frac{1}{3} \times \frac{3}{4}$ means take a third of $\frac{3}{4}$. Oh, gee that's $\frac{1}{4}$!

In the third situation, the child simply needs to remember that to compute products we must use the distributive property. Since

$$2 \times 3 = 6$$
$$2 \times \frac{3}{4} = \frac{6}{4}$$
$$\frac{1}{3} \times 3 = 1$$
$$\frac{1}{3} \times \frac{3}{4} = \frac{1}{4}$$

This is multiplication, so I have to use the distributive property just like I did with whole numbers.

$$2\frac{1}{3} \times 3\frac{3}{4} = 8\frac{3}{4}$$

In the fourth situation, the child simply needs to interpret $\frac{3}{4} \div \frac{1}{4} = \square$ as asking "how many $\frac{1}{4}$s are there in $\frac{3}{4}$?"

$$\frac{3}{4} \div \frac{1}{4} = \square$$
$$\frac{3}{4} \div \frac{1}{4} = 3$$

How many $\frac{1}{4}$s are there in $\frac{3}{4}$? Hmmm. There are 3.

Of course, there are many more errors that children could make when working with fractions. But most of them are related to the preceding difficulties. In any event, we can safely say that a large percentage of errors are made because a child has one or more of these difficulties.

■ Difficulties using decimals

Now, what can we say about learning concepts involving decimal numerals that don't represent whole numbers? Actually, we can't say too much more than what we have already said. The meanings for these numerals depend on place-value and fractional ideas, and when we compute with these numerals we must know and understand whole-number algorithms. The only difference between computing with whole numbers and computing with decimal numbers is that we must make important decisions concerning the placement of the decimal point. When children have difficulties comparing decimal numerals or difficulties computing with them, these difficulties are related to the ones discussed in connection with place-value or whole-number computation or to difficulties listed for fractional concepts. Let's examine these claims.

1. *Difficulty associating meaning with a decimal.* If children confuse the meanings of .3, .03, and .003, then either they can't identify place-value positions in these numerals or they don't understand the meaning of 3 tenths $\left(\frac{3}{10}\right)$, 3 hundredths $\left(\frac{3}{100}\right)$, or three thousandths $\left(\frac{3}{1000}\right)$.

2. *Difficulty with place value and with the equivalent fraction rule.* If children think that .29 > .3, then they do not understand that .29 is 29 hundredths, and .3 is 3 tenths or 30 hundredths. They are unable to think "3 tenths is the same as 30 hundredths, so .3 is greater than .29." What are the difficulties? The difficulties involve appropriate place-value and fractional interpretations, as well as being able to use the equivalent-fraction rule.

3. *Difficulty with place value and with common denominators in addition and subtraction.* If children compute 32.4 + .25 as 3.49, or if they compute 2.5 − .11 as .14, then it is obvious that they don't realize that when we compute sums with decimal numerals, we must observe certain place-value principles. We must add ones to ones, tenths to tenths, and hundredths to hundredths, and we must subtract in this same manner.

4. *Difficulty with the meaning of the operations and in distinguishing between various rules for the operations.*

   a. If children compute as in the following examples, then there is a pretty good chance that they have no notion what the decimal point in a numeral indicates. Obviously, they have heard a rule for multiplication that they are applying when adding and subtracting.

$$
\begin{array}{r}
32.4 \\
+\ \ .25 \\
\hline
3.265
\end{array}
\qquad
\begin{array}{r}
14.32 \\
+\ 1.41 \\
\hline
.1573
\end{array}
$$

   To remediate, the teacher must reteach the meaning of decimal numerals and the associated fractional ideas.

   b. If children compute products as in the following examples, then, again, it is obvious that they don't understand the significance of the decimal point and probably don't recognize that these products can be "thought of" in terms of fractions, which makes determining the position of the decimal point relatively easy.

$$
.3 \times .4 = 1.2
\qquad
\begin{array}{r}
3.4 \\
\times\ \ .2 \\
\hline
6.8
\end{array}
$$

   Children should think

$$\frac{3}{10} \times \frac{4}{10} = \frac{12}{100}$$

c. If children compute quotients as follows, then, again, they're just pushing the decimal point around without understanding why. In some cases, in fact, they're ignoring the decimal point.

$$.05\overline{)\overset{.1}{.50}} \qquad .46\overline{)\overset{2}{92}}$$

Children need to know the equivalent-fraction rule. In each of the above two examples, they can multiply both numbers by 100. They can usually estimate, and they can use money concepts to help learn about decimals. "How many nickels are in fifty cents?" is an exact representation for the first example. "About how many half-dollars would you get for 100 dollars?" provides a reasonable estimate for the second. They need to be able to relate fractional notation and the division operation, and they need to recognize when it's important to record zero as a place holder. There are no shortcuts if we want children to compute efficiently and to apply these skills to problem-solving situations. We cannot depend on routines that are memorized and not understood. We must make sure that students understand the meaning of the numerals with which they compute, understand the interpretations for the arithmetic operations, and understand how the steps in the algorithms are related to these meanings. Sometimes, teaching and learning meaning takes longer than just teaching and learning a rule or a set of steps. But in the long run, it's far more efficient. This kind of teaching prevents errors before they occur.

➡ Reading and writing difficulties

By now, you're certainly aware that there are many mathematical concepts and skills to be acquired during the elementary school years. Acquiring these concepts and skills involves learning to interpret and use symbols, terms, and procedures different from those used in ordinary language. Unfortunately, even when learning conditions are at their very best, many children experience difficulties when reading and writing mathematical ideas. In this section, we will do things a little bit differently. Instead of discussing specific difficulties, we will discuss factors that may result in difficulties as children learn to read and write mathematics. By being aware of these factors, you will be able to anticipate difficulties and help children through periods of confusion.

1. Mathematical terms and symbols are based on precise definitions that are not generally learned in out-of-school environments. Learning to read, write, and use words such as *dog* and *cat* is easier than learning words such as *area, quadrilateral, division,* and so on. In the process of just "growing up," children encounter many examples of "not-dog" and "not-cat." By the time most children are 2 or 3 years old, they can successfully classify both St. Bernards and poodles as dogs, even though these two types are very different. In contrast, learning to read and write terms such as *area* requires structured experiences in which children use appropriate vocabulary to describe actions on objects; for instance, it takes 10 tiles to cover the first rectangle below but only 8 to cover the second, so the area of the first rectangle is greater than the area of the second. The areas of the two figures that take 8 tiles to cover are the same.

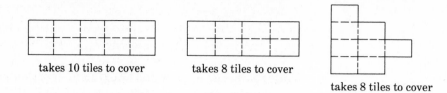

takes 10 tiles to cover   takes 8 tiles to cover

takes 8 tiles to cover

2. Sometimes words that have precise mathematical meanings are used ambiguously in everyday conversation. Such casual expressions as "the dining room area" or "a nice area of town" can certainly interfere with children's development of precise ideas about what "area" means in mathematics. Is area a measure? If so, what does the measure describe? Is area a geometric figure? If so, what kind—a closed region, a space figure, or what?

3. Many mathematical words have other meanings in ordinary usage. Children must be familiar enough with associated concepts to use contextual clues to distinguish among the various meanings of a word (or of words that sound alike, such as "sum" and "some").

a. The *product* is on sale at a 50% discount.
   The *product* of 3 and 5 is 15.
b. *Some* people came to the benefit before dinner, and *some* came after.
   The final *sum* indicated that the number of people who came to the benefit was greater than usual.
c. The *set* of whole numbers is *closed under* the *operation* of addition.
   Please *close* the door *under* the shelf and *set* the table. I'd help, but my arm still hurts from the *operation*.

4. Some words are used inappropriately in mathematical contexts. For example, "To multiply a number such as 34 by 10, just add a zero to the right of 34." What *should* be said is, "*Annex* a zero to the right of the numeral 34." Obviously, if you *add* a zero to 34 (34 + 0), you get 34. How about statements such as "Reduce the fraction $\frac{5}{10}$." Does this mean that you are to make $\frac{5}{10}$ smaller? After all, *people* who reduce get smaller!

5. The names and meanings of many mathematical symbols cannot be determined by looking at the symbols. There are no "attack skills" for decoding symbols such as $+$, $-$, $\sqrt{}$, x, and b^2. Mathematical symbols can be learned only through definition and appropriate usage with correct vocabulary, models, and applications.

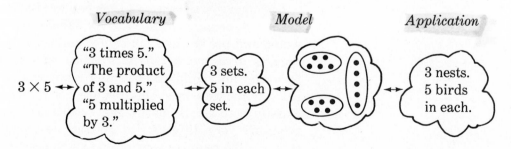

Vocabulary   *Model*   *Application*

$3 \times 5 \leftrightarrow$ "3 times 5."
"The product of 3 and 5."
"5 multiplied by 3."
$\leftrightarrow$ 3 sets. 5 in each set. $\leftrightarrow$ $\leftrightarrow$ 3 nests. 5 birds in each.

6. Sometimes inappropriate or misleading visual models are given to illustrate the meaning of mathematical terms and ideas. Children who are shown only figures such as that in illustration A (below) in connection with the term *parallelogram* get the mistaken idea that rectangles and squares are not parallelograms. But a parallelogram is any four-sided polygon whose opposite sides are parallel. Certainly, squares and rectangles meet this criterion. A more appropriate model for visually illustrating the meaning of *parallelogram* is illustration B.

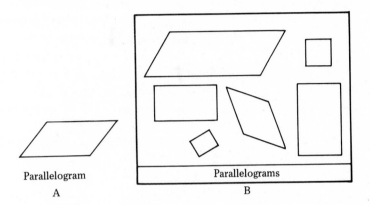

Parallelogram
A

Parallelograms
B

How about the number line for illustrating addition and subtraction ideas? This model is often used in the first and second grades, yet there is evidence that some children cannot conserve length or read a scale until later. More appropriate models for illustrating number concepts in the early grades involve sets of objects, such as beads or tiles.

7. Mathematical language is more concise than ordinary language. To achieve this conciseness, rules for connecting terms, phrases, and symbols are used. The sentence "If m and n are whole numbers, then $(3 \times m) + 1 = n$" cannot be glibly read over if serious interpretations are to be made. It must be understood: first, that m and n are *variables;* second, that the sentence is an *open sentence;* third, that the sentence represents an *equality;* and fourth, that the parentheses indicate the *order* in which calculations are to be made. Even with these understandings, children might limit their interpretation to sentences such as, "The sum of 3 times a certain whole number, m, and 1 is another whole number, n." The payoff comes when they are able to see such a sentence in relationship to models and applications. Children should read mathematical material slowly and carefully, and they should learn a variety of ways for translating these communications.

To help children to more fully appreciate the sentence $3m + 1 = n$, we might do one of the following:

1. *Make a chart.*

m	$3m + 1$, or n
1	$(3 \times 1) + 1$ or 4
2	$(3 \times 2) + 1$ or 7
3	$(3 \times 3) + 1$ or 10
(Etc.)	(Etc.)

2. *Look at some models.*

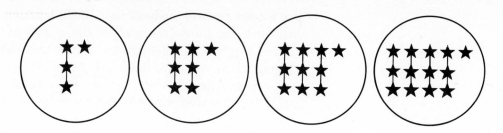

3. *Dream up some applications.* "You get one nickel for good conduct and three nickels for every star received for good work."

4. *Construct a graph.*

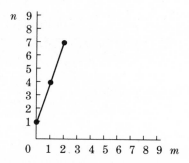

m	n
0	1
1	4
2	7
(Etc.)	(Etc.)

8. *The organization of mathematical communications differs from the organization of ordinary reading materials.* Young children are just learning to read from left to right and top to bottom when they are confronted with numerous patterns for organizing mathematical ideas. Fractions are read from top to bottom.

$$\frac{3}{4} \qquad \frac{4}{8} \qquad \frac{9}{10}$$

Some charts are read from left to right, some from top to bottom, and others from bottom to top.

Name	Marge	Lou	Bill	Sue
Height in inches	56	62	57	62
Weight in pounds	102	104	100	105

Name	Height in inches
Marge	56
Lou	62
Bill	57

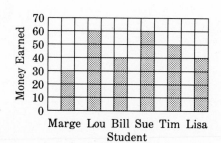

To fill in addition and multiplication tables, children must find the intersection of a row and a column. Needless to say, this is very difficult for children who are not secure in using horizontal and vertical references.

×	0	1	2	3	4	5	6	7	8	9
0										
1						5				
2								16		
3										
4								28		

Number sentences are written horizontally, but most computational forms are written vertically. Also, word problems are read from left to right, but computations are performed from right to left.

$$5 = 2 + 3 \qquad \begin{array}{r} \overset{2\,1}{7.43} \\ \times \quad 6.2 \\ \hline 1486 \\ 4458 \\ \hline 46.066 \end{array}$$

The eight points that we have reviewed touch on some of the critical aspects of learning to read and write mathematical ideas. Thus, they should be considered whenever you are doing any of the following:

1. Reviewing textbooks and other instructional materials for children, including teacher-made materials.
   a. Are the methods for communicating concepts compatible with children's maturity?
   b. Is the development of ideas logical? Are appropriate models and applications given? Is meaningful practice given?
   c. Are pages uncluttered? Do tables, charts, diagrams, and pictures aid the development of concepts or are they distracting?
   d. Is the print legible? Are visual cues given to indicate how the material is to be read (left to right or top to bottom)?
2. Presenting or reinforcing mathematical concepts.
   a. Have children developed the concepts necessary for understanding the language you use?
   b. Can you find intuitive, nontechnical ways of expressing ideas that will lead to clear conceptualization?
   c. Are you sure that you're not using words or models that will mislead children?
3. Evaluating student progress.
   a. Are errors made because children do not understand the concepts, or are they made because children do not understand the *language and symbols* used to express the concepts?

The computer corner

In this chapter, we have covered many important topics on the diagnosis of mathematical difficulties. We have neglected, however, to discuss how computer technology can assume many of the tasks in identifying and analyzing mathematical difficulties that children may encounter. In Chapter 16, we discuss computer technology and its potential for managing the instructional environment. But before we leave this chapter, we feel we should give you a small peek at what is possible on a much grander scale.

Load the MGB Diskette, then run the program DIAGNOSE. This program will ask for your name and the date and then will give you a small inventory on arithmetic skills. You are to make errors that you think children will make. Observe what the program does after you complete the inventory. You should get diagnostic information regarding your errors.

Think about the program provided here and imagine one that is infinitely more powerful. Consider a program that can grade tests and analyze student errors to determine difficulties, or one that contains an inventory of all instructional materials in the environment—a system that can prescribe, deliver, and monitor activities that prevent and remediate difficulties. The teacher's job would be to orchestrate, making sure that substantial materials are keyed to learning objectives and so on. In this setting there might be more time than is usually available for teacher-student interaction.

■ Computer tips

When purchasing computer equipment for a school for instructional purposes, it is generally not wise to invest all the budget on one type of machine. Equipment should be purchased in terms of educational goals. Sophisticated machines should be purchased to handle management tasks, the development of instructional materials, and the delivery of in-depth instruction. Less sophisticated machines should be purchased for executing educational programs with limited objectives. Ideally, it would be best to purchase different machines with differing capabilities and integrate their uses into a substantial instructional program.

In closing . . .

In this chapter, we have discussed difficulties that children may encounter when dealing with whole numbers, fractions, and decimal numerals and the difficulties they may have when reading and writing mathematical ideas. Suggestions for remediation have also been given, and we have encouraged estimation and the use of physical models.

The following chart lists the difficulties discussed in this chapter.

Common mathematical difficulties

Place value	*Addition/subtraction*	*Multiplication/division*
1. Difficulty associating place-value models with numerals. 2. Difficulty using zero to write numerals. 3. Difficulty using regrouping concepts. 4. Difficulty naming place-value positions in a numeral. 5. Difficulty giving nonstandard place-value representations for a numeral.	1. Difficulty identifying addition or subtraction situations. 2. Difficulty using counting to find basic addition facts. 3. Difficulty using zeros when computing. 4. Difficulty using counting to find differences related to addition facts. 5. Difficulty regrouping. 6. Difficulty computing with numerals that don't have the same number of digits. 7. Difficulty when a sum involves several addends or when a sum or difference involves large numbers.	1. Difficulty identifying multiplication or division situations. 2. Difficulty determining the basic multiplication facts. 3. Difficulty using multiplication facts to find related quotients. 4. Difficulty applying place-value concepts and basic facts to find products and quotients of multiples of ten. 5. Difficulty using zeros in a product or a quotient. 6. Difficulty using the distributive property of multiplication over addition. 7. Difficulty in regrouping. 8. Difficulty in aligning partial products.

Common mathematical difficulties

Fractions	*Decimals*
1. Difficulty associating meaning with a fraction.	1. Difficulty associating meaning with a decimal.
2. Difficulty using the equivalent-fraction rule.	2. Difficulty with place value and the equivalent-fraction rule.
3. Difficulty applying appropriate uses of a common denominator.	3. Difficulty with place value and with common denominators in addition and subtraction.
4. Difficulty making appropriate interpretations for mixed numerals.	4. Difficulty with the meaning of the operations and in distinguishing between various rules for the operations.
5. Difficulty with the meaning of the operations.	

Reading and writing difficulties

1. Mathematical terms and symbols are based on definitions that are not generally learned outside of school.

2. Sometimes words with precise mathematical meanings are used ambiguously in everyday conversation.

3. Many mathematical words have other meanings in ordinary usage.

4. Some words are used inappropriately in mathematical contexts.

5. The names and meanings of many mathematical symbols cannot be determined by looking at the symbols.

6. Sometimes inappropriate or misleading visual models are given to illustrate the meaning of mathematical terms and ideas.

7. Mathematical language is more concise than ordinary language.

8. The organization of mathematical communications differs from the organization of ordinary reading materials.

We would like to stress that the best "remediation" is prevention. But whether the teacher is preventing or remediating, the key is to design activities that build sound concepts, rather than encouraging rote memorization. Once children have built adequate concepts, then practice activities are in order. Not only does practice enable children to reinforce these concepts, but practice, supported by understanding, provides opportunities for children to experience success.

It's think-tank time

1. Relate each of the specific errors given below to a mathematical difficulty listed in this chapter.

a.
$$\begin{array}{r} 35 \\ + 29 \\ \hline 514 \end{array}$$

b.
$$\begin{array}{r} 311 \\ 4\,\overline{)1374} \end{array}$$

c.
$$\begin{array}{r} 32 \\ + 5 \\ \hline 87 \end{array}$$

d.
$$\begin{array}{r} 400 \\ - 163 \\ \hline 363 \end{array}$$

e.
$$\begin{array}{r} 48 \\ \times 52 \\ \hline 2016 \end{array}$$

f.
$$\begin{array}{r} 34 \\ 8\,\overline{)2432} \end{array}$$

g.
$$\begin{array}{r} 408 \\ 40\,\overline{)1632} \end{array}$$

h.
$$\begin{array}{r} 37 \\ \times 42 \\ \hline 74 \\ 148 \\ \hline 222 \end{array}$$

i.
$$\begin{array}{r} 2 \\ 17\,\overline{)3473} \\ 34 \\ \hline 0 \end{array}$$

j.
$$\begin{array}{r} 603 \\ \times\ 2 \\ \hline 126 \end{array}$$

k.
$$\begin{array}{r} 718 \\ 9\,\overline{)716} \\ 63 \\ \hline 12 \\ 9 \\ \hline 76 \\ 72 \\ \hline 4 \end{array}$$

l.
$$\begin{array}{r} 235 \\ + 601 \\ \hline 806 \end{array}$$

m. $3.4 + .08 = .42$

n. $2\frac{3}{4} + 4\frac{1}{3} = \frac{9}{12} + \frac{4}{12} =$
$\frac{13}{12} = 1\frac{1}{12}$

o.
$$\begin{array}{r} 3.2 \\ \times\ .5 \\ \hline 16.0 \end{array}$$

p. $\dfrac{2}{3} = \dfrac{2}{6}$ q. $\dfrac{1}{3} + \dfrac{2}{4} = \dfrac{3}{7}$

r. $\dfrac{3}{5} - \dfrac{1}{4} = \dfrac{3}{20} - \dfrac{1}{20} = \dfrac{2}{20}$ or $\dfrac{1}{10}$

s. $\dfrac{5}{10} = 2$ t. $\dfrac{\cancel{2}}{\cancel{3}} \times \dfrac{\cancel{3}}{\cancel{8}} = 4$

u. $.03\,\overline{\smash{\big)}\,9}\;\;^{.3}$ v. $10 \div \dfrac{1}{5} = 2$

w. $\dfrac{3}{4} \div \dfrac{1}{8} = \dfrac{\overset{1}{\cancel{4}}}{3} \times \dfrac{1}{\underset{2}{\cancel{8}}} = \dfrac{1}{6}$

2. For each difficulty listed in this chapter, make a 5 × 8 file card. On each file card, list a brief outline for remediation, then find activities or discussions given in Chapters 2, 3, 4, 6, and 7 that could be used to help you carry out the remediation activities.
3. Represent all the kinds of errors you can think of that involve (a) regrouping and (b) zeros.

Suggested readings

Ashlock, Robert B. *Error Patterns in Computation* (2nd Ed.). Columbus, Oh: Charles E. Merrill, 1976.

Ballew, Hunter, and Cunningham, James W. "Diagnosing Strengths and Weaknesses of Sixth-Grade Students in Solving Word Problems." *Journal for Research in Mathematics Education*, May 1982, pp. 202–210.

Baroody, A. J. "Children's Difficulties in Subtraction: Some Causes and Questions." *Journal for Research in Mathematics Education*, 15(1984):203–213.

Beattie, John, and Algozzine, Bob. "Testing for Teaching." *The Arithmetic Teacher*, September 1982, pp. 47–51.

Driscoll, Mark J. "Unlocking the Mind of a Child: Teaching Remediation in Mathematics." In *Research within Reach, Elementary School Mathematics*. St. Louis:R&D Interpretive Services, Inc., 1981.

Engelhardt, J. M. "Analysis of Children's Computational Errors: A Qualitative Approach." *British Journal of Educational Psychology*, 47(June 1977):149–154.

Engelhardt, J. M. "Using Computational Errors in Diagnostic Teaching." *The Arithmetic Teacher*, April 1982, pp. 16–19.

Hambleton, Ronald K., and Swaminathan, Hariharan. "Criterion-Referenced Testing and Measurement: A Review of Technical Issues and Developments." *Review of Educational Research*, 48(1)(Winter 1978):1–47.

Liedtke, Werner. "Learning Difficulties: Helping Young Children with Mathematics–Subtraction." *The Arithmetic Teacher*, December 1982, pp. 21–23.

Mattair, Judy Elizabeth Moore. "The Use of Error Patterns Analysis in the Diagnosis and Remediation of Whole Number Comprehension Difficulties." (Doctoral dissertation, Texas A & M University, 1981). *Dissertation Abstracts International*, 42A(1982):4343A.

Rosner, Jerome. *Helping Children Overcome Learning Difficulties* (2nd Ed.). New York: Walker and Co., 1979.

Sapir, Selma G., and Nitzburg, Ann C. (Eds.). *Normal and Deviant Learning Patterns in Children with Learning Problems*. New York: Brunner/Mazel, 1973.

Simon, Anita, and Byram, Claudia. *You've Got to Reach Them to Teach Them*. Dallas, Tex: TA Press, 1977.

Smith, Sally L. *No Easy Answers—The Learning Disabled Child*. Rockville, Md.: National Institute for Mental Health, 1978.

Troutman, Andria P. "Error Sources and Diagnosis." In *Diagnosis: An Instructional Aid, Levels A and B*. Chicago, Ill.: Science Research Associates, 1979.

Troutman, Andria P., and Lowe, Kenneth. "Identification and Remediation of Mathematical Difficulties Students Experience as Evidenced by the Florida State Assessment Program." Unpublished paper presented at the Sixth National Conference on Diagnostic and Prescriptive Mathematics, Tampa, Florida, 1979.

Underhill, R., Uprichard, E., and Heddens, J. *Diagnosing Mathematical Difficulties*. Columbus, Oh: Charles E. Merrill, 1979.

Putting It All Together

Problem Solving

■ *In this chapter, we describe a general view of problem solving and identify related abilities. Many of the activities presented in this book could be classified according to the problem-solving abilities that they require. Such a classification would lead to a general scheme for selecting problem-solving activities for all levels of elementary instruction.*

We believe that children can no more be taught to be great "problem solvers" than they can be taught to be great painters or musicians. But problem solving, like other art forms, does require the development of specific abilities, and teachers can provide the experiences necessary for children's development of these abilities. ■

Objectives for the child

1. Indicates the appropriate operation for a word problem.
2. Writes simple number sentences to correspond to specific actions with sets.
3. Selects number sentences for word problems.
4. Uses number sentences to classify simple word problems.
5. Uses a number sentence with two variables to express a simple relationship between two sets of data.
6. Finds ordered pairs of numbers that can be used to replace the variables in a two-variable number sentence, producing a true statement.
7. Determines whether a number sentence is true or false.
8. Uses symbols of inclusion appropriately.
9. Constructs and uses simple graphs.
10. Solves simple number sentences with one variable.
11. Lists attributes of an object or mathematical idea.
12. Translates a mathematical communication into a variety of forms.
13. Identifies similarities and differences among objects and among mathematical ideas.
14. Classifies objects and mathematical ideas according to some attribute.
15. Determines when information given in a problem is sufficient and relevant or irrelevant.
16. Finds relationships and patterns.
17. Systematically determines cases or alternatives.
18. Approximates.
19. Extends given information.
20. Compares objects and mathematical ideas with a set of criteria.

Teacher goals

1. Be able to identify problem-solving abilities that children should develop.

2. Be able to identify and carry out activities that will aid in the development of problem-solving abilities.

What comes next? $\frac{3}{8}, \frac{3}{4}, 1\frac{1}{2}, 3, \ldots$

If 3 doughnuts and 1 fudgecake together weigh 12 cupcakes, and if 1 fudgecake weighs 8 cupcakes, how many cupcakes does a doughnut weigh?

What is the area of the shaded region?

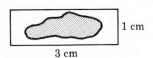

Probably the most talked about but least understood topic in mathematics education is *problem solving*. Most teachers vow that one of their most important tasks is to provide learning experiences that help children develop problem-solving abilities. Most teachers also admit, however, that it's not always easy to know how or where to begin providing these experiences. How do teachers choose content, select and organize materials, and direct questions in order to encourage problem solving? How do they capitalize on problem-solving situations that spontaneously occur in the classroom? How do teachers transform ordinary comments, questions, or tasks into activities requiring children to search deeper into an idea? What behaviors should teachers solicit or encourage?

We can't give complete answers to these questions—no one can. Still, we *can* identify general abilities related to efficient problem solving, and we can come up with ways to help children develop them.

We have drawn heavily on Polya's four-step model for problem solving, first published in 1945. This was very simply stated:

1. Understand the problem
2. Devise a plan
3. Carry out the plan
4. Look back

We have also examined research since then (and there is a lot of it) concerning this topic.

Bloom's Taxonomy of Educational Objectives is another resource that is helpful in providing insight into this topic. Bloom's six levels of thinking skills are knowledge, comprehension, application, analysis, synthesis, and evaluation. We are clearly suggesting that problem solving in mathematics involves the latter, or higher-order, thinking skills.

Actually, this entire book is about problem solving, from the presentation of classification ideas in Chapter 1 to the discussion of ways to use computers in Chapter 16. In fact, all of the abilities that we identify in this chapter have been developed implicitly elsewhere in this book. Here, we focus our attention directly on these abilities, making them explicit and providing a conceptual framework for an understanding of problem solving.

What is problem solving?

Before we identify problem-solving abilities, we need to say what problem solving *is*. There are two kinds of problem solving. The first involves using established procedures to solve routine problems. Related abilities include being familiar with routine problems and knowing how to carry out standard procedures. Understanding what procedure to apply to what routine problem is also handy!

■ Routine problems

What's the area of a 100 meter × 100 meter car lot?
There is a 4% sales tax on the cost of clothing. What's the tax on a
   $25 dress?
An employee makes $8.50 per hour. How much will the employee
   make in 40 hours?

The second kind of problem solving occurs when an individual is confronted with an unusual problem situation and is not aware of a standard procedure for solving it. The individual thus has to create a procedure. To do so, he or she must become familiar with the problem situation, collect appropriate information, identify an efficient strategy, and use the strategy to solve the problem.

■ Unusual problems

What's the distance between your ears?
Approximately how many hairs are there on your head?
How long does it take a person to react to a certain signal?
Separate this region into four congruent regions.

Both kinds of problem solving are important and should be practiced at every level in the elementary school. After all, that's what mathematics is all about—solving problems! The routine kind should be emphasized because of the many everyday decisions that involve routine problems and standard procedures. The unusual kind should be emphasized because as children grow into adulthood, they will constantly be required to solve problems for which there will be no ready-made procedures.

The following table illustrates a model for the problem-solving process discussed in this chapter. The model has two main parts. Part 1 indicates a set of steps for solving routine problems. Part 2 gives steps for solving nonroutine or unusual problems.

The problem-solving process

Part 1: Routine or familiar problems	*Part 2: Nonroutine or unusual problems*
1. Select established procedures.	1. Become familiar with the problem.
2. Carry out procedures and find solutions.	2. Collect information related to the problem.
3. Evaluate the solutions.	3. Devise strategies for solving the problem. Evaluate the strategies.
	4. Select a strategy and carry it out to find solutions. Evaluate the solutions.

Solving routine problems

There are many kinds of routine problems that students should learn to solve before they leave secondary school. But in the elementary school, the focus should be on solving word problems involving one variable and the four basic operations, as well as learning how to characterize problem situations involving two variables with appropriate number sentences. Most of these experiences should involve whole numbers. Situations involving fractions and decimals should be used primarily to extend problem-solving ideas and to demonstrate the connection between computation and applications involving rational numbers.

To solve word problems, children must learn to translate a problem situation into a number sentence or number expression that compactly represents the problem situation and indicates which operations are to be performed and in what order.

Children must also learn to solve number sentences and evaluate number expressions. These activities involve learning to compute efficiently. In many real-life situations, it is not convenient to compute, so a good problem solver must be able to estimate. Sometimes problems involve two or more variables. Solving these problem situations involves writing the appropriate number sentence and identifying number pairs that make the sentence true.

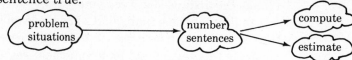

■ Translating word problems involving one operation

To translate a problem situation involving one operation into a number sentence requires children to be familiar with the different interpretations for each operation. For example, children must recognize that different types of problems can be solved using this number sentence: $7 - 3 = 4$.

Take away:	7 birds on a fence, 3 fly away. How many are left?	$7 - 3 = 4$
Comparison:	Sam has 7 hats, his brother has 3. How many more does Sam have than his brother?	$7 - 3 = 4$
Missing addend:	Jan needs 7 dollars, she has 3. How many does she need?	$7 - 3 = 4$
Partition:	Marion has 7 toy trucks, 3 are red, the rest are blue. How many are blue?	$7 - 3 = 4$
Decreasing:	A punchbowl has 7 liters of punch. We drank 3 liters. How much is left?	$7 - 3 = 4$

In Chapters 3 and 4, we discussed in detail the interpretations for each whole-number operation. We also described detailed methods for teaching children to write number sentences for each interpretation. In Chapters 6 and 7, we did the same for rational numbers. If you remember these interpretations, read on! If not, you should review before you read the rest of this chapter. In any event, the following chart summarizes these interpretations. See if you can identify a word problem for each.

Addition	Subtraction	Multiplication	Division
Combination	Take away	Additive	Distributive
Static	Comparison	Row by column	Subtractive
	Missing addend	Combination	
	Partition		
Increasing . . .	Decreasing		
(These situations involve measurement ideas such as gaining or losing weight instead of sets of objects.)			

Now assume that children at the appointed time learn the interpretations for the given operations; that is, in the first grade they begin to learn the interpretations for addition and subtraction, and in the third grade they do likewise for multiplication and division. Is this enough? No! Children need many learning experiences that challenge their abilities to recognize and use these interpretations.

1. Children should learn to pick appropriate information from a problem situation. Instructionally, we tend to compartmentalize the teaching of arithmetic. First we teach addition, then subtraction, and so on. Consequently, children often learn to solve addition problems only after spending days on addition computation. The same is often true for the other operations. So, many children who have been computing sums are "programmed" to add. When confronting word problems on paper, therefore, they do not bother to read. When given word problems orally, they do not listen carefully or analyze them. They simply pick out the numbers and add. To make sure that children are reasoning, we need to provide learning experiences where children solve problems that include extra information. Of course, this type of learning experience should occur after children have had sufficient experiences solving problems that include information that is just adequate.

Here is an example activity in which children concentrate on finding the right information in problem situations. Notice that the numbers used are not large and that students are not asked to compute. The emphasis is entirely on choosing the right information and writing the correct number sentence.

Example 9-1   **C-1, 2, 3**
Eliminating extra information in a word problem

Materials: Cards with simple word problems involving the arithmetic operation being studied. Each word problem includes extra information. Numbers used are small.

> Shane is 17 years old. His brother is 15, and his sister is 13. How much older is Shane than his sister?

Teacher directions: Draw a card. Read the problem on the card (children may need teacher assistance). Tell what information is necessary and what is not necessary. Write a number sentence.

---

2. Children should learn to identify problem situations containing insufficient information. They should develop a sense of what is realistic and what is practical in a problem situation. They should be able to identify problem situations where insufficient information is given and should learn to supply information that is plausible. The following is a sample activity that illustrates this idea. Notice that this example, like Example 9-1, does not use large numbers and does not require computation.

Example 9-2
Identifying needed information in a word problem

C-1, 2, 3

Materials: Teacher-made ditto with simple word problems involving arithmetic operation being studied. Each word problem has missing information. Numbers used are small.

Sample problem:
There are 5 Easter baskets.
Each basket has the same number of eggs.
How many eggs are there in all?

Teacher directions: Read each problem (teacher may have to read problems with children). For each problem, tell what information is missing. Give information that you think would work, then write a number sentence. (Teacher should discuss student responses.)

---

3. *Children should learn to distinguish among the operations.* As children learn the different operations, they should have experiences dealing with sets of word problems that involve different operations. These experiences should vary in levels of difficulty, proceeding from simple tasks to more difficult ones. Children should first learn to identify the needed operation, then learn to select the number sentence, and finally learn to solve the number sentence. The following activities illustrate these ideas.

Example 9-3
An activity for learning to distinguish operations

C-1, 2, 3

Materials: Prepare a set of word problems. For each child, prepare cards labeled with operation symbols. (Symbols should be restricted to the operations having been studied.)

Teacher directions: I am going to read a problem to you. You choose the card that shows the right operation for the problem. (Teacher reads each of the problems.)

$$\boxed{+} \quad \boxed{-} \quad \boxed{\times} \quad \boxed{\div}$$

Example 9-4
Matching a word problem and an appropriate number sentence

C-1, 2, 3

Materials: Cards with word problems involving different arithmetic operations. (Operations should be limited to ones that have been studied.) Cards labeled with number sentences. There should be two number sentences—a correct one and an incorrect one—for each word problem.

Teacher directions: Read the problems on the cards. Go through the number sentences and find the correct one for each problem.

Word problem card

Sam has to work 14 days. He has already worked 5 days. How many more days must he work?

Number sentence cards

$$\boxed{14 + 5 = \square}$$

$$\boxed{14 - 5 = \square}$$

4. Children should learn to think of a number sentence as representing a class of problem situations. Number sentences often are too narrowly conceived. The number sentence $3 + 2 = 5$ is more than a translation for a story like this one:

Abbey has 3 pennies in her left pocket and 2 in her right pocket.
   Altogether, she has 5 pennies.

Actually, $3 + 2 = 5$ represents the *class* of all situations "like" the one above. Other number sentences represent other classes. We can use the number sentence $a = b \times c$, for instance, in countless different situations.

We use the number sentence $a = b \times c$ to solve area problems.

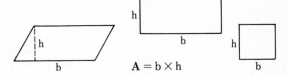

$$A = b \times h$$

We use the number sentence $a = b \times c$ to solve percentage problems.

amount off = rate × original cost

We use the number sentence $a = b \times c$ to solve distance problems.

distance = rate × time

We use the number sentence $a = b \times c$ to solve force problems.

force = mass × acceleration

Of course, children can't be expected to understand this global idea about classes of situations the first time they see a number sentence, and they won't learn to appreciate it by translating one story problem after another into number sentences. Activities have to be specially designed that illustrate this idea, such as the following: "Draw pictures to show a problem that fits $3 + 5 = 8$," or "Write a story problem that fits $4 = 7 - 3$," or, more generally, "Write as many different story problems as you can that illustrate the number sentence $\triangle \times \bigcirc = \square$." The following is a specific activity requiring children to classify story problems in terms of corresponding number sentences.

☞

Example 9-5

C-2, 3

Classifying story problems in terms of number sentences

Materials: Index cards with appropriate story problems. Cards labeled with number sentences involving different operations.

Teacher directions: Read the story problems. Put each problem with the right sentence.

New teacher directions: Write your own story problem for each of the number sentences.

Example:

3 dogs and 6 cats. How many pets?	3 rows, and 6 children in each row. How many children in all?
$3 + 6 = \square$	$3 \times 6 = \square$

Okay, so we have talked about four problem-solving abilities children should acquire, and we've suggested activities for developing and practicing each of these abilities. When children gain security with each ability, they should be involved in continuous learning activities requiring them to use a combination of these abilities.

■ Solving number sentences

When we talk about solving number sentences, we are talking about open sentences. A number sentence is called "open" when we cannot establish whether it is true or false. Here are some examples:

$$2 + \square = 5 \qquad (3 \times \square) + 2 = 14 \qquad \square + 4 < 8$$
$$2 \times m = n \qquad (4 \times m) + 1 = n \qquad n + m < 10$$

The first three sentences each have one variable, $\square$, which can be replaced with different numbers. Sometimes the replacement will result in a true statement; sometimes it will result in a false statement.

For $2 + \square = 5$, replace $\square$ with 3 to get a true statement $(2 + 3 = 5)$;
replace $\square$ with 4 to get a false statement $(2 + 4 = 5)$.

The other three sentences contain two variables, m and n. Again, replacing the variables with numbers will produce either false or true statements.

If children can write number sentences such as $3 \times 15 = \square$, $17 - 5 = \square$, $14 + 24 = \square$, or $16 \div 4 = \square$ for problem situations, then they can learn to solve them easily since each sentence involves direct computation. Notice that each of these number sentences has one variable and one operation.

But to learn how to solve number sentences such as $3 \times \square = 15$ and $\square + 4 = 76$, children must understand the inverse relationships between addition and subtraction and between multiplication and division. They must be aware that because of these relationships, the sentences can be rewritten as equivalent sentences that are more convenient: $\square = 15 \div 3$ and $\square = 76 - 4$. Whether children acquire this understanding depends for the most part on how the operations are developed in the first place. (Appropriate development is discussed in Chapters 3 and 4.) This understanding also depends on using informal models that illustrate these inverse relationships.

Here are two sample activities. The first emphasizes inverse relationships between operations; the second emphasizes the use of appropriate models for teaching ways to solve simple open sentences.

Example 9-6

C-3

Finding equivalent number sentences

Materials: Paper strips of various lengths. The strips are measured to the nearest centimeter and labeled similarly to the one shown below.

Teacher directions: For each paper strip, you can write several number sentences. Here is an example.

7	
3	n

This strip illustrates these number sentences:

$n + 3 = 7$	$n = 7 - 3$
$7 = n + 3$	$7 - n = 3$
$7 - 3 = n$	$3 = 7 - n$

The number that we can use for n to make all these sentences true is 4. Check this out. For each of the strips, write a set of number sentences. For each set of sentences, find a number replacement for n that will make all the sentences in the set true.

Example 9-7

C-3

Using models to solve open sentences

Materials: Centimeter ruler. Teacher-made ditto showing line segments.

Teacher directions: This line segment is $n + 3$ cm long. We can measure it to find that $n + 3 = 8$.

To find n, we need to subtract. First we rewrite $n + 3 = 8$ as $n = 8 - 3$. In this way, we find that $n = 5$.

For each line segment, measure the segment and write an addition sentence. Then rewrite the sentence as a subtraction sentence. Find a value for n that makes the sentence true.

Example:

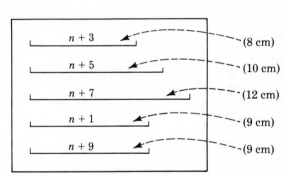

■ Dealing with more than one operation

To solve open sentences such as $(3 \times n) + 5 = 23$, children need a greater number of skills than they need to solve open sentences that involve only one operation. The ability to solve these sentences depends on an understanding of an ability to evaluate number expressions such as $(3 \times 6) + 5$ and $3 \times (6 + 5)$. Each set of parentheses in these expressions not only indicates which operation is to be performed but also marks off a nonstandard representation for a particular number.

The number $(3 \times 6) + 5$ is *not* the same as $3 \times (6 + 5)$.

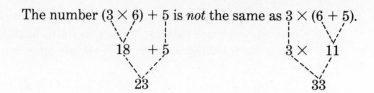

Conveniently for us, the expressions themselves suggest concrete models. By studying these models in conjunction with corresponding number sentences, children will acquire skills needed for translating word problems into number sentences when appropriate.

$(3 \times 6) + 5$

$3 \times (6 + 5)$

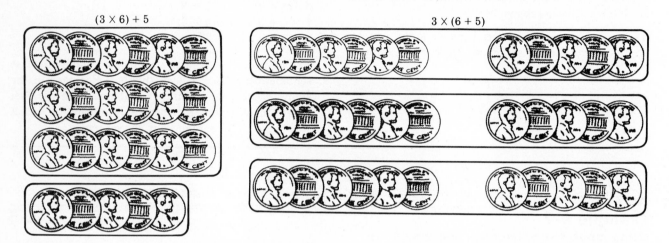

Three candies at 6¢ each.
One jaw breaker for a
  nickel.
How much do they cost
  altogether?

Six pennies and a nickel's worth of
  pennies for each of three boys.
How much money altogether?

To develop skills necessary for solving problems, children should have experiences comparing number expressions to models and applications. They should also have different kinds of practice with such expressions. Here are a couple of sample activities that illustrate how appropriate experiences can be provided.

Example 9-8
Matching models to number expressions with parentheses

C-3

Materials: Teacher-made ditto, and squared paper (see Material Sheet 17A).

Teacher directions: Replace the △s, ○s, and □s with numbers so that the expressions describe the pictures.

New teacher directions: Draw a picture for each of these expressions.

Sample
Ditto

$(\square \times \triangle) + \bigcirc$ $(4 \times \triangle) - (2 \times \square)$

$(7 \times 2) - 1$ $(2 \times 4) + 1$ $3 + (2 \times 5)$
$(5 \times 5) - (3 \times 3)$ (Etc.)

Example 9-9
Using parentheses to make number sentences true

C-3

Materials: A teacher-made ditto showing number sentences with the parentheses left out.

$3 \times 5 + 4 = 19$ $6 = 3 + 3 \times 1$
$5 \times 2 + 3 = 25$ $4 + 3 \times 2 = 10$
$8 \div 2 + 1 = 5$ $11 = 7 \times 2 - 3$
$9 + 1 \times 3 = 30$

Teacher directions: Expressions such as $3 \times 5 + 4$ don't make sense, because we don't know whether to multiply or to add first. If we multiply first, we get 19. If we add first, we get 27. To make the meaning of such expressions clear, we use parentheses. For instance, $(3 \times 5) + 4$ tells us to compute (3×5) first and then add the results to 4. The expression $3 \times (5 + 4)$ tells us to add first, and then multiply the result by 3. Use parentheses to fix the sentences on the ditto so that they will be true statements.

Children who can evaluate the kinds of number expressions discussed above and identify appropriate models and applications have the background skills necessary for learning to solve number sentences involving more than one operation. For elementary children, the "cover-up method" is suitable. You need a small piece of construction paper to use this method.

Example: Find the correct replacement for n: $(3 \times n) + 5 = 23$.

To find the value of n, we must find the value of $3 \times n$. How can we do this? Let's cover up $3 \times n$ and think about it.

$\boxed{} + 5 = 23$

Our new sentence asks "What number added to 5 is 23?" We can think: $\boxed{} = 23 - 5 = 18$. So we get a new sentence.

$(3 \times n) = 23 - 5 = 18$
or
$3 \times n = 18$

Our new sentence asks "What number multiplied by 3 is 18?" We can think: $\boxed{} = 18 \div 3$. So, we get another new sentence.

$$n = 18 \div 3$$

And now we can compute!

$$n = 6$$

This tells us that the replacement for n that makes $(3 \times n) + 5 = 23$ a true statement is 6.

■ Dealing with problems involving two variables

Elementary school children should also learn to write and interpret simple number sentences having two variables. Learning these skills involves comparing sets of objects, collecting and organizing information, and looking for relationships. Activities like "What's My Secret?" (Example 9-10, below) can engender a lot of enthusiasm while helping children to develop some of the basic skills associated with finding relationships.

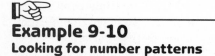

Example 9-10
Looking for number patterns

C-3

Materials: Tiles.

Teacher directions: Choose a partner. One person puts out so many tiles. Then the partner puts out more tiles, using a secret plan. Try to guess your partner's secret.

Example:

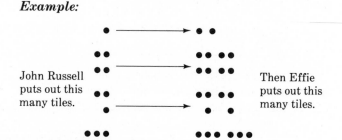

John Russell puts out this many tiles.

Then Effie puts out this many tiles.

What's Effie's secret?

Children should learn the following characteristics of number sentences having two variables.

1. A number sentence with two variables describes a relationship between two sets of data. Sometimes the relationship can be discovered by examining simple instances of the data.

Example: Suppose that we glued a row of cubic blocks together and then dipped the row in a bucket of paint. How many faces of the blocks would be painted. If the row had only one block, all six faces would be painted. If the row had two blocks, ten faces would be painted. If the row had three blocks, 14 faces would be painted. We don't have to continue this process for very long before realizing that the number sentence $f = (4 \times b) + 2$ describes the relationship between blocks and painted faces.

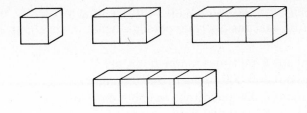

Number of blocks *(b)*	Painted faces *(f)*
1	6
2	10
3	14
4	18
.	.
.	.
.	.

$f = (4 \times b) + 2$

$\downarrow$ number of blocks

$\downarrow$ number of painted faces

2. A number sentence with two variables can be used to find the values for one of the variables when a value for the other variable is given.
Example: Suppose we made and painted a row of 999 blocks. Would we have to count the painted faces? Of course not! We could use the number sentence we discovered above.

$$f = (4 \times 999) + 2$$
$$= 3996 + 2$$
$$= 3998$$

The beauty of these first two attributes of number sentences having two variables is that we can study the problem situation in which values of the variables can be determined by direct methods such as counting, measuring, and so on. The use of a number sentence is not necessary. Sometimes this allows us to *discover* a number sentence. The number sentence can then be used to determine values of the variables that otherwise might be too difficult to find.

3. It's not always possible to find number sentences that will describe situations perfectly. When we derive a number sentence from mathematical theorems and definitions, then the number sentence describes a mathematical relationship exactly. For example, $P = 4 \times s$ describes the exact relationship between the length of a side of a square and its perimeter. (But when we calculate the perimeter of a square object, our calculation *does* involve measurement errors.) On the other hand, suppose we wanted to find a number sentence expressing the relationship between a person's age and total number of heartbeats. We could experiment and find that the average person's heart beats about 72 times in one minute, or

about 4300 times in one hour, or
about 100 000 times in one day, or
about 3 000 000 times in one month, or
about 36 000 000 times in one year.

So, a number sentence that relates age in years and number of heartbeats is

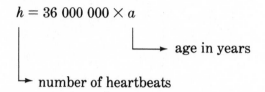

$h = 36\ 000\ 000 \times a$

        → age in years

→ number of heartbeats

Example: To find how many times your heart has beaten—give or take 100 000 beats—as of your 21st birthday, just solve the number sentence for $a = 21$.

$h = 36\ 000\ 000 \times 21$
$= 756\ 000\ 000$

→ number of heartbeats

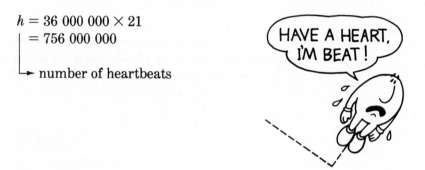

This number sentence gives a crude approximation at best. It does not account for differences in people's metabolic rates, ages, occupations, degrees of stress, and so on. By more careful experimentation, we could come up with a number sentence that gives more accurate results than this one does, but we could never describe the relationship perfectly.

4. There are geometric interpretations for number sentences having two variables. These interpretations are called *graphs.* Sometimes we must use a number sentence so often that the computation becomes too tedious and repetitious. To avoid having to compute each time that we replace a variable with a new number, we use a coordinate system to make a graph. Let's go back to the painted-block problem. The chart looks like this:

Number of blocks (b)	Painted faces (f)
1	6
2	10
3	14
4	18
.	.
.	.
.	.

We can use the values of *b* and *f* to obtain *ordered pairs* of numbers: (1, 6), (2, 10), (3, 14), and so on. The first number in each pair indicates the value of *b*, and the second number indicates the value of *f*. The ordered pairs are used to find points on a rectangular coordinate system. By finding the line through these points, we obtain the graph of our number sentence. Of course, in this case, only whole-number values for *b* and *f* make any sense. (Why?)

By using the graph, we can now find values for *f* when we are given values for *b*.

Example: Use the graph to find the value of *f* when *b* = 5. First find *b* = 5 on the horizontal axis. Move vertically up the line of the graph and then horizontally over to the *f*-axis to find that the corresponding value of *f* is 22.

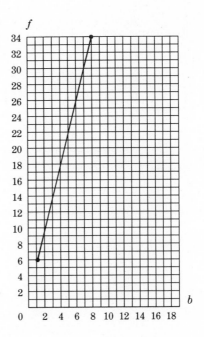

Devising situations that allow children to discover number sentences having two variables is easy. Here are some examples:

If you drop a golf ball from a distance *(d)*, how high *(h)* will the first bounce be?

What is the relationship between quarters *(q)* and dollars *(d)*?

How high *(h)* will this plant grow in *d* days?

How much *(m)* of a candle will melt in *h* hours?

How many tires *(t)* are needed for the production of *c* cars?

If it costs 15¢ just to get onto the freeway and 1¢ for every mile traveled, what is the relationship between miles *(m)* and cost *(c)*?

How does the distance around your wrist *(w)* compare to the distance around your neck *(n)*?

These situations can be used to plan activities that emphasize different attributes of number sentences. If the activities are appropriately sequenced, children can learn how these attributes complement one another. The following are examples of appropriate activities.

Example 9-11
Writing number sentences for data in charts

Materials: Teacher-made dittos.

Teacher directions: We can use charts to study relationships. For example, this chart shows us a relationship between nickels and quarters. We can write a number sentence to describe this relationship.

Study these charts. Write a number sentence for each one. Make other charts that involve situations such as these: days and weeks, hotdogs and packages, drumsticks and fried chickens.

Chart

Nickels (n)	Quarters (q)
5	1
10	2
15	3
20	4
.	.
.	.
.	.

Pennies (p)	dimes (d)
10	1
20	2
30	3
.	.
.	.
.	.

Number sentence

$n = 5 \times q$

⌐→ number of quarters

└→ number of nickels

Example 9-12   C-3
Collecting data, preparing charts, and drawing graphs

Materials: Ten willing friends. Pencil and paper. Centimeter tape or ruler.

Teacher directions: Choose ten friends. For each friend, do these things:

1. Measure the length of the person's foot, to the nearest centimeter.
2. Measure the distance from the knee to the bottom of the heel.
3. Record the information in a chart like this:

Questions:

1. About how many times longer is the knee/heel distance than the length of the foot?
2. Can you give a number sentence that "approximates" the relationship?
3. Construct a graph for your data. Use a coordinate system labeled like this one.

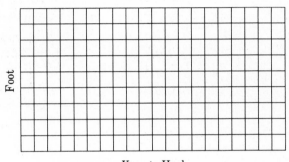

Person	Knee to heel	Foot
1		
2		
3		
4		
5		
6		
7		
8		
9		
10		

Solving unusual problems

As we said earlier, many problems that people encounter are nonroutine or unusual, and there are no standard methods of solving them. So, we can't teach children specific procedures for solving unusual problems; we can only expose them to unusual problems and help them develop problem-solving abilities that may help them develop their own strategies.

In the discussion that follows, these problem-solving abilities are identified and associated with the steps in Part 2 of the problem-solving model given on page 283.

1. Becoming familiar with the problem. Sometimes, half the battle in solving an unusual problem is just becoming familiar with it. There are two abilities that can be especially helpful in this connection.

a. The ability to recognize attributes that an object or mathematical concept may have. The more capable that children become at discovering attributes of an object or concept, the better their chances will be of seeing what objects or ideas are related.

Examples:

1. "Tell me what you can about this cube."
   Response: "It has six square faces.
   It has square corners (right angles).
   It has . . . ."

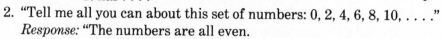

2. "Tell me all you can about this set of numbers: 0, 2, 4, 6, 8, 10, ..."
   Response: "The numbers are all even.
   The set contains all the (nonnegative) multiples of 4
   but not all the multiples of 3.
   The 'ones' digits appear in this order:
   0, 2, 4, 6, 8, 0, 2, 4, ...
   Each number is 2 more than the preceding one.
   Any number in the set can be found by multiplying
   some whole numbers by 2.
   We can use pairs of dots to illustrate the set.———→
   . . ."

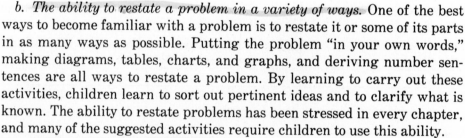

b. The ability to restate a problem in a variety of ways. One of the best ways to become familiar with a problem is to restate it or some of its parts in as many ways as possible. Putting the problem "in your own words," making diagrams, tables, charts, and graphs, and deriving number sentences are all ways to restate a problem. By learning to carry out these activities, children learn to sort out pertinent ideas and to clarify what is known. The ability to restate problems has been stressed in every chapter, and many of the suggested activities require children to use this ability.

2. Collecting information. The next thing to do after clearly interpreting the problem is to collect information, look for patterns, and identify relationships. The following abilities are needed for this step.

a. The ability to find similarities and differences. Comparing objects or ideas to find attributes they share or do not share allows us to classify them.

Example:

"How are these two figures alike? How are they different?"

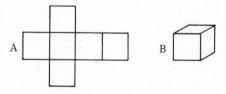

Response:

"A is flat.	"B is not flat.
It has 6 squares.	It has 6 faces that are square.
A can be folded to form a figure like B."	It has 8 corners.
	B can be unfolded to form a figure like A."

b. The ability to classify objects and mathematical ideas. This ability is described in Chapter 1, and activities are presented in almost every chapter that require children to use it. Here is another example.

Example: Think of as many ways as you can to classify these
numbers: 0, 2, 4, 5, 6, 9, 13, 15, 16, 25."

Response:

"Even: 0, 2, 4, 6, 16

"Odd: 5, 9, 13, 15, 25

Prime: 2, 5, 13

Not Prime: 0, 4, 6, 9, 15, 16, 25

Perfect Squares: 0, 4, 9, 16, 25 . . ."

Not Perfect Squares: 2, 5, 6, 13, 15 . . ."

c. The ability to determine when information is sufficient and to eliminate irrelevant information. This ability allows us to distinguish between useful and useless information in a problem situation and to determine the adequacy of the information given. Despite the importance of this ability for the problem solver, its development is rarely encouraged in school. Most textbook exercises include only the necessary information in problem situations—neither too much nor too little is given. Unfortunately, real-life problems are rarely so orderly.

Example: "Read these problems. Decide whether you have enough information to solve the problem. Indicate information that is not needed to solve the problem. Tell what other information might be needed. Tell how you would solve the problem if you had the correct information."

1) Three nests. Some birds in each nest. How many birds?
2) Alice is three years older than Dawn. She is also 10 centimeters taller. How old is Dawn?

d. The ability to find relationships or patterns. This ability is extremely important, because it enables us to tie mathematical ideas together.

Examples:
1) "Match up the rest of the numbers."

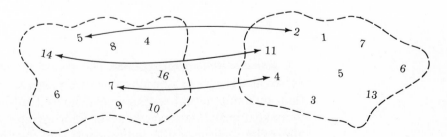

2) "Find the next number."

3 428 915

3 528 915

3 628 915

3) "Give the largest number you can if you use each of these numerals once for digits."

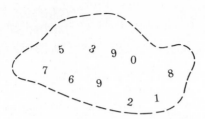

e. The ability to systematically determine cases or alternatives. Some problems must be considered under different sets of circumstances or for different cases. It's very helpful to the problem solver when this is systematically done instead of by trial and error, although finding a system is not always easy.

Examples:

1) "Describe what can happen if we compute the quotient of two whole numbers (divisor is not zero)."
   Response: "If the divisor is smaller than the dividend, the quotient will be larger than 1. If the divisor is larger than . . . ."

2) "Every two-digit numeral can be associated with a sum—for example, $21 \rightarrow 2 + 1 \rightarrow 3$. Find all the two-digit numerals associated with 8."
   Response: "The sums for 8 are: $0 + 8, 1 + 7, 2 + 6, 3 + 5, \ldots$, so the number associations are 8, 17, 26, 35 . . . ."

3) "Find at least six figures that each have an area of 24 square units. No more than two can be rectangles. Tell the perimeter of each."
   Response:

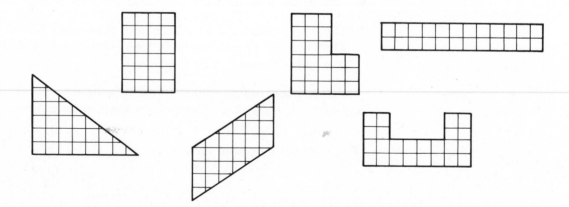

f. The ability to approximate. The ability to approximate is the ability to arrive at answers to a problem by using logical estimates or guesses rather than by using actual counts or measures. Children should be encouraged to make approximations whenever appropriate, and they should base their approximations on skills and concepts that they have previously learned.

Example: "Here is a cola can. Which is greater, the distance around the can or its height? Draw line segments to show the height and the distance around."

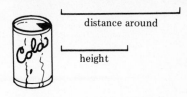

Perception usually interferes with the child's response to this question; that is, it looks as though the height of the can is greater than the distance around it. Children who are aware that the circle's diameter is about $\frac{1}{3}$ its circumference and who are reminded to use the facts they know when solving problems have a better chance of making appropriate approximations.

g. The ability to extend given information. This ability allows children to take what is given in a mathematical situation and use it to generate more information or to draw conclusions.

Example: "What will happen to the area of a square if you increase the side's length?"

Response: "I noticed that when I doubled the side of a square, the area was multiplied by 4. When the side's length was multiplied by 3, the area was multiplied by 9. When the side's length was multiplied by 4, the area was multiplied by 16. So it seems that if you multiplied the side's length by 5, then the area would be multiplied by 5×5."

h. The ability to compare objects or ideas with a set of criteria. One of the most fundamental tasks in mathematics is making appropriate use of mathematical definitions. Children should be given many opportunities to develop this ability. They should be required to use, construct, and learn certain mathematical definitions.

Example: Make up a statement or set of statements that will define the set of figures that includes these figures and only figures like these."

Response: "They have four sides.
They are flat (lie in a plane).
Opposite sides have the same length."

3. Devising and evaluating strategies. After children have formulated the problem and have collected what seems to be a sufficient amount of information, they must begin to look for strategies that might provide solutions. Actually, the mathematical thinking abilities needed in this step are the same abilities needed for previous steps. Students will have to set up the problem parts so that they resemble some arithmetic model or mathematical procedure. If children have carefully carried out the first two steps, however, it should be easy for them to see similarities between the problem and the mathematical procedures needed to solve it. For example, suppose children are asked to find how many grains of rice are in a certain bag. They might decide on any of these strategies:

a. Count the grains.
b. Count the number of grains required to fill a small container and then multiply by the number of containers that could be filled by the grains in the bag.
c. Find the number of grains required to make a weight of one gram, weigh the rice in the bag, and multiply by the number of grams.

Once children have devised a strategy, they must evaluate whether it leads efficiently to an acceptable solution.

4. *Using a strategy to find solutions.* Using a strategy to find a solution to a problem generally involves knowing mathematical information and being able to use mathematical skills. Thus, as we have discussed many times, teachers should provide situations in which children practice using mathematical information and skills. Children should also be encouraged to evaluate strategies they use by doing the following:

a. approximating in advance to see whether solutions are sensible
b. verifying solutions by using more than one strategy
c. checking calculations, conclusions, and so on

→ Encouraging problem solving in the classroom

What teachers do in the classroom and how they do it influences whether children feel secure enough to venture into problem situations, particularly children who haven't been successful at problem-solving activities. The following suggestions might enable you to develop an atmosphere in which students feel the freedom and the inclination to solve problems.

1. *Let children use their own language to express mathematical ideas.* Don't use technical language before children have developed the concepts necessary for handling the language.

Much mathematics teaching is based upon the faulty idea that language automatically *communicates.* Not necessarily! The language teachers use can only bring out in children their own private meanings, images, and feelings. Thus, communicating a problem situation to children involves finding words to which both the teacher and the children give similar meaning, imagery, and feeling. By using these words, along with appropriate learning activities, teachers can help children to bridge gaps between vague conceptions of mathematical ideas and clearer, more refined ones.

2. *Let children work together.* Encourage them to share and debate ideas and to teach each other.

Quite often, a good way to provide problem-solving experiences for children is to provide activities in which they learn while interacting with one another. Children involved cooperatively in appropriate problem-solving activities have the opportunity to discuss problems and to debate cases, alternatives, and solutions. In doing so, they modify each other's mathematical perceptions and help to refine and reinforce their individual concepts.

3. Select learning aids that are compatible with the children's level of development.

When teachers plan problem-solving activities for children, they should consider carefully the methods they use for representing ideas. Problem situations must be represented in ways that correspond to the children's mathematical abilities. For example, a teacher wanting to introduce volume ideas to young children should not begin by discussing abstract formulas. Instead, the teacher should provide activities in which children count the "units" (beans, blocks, washers) that are needed to fill certain containers. Activities such as these provide legitimate ways to informally introduce the problem situations that will be formalized later.

4. Construct learning experiences in such a way that central ideas are not obscured by related ones. Break complicated problems into manageable parts.

Children's early problem-solving experiences should be manageable. The situations should not involve several ideas that have to be considered simultaneously; often a difficult problem can be broken into more manageable problems. The following example illustrates this idea for a problem that might be used in fifth or sixth grade. The problem involves multiplication with decimals. In such situations, children sometimes have trouble deciding whether to multiply or divide.

> Difficult problem: Suppose you know that it rained .7 of an inch in one hour. At this rate, how much would it rain in .5 of an hour?
>
> Problem made more manageable: Boys and girls, let's substitute whole numbers in the problem, decide what operation is to be performed, then switch back to the numbers we had to start with. Suppose you know that it rained 2 inches in one hour. At this rate, how much would it rain in 3 hours?
>
> Solution: It's easy! We multiply 2×3 to get 6 inches, so, in the problem we started with, we must multiply $.7 \times .5$ to get .35 inches.

5. Ask thought-provoking questions. Allow children time to find careful answers.

Asking thought-provoking questions is one of the most powerful ways in which the teacher can help children to develop problem-solving skills. These questions involve more than soliciting specific facts or information. Using such questions in conjunction with appropriate instructional materials, teachers can guide children through many types of problem activities. For example, teachers can pose questions that arouse curiosity, and questions can then be asked that enable children to discover mathematical strategies, alternatives, and solutions. The right question asked at the opportune time can help children to overcome obstacles or to search deeper. We've provided many examples of thought-provoking questioning techniques throughout this book.

6. Provide problems that have no answers or that have many answers.

Often instructional materials used for developing problem-solving abilities involve contrived problem situations that leave children with the impression that "every math problem has exactly one answer." Children should also deal with problems having no answers and with problems having many answers. Here are two examples. The first is suitable for bright sixth-graders; the second can be used with children in third grade or above.

Examples:

a. "Find the number that is smaller than $\frac{1}{2}$ but is right next to $\frac{1}{2}$. Could it be $\frac{1}{3}$? Why not?"

   Response: "$\frac{1}{3} = \frac{4}{12}$, and $\frac{1}{2} = \frac{6}{12}$, so $\frac{5}{12}$ is closer to $\frac{1}{2}$ than $\frac{1}{3}$ is.

          But $\frac{5}{12} = \frac{10}{24}$, and $\frac{1}{2} = \frac{12}{24}$, so $\frac{11}{24}$ is closer to $\frac{1}{2}$ than $\frac{5}{12}$ is.

          [Etc.]"

"Do you think that we can ever find a number that is right next to $\frac{1}{2}$?"

b. "In how many different ways can a bus driver get from City A to City B if the driver always moves toward B?"

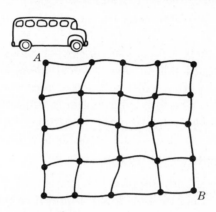

   7. Encourage children NOT to make superficial interpretations of words such as "and," "of," "more," and so on.

   Often children associate words such as "and" and "of" with certain operations. Rote association causes children to make errors when translating word problems into mathematical sentences. Children should be encouraged to interpret word problems carefully to see whether such associations are valid. For example, consider the following word problems and their corresponding number sentences. How are they different? What difficulties might children encounter in trying to solve them?

a. "Glenn had $\frac{3}{4}$ of a bar of candy. He gave his friend $\frac{1}{4}$ of a bar. How much did he have left?"

$$\frac{3}{4} - \frac{1}{4} = \square$$

b. "Glenn had $\frac{3}{4}$ of a bar of candy. He gave his friend $\frac{1}{4}$ of his piece. How much did he have left?"

$$\frac{3}{4} - \left(\frac{3}{4} \times \frac{1}{4}\right) = \square$$

The computer corner

In this computer corner, we highlight the program titled ITEM on your MGB diskette. This program was prepared for this chapter and Chapter 16. It is a small database that allows teachers to store and retrieve problems, test items, or brain teasers. Once problems have been stored, the program allows users to search for problems and to use selected ones to prepare lessons. We have entered several word problems, some of which have sufficient information. We want you to enter more problems using the codes we have established. To use this program successfully, you must first copy it to another disk, otherwise you may destroy data on the original MGB.

After you have made a copy of the necessary programs, load the program ITEM. When the MAIN MENU appears select the option to ADD ITEMS. Follow the directions given on the screen to begin adding items to the database. Use the codes indicated below:

For Subject, use one of the following codes:

ADD, SUB, MUL, or DIV

For Objective Number or Code, use one of these codes:

INSUFFICIENT, SUFFICIENT, or EXTRANEOUS

For Learning Objective, enter the problem type:

SUBTRACTION, TAKE-AWAY,
DISTRIBUTIVE DIVISION, etc.

Then use lines 1–5 to enter the text of the problem. Use line 6 to enter the answer.

Example:

```
Item Number: 25
Subject: SUB
Learning Objective: SUBTRACTION PARTITION
Objective Number or Code: INSUF
Line 1: The lazy dog had 1001 fleas.
Line 2: Some of the fleas were big.
Line 3: Some of the fleas were small.
Line 4: Some of the fleas were gigantic.
Line 5: How many of the fleas were small?
Answer: n/a
```

Don't worry about entering the Item Number—the program will do that.

After at least 40 problems have been entered, use the database to prepare a lesson. To do this you will have to use the SEARCH option of the program. When you search for items, be sure you enter search criteria exactly as you entered information for problems. In this program, you can search by SUBJECT and OBJECTIVE CODE, which means that the search criteria must be consistent with the codes we have established. Here is an example of how to search:

The program asks:

ENTER SEARCH CRITERION FOR SUBJECT:

You enter:

ADD

The program asks:

ENTER SEARCH CRITERION FOR OBJECTIVE CODE:

You enter:

INSUFFICIENT

With these search criteria, the program will look for all *addition* word problems having *insufficient* data. Now suppose you do not enter any search criterion for SUBJECT. Instead you simply press the enter key. In this case, the program will look for all word problems in the database that have insufficient data.

When your search is complete, return to the MAIN MENU and select the option to PREPARE A LESSON. You will be able to scan and select problems found in the search for inclusion in a lesson. Follow directions. See what happens.

■ Computer tips

Let children plan criteria for developing their own database of brain teasers. When their plan is complete and has been checked for consistency, let them enter problems. Once a sufficient number of problems have been entered, search for problems to use for small competitions. Many cognitive abilities are strengthened in an activity such as this. At the same time, children acquire realistic computer skills.

 ───────────────────────────────────────

In closing . . .

In this chapter, we've provided a general description of problem solving. We have identified methods for solving routine problems and methods for solving unusual problems. We've also given suggestions for encouraging problem-solving activities in the classroom. Actually, this chapter brings into focus many ideas that have been extensively discussed in this book; that is, it provides a *guide* for selecting and organizing activities and emphasizes that the development of skills and concepts is meaningless without the development of related problem-solving abilities.

The set of strategies presented in this chapter is by no means intended to be a formula for the teaching of problem solving. We recognize that capable problem solving involves the acquisition of diverse and ever-changing abilities. But we also recognize that children cannot learn to problem solve if they don't practice problem-solving tasks. So, we have tried to present realistic tasks that will promote this development in the elementary grades.

It's think-tank time

1. Prepare the materials for the activities in Examples 9-1 through 9-5, then carry out each activity.
2. Sketch a model for each of these number expressions:
   a. $(4 \times 1) + 2$ b. $(4 \times 2) - 2$ c. $3 \times (6 + 4)$
3. Write number sentences for these models, using symbols of inclusion appropriately.
   Example: $1 + 3 + 5 + 7 + 9 = 25$

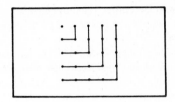

a.

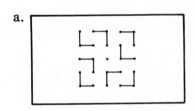

b.

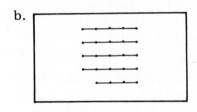

c.

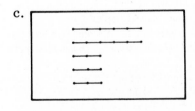

d.

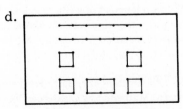

4. Sometimes we must use more than one set of inclusion symbols to write a number expression. For example, we need to distinguish $[3 \times (5 - 2)] + 1$ from both $3 \times [5 - (2 + 1)]$ and $(3 \times 5) - (2 + 1)$. Here's how we evaluate the first expression.

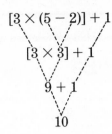

Start with the innermost set of enclosures (parentheses, braces, brackets, etc.) and "work out."

$$[3 \times (5 - 2)] + 1$$
$$[3 \times 3] + 1$$
$$9 + 1$$
$$10$$

Evaluate the other two expressions to show that neither is equivalent to the first.

5. A number can be represented by many number expressions. Here are examples for the number 17.

$$[3 \times (2 + 4)] - 1 \qquad (3 \times 2) + (2 \times 2) + 7$$
$$(2 \times 10) - 3 \qquad (3 \times 5) + 2 \qquad \left(\frac{1}{2} \times 40\right) - 3$$
$$(21 \div 3) + 10 \qquad (4.2 \div .6) + 10$$

Write at least ten different expressions for the number 23. Use each of the four operations at least once. Use fractions; use decimals.

6. Study these situations and write number expressions for each one. Use inclusion symbols when necessary. Do not compute.
   a. Three outfits. Each consists of two hats costing $3 each and one dress costing $8. What is the total cost?
   b. Two children plan a biking trip. They buy two candy bars for 15¢ each, three sandwiches for 50¢ each, and one large bottle of cola for 60¢. If the children share the cost, how much does each spend?
   c. A lady spends $\frac{1}{3}$ of her money in one store and $\frac{1}{2}$ of what is left in another store. How much did she spend in the second store?

7. Describe situations for each of these number expressions.
   a. $3 \times (2 + 5)$ b. $(6 \times 12) - 2$ c. $(6 \times 5) - 3$

8. Using the cover-up method, solve each of these number sentences. Show each step.
   a. $(4 \times n) + 3 = 23$ b. $n \div 6 = 5$
   c. $(3 \times n) - 4 = 16$

9. Describe each of the problem-solving abilities identified in this chapter. Then go back through the book and, for each of these abilities, find two activities that emphasize it.

10. Make up three original activities (or find them in other sources) that emphasize each of the problem-solving abilities discussed in this chapter at these three levels of development:

a. Grades K–2 b. Grades 2–4 c. Grades 4–6

Suggested readings

Bartalo, Donald B. "Calculators and Problem Solving Instruction: They Were Made for Each Other." *The Arithmetic Teacher*, January 1983, pp. 18–21.

Brownell, William A. "Problem Solving." In *The Psychology of Learning*, Forty-first Yearbook, Part II. National Society for the Study of Education. Chicago: University of Chicago Press, 1942. (Pp. 415–443.)

Campbell, Patricia F. "Using a Problem Solving Approach in the Primary Grades." *The Arithmetic Teacher*, December 1984, pp. 11–14.

Charles, Randall I. "The Role of Problem Solving." *The Arithmetic Teacher*, February 1985, pp. 48–50.

Davidson, James E. "The Language Experience Approach to Story Problems." *The Arithmetic Teacher*, October 1977, pp. 28–29.

Dewey, John. *How We Think* (Rev. Ed.). Boston: D.C. Heath, 1933.

Easterday, Kenneth E., and Clothiaux, Clara A. "Problem Solving Opportunity." *The Arithmetic Teacher*, January 1985, pp. 18–20.

Fennell, Francis (Skip). "Focusing on Problem Solving in the Primary Grades." In *The Agenda in Action*. 1983 Yearbook of the National Council of Teachers of Mathematics. Reston, Va.: The Council, 1983.

Henderson, Kenneth B., and Pingry, Robert E. "Problem Solving in Mathematics." In *The Learning of Mathematics: Its Theory and Practice*. Twenty-first Yearbook of the National Council of Teachers of Mathematics. Washington, D.C.: The Council, 1953.

Henney, M. "Improving Mathematics Verbal Problem-Solving Ability through Reading Instruction." *The Arithmetic Teacher*, April 1971, pp. 223–229.

Kilpatrick, Jeremy. "Problem Solving in Mathematics." *Review of Educational Research*, October 1969, pp. 523–534.

Krulik, Stephen. "Problem Solving: Some Considerations." *The Arithmetic Teacher*, December 1977, pp. 51–52.

Lester, Frank K. "Ideas about Problem Solving: A Look at Some Psychological Research." *The Arithmetic Teacher*, November 1977, pp. 12–14.

Lichtenberg, Don. R. "The Difference Between Problems and Answers." *The Arithmetic Teacher*, March 1984, pp. 44–45.

Lodholz, Richard David. "The Effect of Student Composition of Mathematical Verbal Problems on Student Problem Solving Performance." (Doctoral dissertation, University of Missouri–Columbia, 1980.) *Dissertation Abstracts International*, *42A* (1982):3483A.

Marcucci, Robert G. "Problem-Solving Activities for Prospective Elementary School Teachers." In *The Agenda in Action*. The 1983 Yearbook of The National Council of Teachers of Mathematics. Reston, Va.: The Council, 1983.

Morris, Janet P. "Problem Solving with Calculators." *The Arithmetic Teacher*, April 1978, pp. 24–25.

National Council of Teachers of Mathematics. *Topics in Mathematics for Elementary Teachers: Hints for Problem Solving*. (Booklet No. 17) Washington, D.C.: Author, 1969.

O'Daffer, Phares G. "Problem Solving Tips for Teachers." *The Arithmetic Teacher*, December 1984, pp. 30–31.

Polya, George. *How to Solve It*. Princeton, N.J.: Princeton University Press, 1945.

"Problem Solving." In *Psychology Today: An Introduction*. Del Mar, Calif.: Communications/Research/Machines, Inc., 1970. (Chapter 20.)

Reys, Robert E. "Estimation." *The Arithmetic Teacher*, February 1985, pp. 37–41.

Robinson, Edith. "On the Uniqueness of Problems in Mathematics." *The Arithmetic Teacher*, November 1977, pp. 22–26.

Russell, David H., and Spencer, Peter L. "Reading in Arithmetic." In *Instruction in Arithmetic*. Twenty-fifth Yearbook of the National Council of Teachers of Mathematics. Washington, D.C.: The Council, 1960.

Sairafi, Adnan Abdulghani. "An Investigation of Procedures for Improving Prospective Elementary School Teachers' Problem Solving Abilities." (Doctoral dissertation, Florida State University, 1983.) *Dissertation Abstracts International*, *44A* (1984):3312A.

Shulman, Lee S., and Elstein, Arthur S. "Studies of Problem Solving Judgment and Decision Making: Implications for Educational Research." In *Review of Research in Education* (Vol. 3). Washington, D.C.: American Education Research Association, 1975. (Chapter 1.)

Spencer, P. J., and Lester, F. K. "Second Graders Can Be Problem Solvers." *The Arithmetic Teacher*, September 1981, pp. 15–17.

Spitzer, Herbert F., and Flournoy, Frances M. "Developing Facility Solving Verbal Problems." *The Arithmetic Teacher*, November 1956, pp. 177–182.

Troutman, Andria P. "Development of a Specific Cognitive Observation System for the Analysis of Mathematics Teaching." (Doctoral dissertation, University of Florida, 1971.) *Dissertation Abstracts International*, *33*(1972): 668A.

Troutman, Andria P., and Lichtenberg, Betty K. "Problem Solving in the General Mathematics Classroom." *Mathematics Teacher*, November 1974, pp. 590–597.

Underhill, Robert G. "Teaching Word Problems to First Graders." *The Arithmetic Teacher*, November 1977, pp. 54–56.

Wheatley, Charlotte, and Wheatley, Grayson H. "Problem Solving in the Primary Grades." *The Arithmetic Teacher*, April 1984, pp. 22–25.

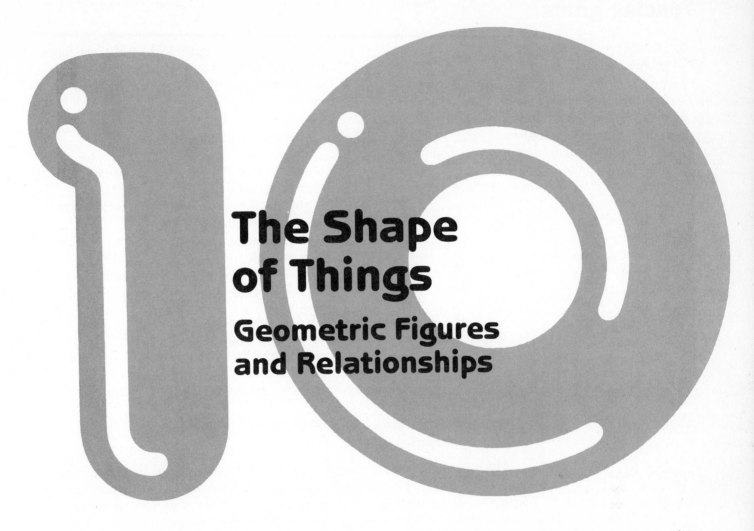

The Shape of Things
Geometric Figures and Relationships

■ This chapter summarizes the basic geometric ideas that teachers in the elementary school should know. The purposes of this chapter include (1) helping you to pinpoint and review specific ideas that often get lost in more rigorous mathematics courses concerned with precise terminology and deductive methods; (2) providing the background for the discussions in Chapters 11, 12 and 14 of children's development of geometric ideas; and (3) providing a concise but comprehensive reference for classroom teachers.

This chapter presents many ideas in relatively few pages. A number of exercises are given at the end of the chapter to help reinforce and refine these ideas and also to introduce you to a variety of intuitive ways for dealing with certain concepts. We do not suggest that all of the concepts presented here can or should be acquired in a single methods course that also considers arithmetic. For such a course, priorities would have to be established among the arithmetic, geometry, and measurement ideas discussed in this book. Actually, Chapters 10 through 14 provide the basis of a methods course for elementary school geometry.

We deliberately postponed the activities for children until the next chapter. This chapter is for you, the teacher. ■

Teacher goals

1. Be able to identify and classify familiar geometric figures.
2. Be able to use appropriate symbols to name familiar geometric figures.
3. Be able to select appropriate definitions for familiar geometric figures.
4. Be able to construct appropriate models for familiar geometric figures.
5. Be able to give definitions for these terms:
   a. congruence
   b. turns
   c. similarity
   d. turning symmetry
   e. slides
   f. flipping symmetry
   g. flips
6. Be able to specify sufficient conditions for congruence of familiar figures.
7. Be able to construct geometric figures to illustrate these relationships:
   a. same size and same shape
   b. same shape but not same size
   c. same size but not same shape
8. Be able to identify physical situations that illustrate specific geometric concepts.
9. Be able to locate points on a rectangular coordinate system.
10. Be able to make scale drawings for simple line drawings.
11. Be able to use a point p and a scale factor to produce an enlargement or reproduction of a figure.

Why is a manhole cover round instead of square?

Why doesn't a three-legged stool wobble?

What happens when you bounce a football?

Which contains more money, a milk carton filled with dimes or a milk carton filled with half-dollars?

Do objects of the same size necessarily have the same shape?

Why is the earth a sphere?

Mathematics can be thought of as a *system* of generalities that are based originally on experiences that human beings have had with their environment. Loosely speaking, these generalities can be separated into three categories. The first category deals with *number concepts* and stems from a need to tell how many and how much. People first wrestled with the problem of finding ways to keep track of separate objects—such as sheep in a flock—and eventually developed systems for representing counting numbers and their operations. Once counting caught on, mathematical techniques were extended so that phenomena other than "separate" objects could be quantified—for example, parts of a single pie. This first category, then, contains generalities from *arithmetic*.

The second category of generalities concerns *geometry*. These generalities were developed because of a need to describe, compare, represent, and relate objects in the environment. Using ideas from geometry, we can answer questions and solve problems that involve the *shape of things* rather than the number of things. The questions at the beginning of this chapter are examples from this category. Try to answer them. Think "shape" and try to find relationships between shapes.

When we use numbers to describe certain *attributes of geometric figures*, we use generalities from both geometry and arithmetic. These generalities form the third category, *measurement*.

So far, this book has presented only arithmetic concepts and ways in which they should be developed in the elementary school. This and the next three chapters cover concepts from geometry and measurement.

The next few sections present a number of concepts in condensed form. Just reading will not enable you to fully appreciate these concepts, even though most of them are familiar to you. It's important that you work the problems at the end of the chapter. Doing so will help you to develop flexibility in dealing with the geometric ideas we present and will also expose you to a variety of materials and ideas that can be used with children.

Geometric figures

Geometry, we've said, involves describing, comparing, representing, and relating objects in the environment by using *geometric figures*. These are not real objects; rather, they are generalized notions for real objects. These figures can be classified into four broad categories: (1) points; (2) lines, line segments, and rays; (3) plane figures; and (4) space figures.

■ Points, lines, segments, and rays

The simplest of all geometric figures is the *point*. A point can describe objects or locations. For example, a point might describe the tip of an ice-cream cone, the place in a room where the ceiling and two walls meet, or the place where two roads intersect on a map. A point has no dimensions or shape.

A second basic geometric figure is the *line*. Two points determine one and only one line.

The line above can be called line AB or line BA, and is symbolized by $\overleftrightarrow{AB}$ or $\overleftrightarrow{BA}$. The arrows indicate that the line extends infinitely in both directions.

Segments and *rays* are geometric figures that lie on a line. Thus, they too can be indicated by two points. A ray has only one end-point, and a segment has two end-points.

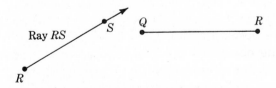

We write $\overrightarrow{RS}$ to designate this ray. (Is $\overrightarrow{RS}$ the same as $\overrightarrow{SR}$? Why not?)

This can be called segment RQ or segment QR. We can write $\overline{QR}$ or $\overline{RQ}$.

Since both of these figures lie on a line, they are called *linear figures*.

■ Plane figures

What kind of figures are these? Each figure contains at least three points that are not all on the same line. Can you identify a set of three such points for each figure?

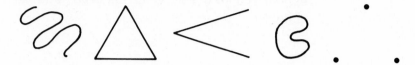

Figures such as these are called *plane figures*. A plane is determined by three points that do not lie all on the same line; it can be thought of intuitively as a region that extends infinitely in all directions.

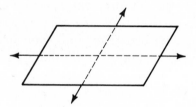

Two lines that lie in the same plane either intersect or don't intersect. If they do, they create *angles;* if they don't intersect, they are *parallel.*

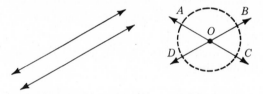

This picture shows angle *BOC*, angle *AOB*, angle *AOD*, and angle *DOC*. We can write "angle *AOB*" as "∠*AOB*." Point *O* is the *vertex* for each of these angles.

When two lines intersect to form four angles having the same size, the two lines are *perpendicular*, and the angles are called *right angles*. Angles smaller than a right angle are *acute;* angles larger than a right angle are *obtuse.*

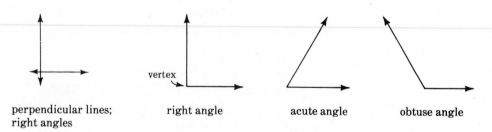

perpendicular lines;   right angle   acute angle   obtuse angle
right angles

Two planes either intersect in a line, or they are parallel.

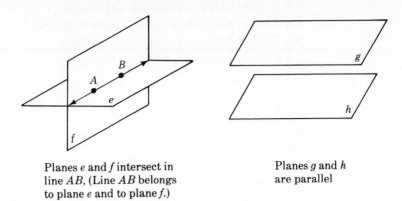

Planes *e* and *f* intersect in
line *AB*, (Line *AB* belongs
to plane *e* and to plane *f*.)

Planes *g* and *h*
are parallel

Two planes can also be perpendicular. The intersection of two planes
creates what can be called four *space angles*. When these angles are the
same size, the two planes are perpendicular, and the angles can be called
right space angles.

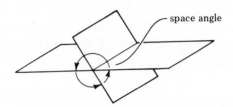

space angle

Rectangular coordinate systems

The concepts of parallel and perpendicular lines enable us to devise a
coordinate system for locating points in the plane. For example, the
coordinate system below consists of vertical and horizontal lines, or *axes*.
Points on the axes are associated with numbers, and these associations
give us what can be thought of as "rulers." With a number from each ruler
we are able to locate points in the plane. Notice that points on the graph
are indicated by *ordered pairs* of numbers. The first number in the pair is
the value for *x;* the second number is the value for *y*. (Make sure you
understand why it is important for these pairs of numbers to be ordered.)

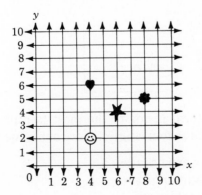

The horizontal line that goes through
the point associated with 0 is called the
x-axis.

The vertical line that goes through the
point associated with 0 is called the
y-axis.

The flower is located at the point (8, 5),
the Happy Face at (4, 2), the heart at
(4, 6), and the star at (6, 4).

A *polygon* is a special kind of plane figure, and is made of line segments.

A 3-sided polygon is called a triangle.
A 4-sided polygon is called a quadrilateral.
A 5-sided polygon is called a pentagon.
A 6-sided polygon is called a hexagon.
(Etc.)

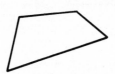

triangle   quadrilateral   pentagon   hexagon
(3-sided)   (4-sided)   (5-sided)   (6-sided)

Whenever the sides and the angles of a polygon both have equal measure, the polygon is called a *regular polygon*.

The triangle is the most basic polygon. It has three sides and can be classified by the characteristics of these sides or by characteristics of the angles created by the sides. If at least two of the sides have the same length, it is an *isosceles triangle*. If all sides have the same length, it is an *equilateral triangle*. (If none of the sides share the same length, it is a *scalene triangle*.) If it has a right angle, then it is a *right angle*.

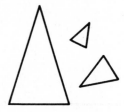

right triangles   equilaterial triangles   isosceles triangles

Here is an inclusion diagram that illustrates some of the relationships among triangles.

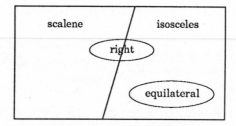

The quadrilateral is one of the most familiar types of polygons. It has four sides that form four angles. Here are some attributes it may have:

1. Opposite sides are parallel.
2. Angles are right angles.
3. Sides have equal length.

The most important quadrilaterals can be classified using the following characteristics.

	Opposite sides parallel	*All right angles*	*Sides of equal length*
Parallelogram	yes	maybe	maybe
Rectangle	yes	yes	maybe
Square	yes	yes	yes

Let's look more closely at these familiar figures.

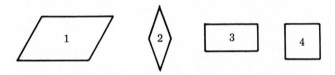

According to the chart, the quadrilaterals above are all parallelograms, since each has parallel opposite sides. Figures 3 and 4 are both rectangles, because their angles are right angles. Figures 1 and 2 are nonrectangular parallelograms, since neither contains right angles. Of the figures above, figure 4 is the only square. It is also the only quadrilateral in the set having all three characteristics—opposite sides parallel, all right angles, and sides of equal length. The square is a regular polygon. As you can see, some classes of quadrilaterals are included in other classes.

Here is an inclusion diagram that illustrates some of the relationships among quadrilaterals.

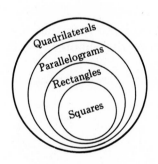

Two simple closed curves are the *circle* and the *ellipse*. A circle can be thought of as the set of all points in a plane that are a given distance, r, from a given point, O. The point O is the *center* of the circle. Any segment that joins O with a point on the circle is called a *radius* (plural: radii) of the circle. A *diameter* of a circle is any line segment that goes through the point O and joins two points on the circle. The "distance around" the circle is called the *circumference*, which is usually referred to as C.

■ Space figures

Space figures are those that represent shapes of real objects. The points of these figures do not lie in the same plane. Thus, although we can draw a two-dimensional figure to represent a cracker box, a figure that truly described it would require points that are not on the page. Too bad! We're stuck with the points on this page. You'll have to use your imagination to identify the points of these common space figures that do not lie on the plane for this page. These drawings are called *graphic representations* for the space figures.

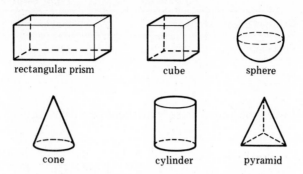

rectangular prism   cube   sphere

cone   cylinder   pyramid

There are many kinds of space figures, but for your purposes, we need only be familiar with *polyhedra* (plural for *polyhedron*), *cones*, *cylinders*, and *spheres*.

A polyhedron has *faces*, *edges*, and *vertices*. All of the faces are polygonal regions.

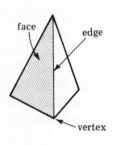

face   edge

vertex

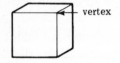

vertex

A polyhedron is called regular when it meets three conditions: (1) the faces of the polyhedron are regular, (2) the faces have the same shape and size, and (3) each vertex joins the same number of edges. For example, the *cube* is a regular polyhedron. All of the faces are *square regions* of the same size. Every vertex joins exactly three edges.

There are only five regular polyhedra, the cube, and the four that are pictured below. In doing the exercises at the end of the chapter, you'll learn to construct models for these space figures.

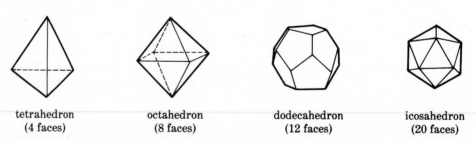

tetrahedron   octahedron   dodecahedron   icosahedron
(4 faces)   (8 faces)   (12 faces)   (20 faces)

The *prism* is a polyhedron having at least one pair of faces that belong to parallel planes. In fact, a prism can be thought of as the space figure "swept out" by moving a polygon through a set of parallel planes.

Prisms

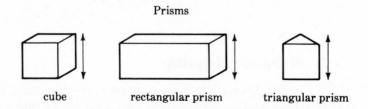

cube   rectangular prism   triangular prism

The *pyramid* can be thought of as the space figure obtained by using line segments to join all of the points of a polygonal region to a point that is not in the plane of the region. There are many different kinds of pyramids. The one below is a pentagonal pyramid.

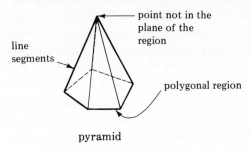

line segments

point not in the plane of the region

polygonal region

pyramid

We can describe the *cylinder* in much the same way as we described the prism. That is, the cylinder is the space figure "swept out" by moving a circle through a set of parallel planes.

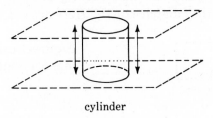

cylinder

Two more space figures merit attention. A *sphere* is the set of all points in space a given distance, *r*, from a given point, *O*. Like the circle, the sphere has a *radius* of length *r*. A radius of the sphere is any line segment that joins the center with a point on the sphere.

sphere

A *cone* can be described in a way similar to the way in which we described a pyramid. See if you can describe a cone.

cone

Congruence and similarity

■ Congruence

So far, we've described certain geometric figures that are relevant to the elementary school curriculum. Many attributes of these figures can be studied without considering size, shape, direction, and angle. For example, we don't have to consider these ideas when we want to establish whether a curve is closed. Other information attributes, however, *are* related to size, shape, direction, or angle. Often these attributes can be associated with numbers in a way that simplifies the comparison of objects. For example, a foot with a shoe size of 7 is smaller than one with a shoe size of 9, and we don't have to play "toesies" to find out!

One key to many attribute-number associations for geometric figures is *congruence. Two geometric figures are congruent if they have the same size and shape.* This definition is easy enough to understand, and, to make it operational, we can say that two figures are congruent if they can be moved so that they match exactly. Here are examples of congruent figures.

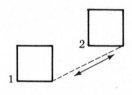

Figure 1 can be moved along a line to make a match with Figure 2.

Figure 1 can be turned about point O to make a match with Figure 2. Figure 2 can be turned about point O to make a match with Figure 1.

Figure 1 can be moved along a line and then flipped to make a match with Figure 2.

Although the operational definition should be used to help young children develop ideas about congruence, it is too broad to be helpful in many specific cases. If you had to construct a rectangular prism congruent to a given one, for instance, you would have to consider such specific attributes as width and height. It's helpful, therefore, to consider a *class* of figures and to focus on the conditions that make two or more figures in the class congruent; for example, under what conditions will triangles be congruent? Under what conditions will rectangles be congruent?

Luckily for us, the congruence of figures can usually be established in terms of the congruence of line segments and angles. We can see whether two line segments are congruent by checking whether a tracing of one can be moved to match the other. And we can check the congruence of angles in a similar way. With these tools and the definitions for perpendicularity and parallelism, we can study classes of figures to find what conditions lead to congruence.

Let's study a few examples. We've said that a circle is the set of all points in a plane that are the same distance from a given point, and we've noted that the radius is a line segment joining the center of the circle to one of the points on the circle. So, to find out whether two circles are congruent, we need only find out whether their radii are congruent. We can check the congruences of spheres in the same way.

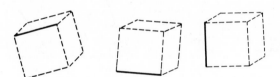

The radii are congruent.
so the circles are congruent.

The radii are congruent,
so the spheres are congruent.

How can we check for the congruence of squares? Easy! Suppose we try to construct different squares with a side congruent to this segment: _____.

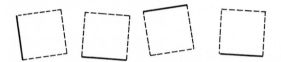

It turns out that the best we can do is to obtain a set of congruent squares in different "positions." So, we can say that if a side of a square is congruent to the side of another square, then the squares are congruent. Of course, an analogous thing can be said about the cube. If an edge of a cube is congruent to the edge of another cube, then the two cubes are congruent.

What about triangles? If we're given three line segments, how many different triangles can we construct? It's possible that we won't be able to construct any. Given line segments congruent to the ones below, for example, we cannot construct a triangle.

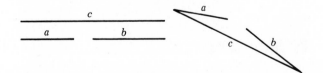

The sides a and b aren't long enough!

What's the trouble? You can see that the sum of the lengths of two sides of a triangle must be greater than the third side.

Given line segments congruent to these, however, we can construct triangles.

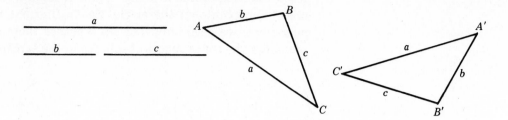

Using the same set of line segments, can we construct triangles that are different in size and shape? As you can see, the answer is no. Try constructing triangles with any set of three pencils, straws, or paper strips. You can construct triangles in such a way that their positions are different, but you cannot change the size or shape of the resulting triangle.

Let's look at one more situation. Suppose we were to construct triangles by using line segments congruent to the ones below.

How many triangles could we construct that are not congruent? (The third side is not given and is dependent on the angle chosen.)

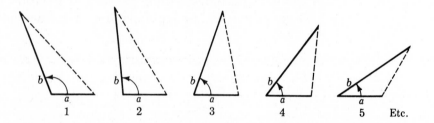

This shows that in order to construct congruent triangles from sides a and b, we must specify the angle included between sides a and b. Other useful questions about congruence include: When will two angles in a triangle be congruent? Three angles? When will two sides be congruent? Three sides? When will two polygons be congruent?

■ Similarity

As we've seen, congruent figures have the same size and shape. Can different figures have the same area and not have the same shape? Of course they can, as the regions below demonstrate.

Can figures have the same shape and not the same size? Sure! (Consider spheres, for example, or cubes.) Figures that have the same shape, regardless of their size, are called *similar* figures. In many practical situations, just looking at the figures is enough to determine their similarity.

These figures are similar.   So are these.

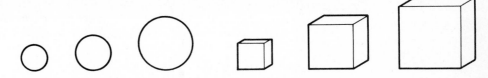

But looks can be deceiving. It may look as if the two superimposed crosses at left are similar, but they're not. The inner cross is longer with respect to its width than is the outer one. So, we have to have more accurate checks for similarity. To illustrate, let's consider triangles.

Two triangles are similar if one can be moved to fit into the other so that two angles match exactly *and* the sides opposite those angles are parallel.

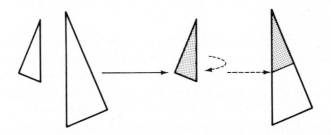

We can use this fact about triangles to devise a method for checking the similarity of polygons. The method depends on the fact that a polygon can be separated into a finite number of triangles.

1. Set up a correspondence between the parts of the polygons.

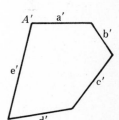

2. Use diagonals from A and A' to separate the polygons into triangles.

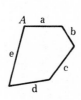

3. If $\angle A$ and $\angle A'$ match exactly, and if corresponding triangles are similar, then the polygons are similar.

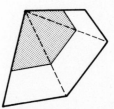
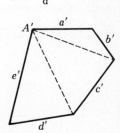

Of course, the method just given will not help you to verify similarity for the crosses shown earlier, and it will not help with other complicated curves. But it will help you verify the similarity of those polygons that receive the most attention in the elementary school.

One of the best ways for elementary school children to learn about similarity is by constructing similar figures and studying the results. One method children can use to construct similar figures involves using coordinate systems. Here we've taken the figure in the first graph and begun a reproduction of it in the second graph. Whenever a plane figure is reproduced in this way, the result is a pair of similar figures.

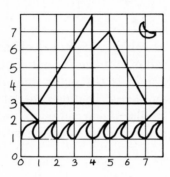

Another method for constructing similar figures involves choosing a point p and a number to indicate the *scale factor* of enlargement or reduction. Below, the scale factor is 2, because the line segment connecting p and a point on the reproduction is twice as long as the line segment connecting p and the corresponding point on the original figure. (See Exercise 10 in this chapter's Think-Tank section.)

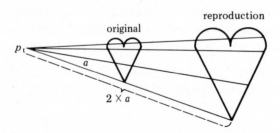

Motions and symmetry

■ Motions

Many "motions" can be performed on a geometric figure.

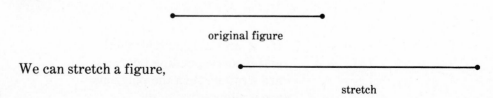

original figure

We can stretch a figure,

stretch

bend a figure,

bend

twist or knot a figure,

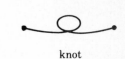

knot

or simply move the figure
without stretching, bending,
twisting, or knotting it.

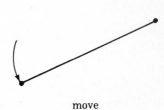

move

In geometry, we're interested in what happens when we perform such
motions. We're particularly interested in motions that do not change the
size and shape of figures. Basically, there are three such motions: the
slide, the *turn*, and the *flip*. Let's consider these motions for plane figures.

Slide. The ray OA indicates
the direction of the slide.

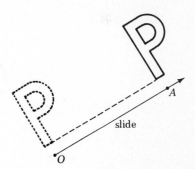

slide

Turn. The center of the turn
in this case is C. The direction is counterclockwise, and
the amount of turn is $\frac{1}{4}$ of a
complete turn.

Flip. This figure has been
flipped about line AB from
left to right.

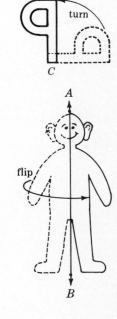

We can also perform a combination of these moves, and the resulting figure will be congruent to the original one.

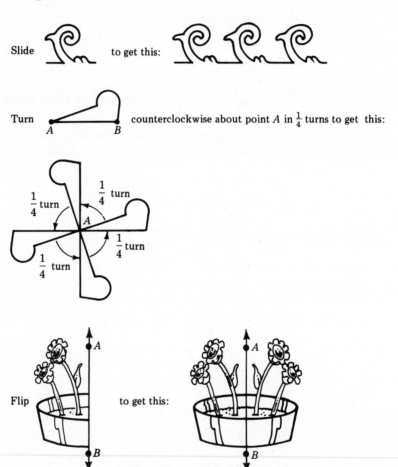

Performing these motions repeatedly can lead to some very interesting designs.

Slide   to get this:

Turn   counterclockwise about point A in $\frac{1}{4}$ turns to get this:

Flip   to get this:

■ Symmetry

People have long been interested in the designs resulting from slides, flips, and turns, and that interest has led to the study of *symmetry*. A figure has symmetry if it can be moved so that the resulting figure exactly coincides with the original figure. There are different kinds of symmetry, depending on the motion. For example, the figure following can be turned counterclockwise so that the resulting figure coincides with the original four times through one full turn. Here's how it will look for all four coincidences.

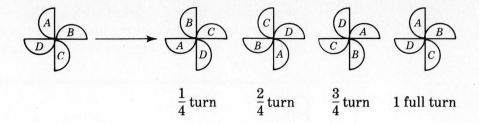

$$\frac{1}{4} \text{ turn} \qquad \frac{2}{4} \text{ turn} \qquad \frac{3}{4} \text{ turn} \qquad 1 \text{ full turn}$$

This figure is said to have *turning symmetry* of order 4.

This figure can be flipped about line $\overleftrightarrow{AB}$ so that the resulting figure coincides with the original one. The figure is said to have *line symmetry*. Line $\overleftrightarrow{AB}$ is the only line it can be flipped about, so it has only one line of symmetry.

This figure has two lines of symmetry. Can you tell where they are?

Now you may be wondering whether there are figures with slide symmetry. Technically speaking, no "finite" figure can have slide symmetry, but an "infinite" one can. Imagine an endless row of C's. We could slide one C onto another and get the same figure back.

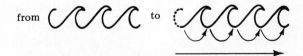

But if the row of C's were finite, such a move would result only in a change of position—that is,

from  to

For our purposes, then, figures can have only turning symmetry or line symmetry. In the case of turning symmetry, we're concerned with a direction (for example, counterclockwise) and a center of turn. In the case of line symmetry, we're concerned with any line having this attribute: The figure can be flipped about the line so that the resulting figure coincides with the original. The examples below will clarify some of the ideas presented here.

No line of symmetry.

One line of symmetry.

Four lines of symmetry; turning symmetry.

Two lines of symmetry; turning symmetry.

In closing . . .

This chapter summarizes many geometric ideas that are important for the elementary school teacher to know. For many of you, it was a thorough review. The method of presentation, however, is not suitable for children. For them, the pace should be slower and the style more discovery oriented. And, just as important, children's discoveries should be the result of working and playing with real objects. Children must acquire many informal concepts before being introduced to such abstractions as points, lines, planes, and so forth. More about this in Chapter 11.

The exercises that follow in "It's Think-Tank Time" are a very important part of this chapter. Doing them will reinforce the ideas we've discussed and will introduce you to a wide variety of materials that can be used in the elementary classroom.

It's think-tank time

1. Use toothpicks to make the following specified figures (you cannot break the toothpicks). Sketch the results.
   a. Two quadrilaterals that each have exactly one pair of parallel sides and are not similar.
   b. A triangle having congruent angles. (Can you use 3 picks to make the triangle? 6 picks? 9 picks? What can you say about the sides of each triangle?)
   c. Two nonrectangular parallelograms that have different shapes. Use the same number of picks for each. (How *must* the two parallelograms be different?)
   d. A parallelogram with exactly one right angle. (Can you do it?)
   e. A pair of rectangles that are similar but not congruent; a pair of triangles that are similar but not congruent.
   f. Regular polygons with 3 sides, 4 sides, 5 sides, 6 sides, 8 sides.
   g. Nonregular polygons with 3 sides, 4 sides, 5 sides. 6 sides. 8 sides.

2. Cut and label polygons like the ones shown in Material Sheet 260. These polygons consist of five triangles, a square, and a parallelogram and are called *tangram pieces*. Use your tangram pieces to do the following, and trace the results.

a. Use two pieces to make a piece congruent to piece F. Then use two pieces to make a piece similar to F but not congruent. Use two pieces to make a piece that is the same size as F but not similar.

b. Make a quadrilateral that does not have right angles. Use two pieces; use all the pieces.

c. Use three pieces to make a square. Use two pieces; use all the pieces.

d. Use two pieces to make a pentagon.

e. Cover piece D with three pieces.

f. Make a hexagon with pieces A, B, C, F, and G.

g. Make some designs that look like real objects. Such designs are called *tangrams*. Trace your designs, and challenge a friend to duplicate them. For example,

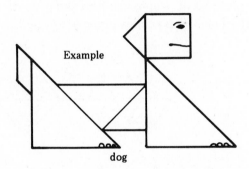

dog

3. Use triangles to help you decide which of the rectangles below are similar. (Hint: Separate the rectangles into two triangles.) Then find the conditions under which two rectangles will be similar.

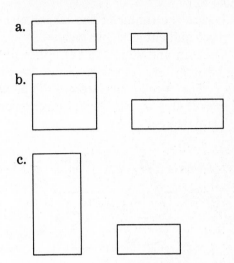

4. The first design below is called a *cover* for a cube, because it represents a plane figure that can be transformed (folded) to represent a cube. Which of the following are covers? Cut them from squared paper (Material Sheet 17C), and try to make cubes.

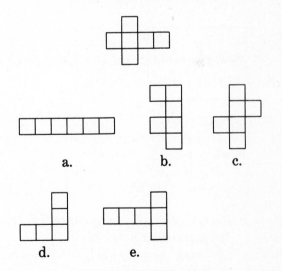

Can you make any other covers?

5. Below are directions for making covers for space figures. Make each cover, and then fold and fit it together with tape to make the corresponding figure.

a. *Cone.* Use two circle-pieces (Material Sheet 22).

  1) Locate the center of the larger circle-piece and cut as shown below.

  2) Overlap and slide the cut edges until the open end forms a circle the size of your second circle-piece. Tape the edges; tape on the second circle-piece. Trim excess.

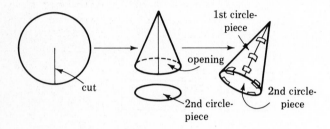

b. *Cylinder*

  1) Make and cut out any rectangle-piece.

  2) Tape two opposite sides together.

  3) Make two tracings of the circular opening. Cut out these pieces and tape them on.

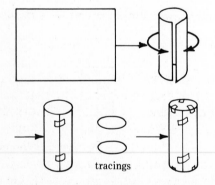

tracings

c. *Tetrahedron.* Use any isosceles triangle from Material Sheet 23.

  1) Trace the triangle three times to form four triangular regions.

  2) Cut, fold, and tape.

tracings

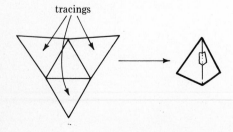

d. *Rectangular prism.* Use any rectangle pieces such that opposite faces are congruent and edges that join are congruent (see illustration below).

  1) Make tracings to form the cover.

  2) Cut, fold, and tape.

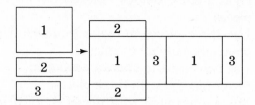

6. Here are covers for polyhedra that you haven't constructed so far. The faces are regular polygons. Use an equilateral triangle-piece and a pentagon-piece to make covers like these, and construct models for the polyhedra. To make them really nice, cut out pictures of people you like and glue them on the faces. (You could make a mobile.)

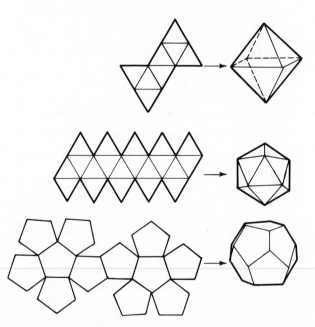

7. Polygons that can be used to fill the plane are ones whose vertices can be fitted together so that there are no holes or overlap. Such polygons are said to *tessellate* the plane. The square is such a polygon.

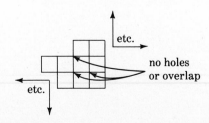

a. Tell which of these regular polygons tessellate the plane and which do not.

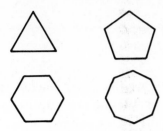

b. We can use different polygons to tessellate the plane and create interesting patterns. Cut out polygon-pieces of different colors from construction paper. Make three designs. (Use the one shown at left as a hint to get you started.)

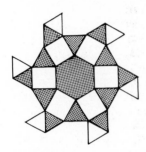

8. Classify the capital letters of the alphabet in terms of these classes.
   a. exactly no lines of symmetry
   b. exactly one line of symmetry
   c. exactly two lines of symmetry
   d. exactly four lines of symmetry
   e. more than four lines of symmetry
   How many letters are in each of the classes?
9. You can fold and cut paper to get interesting designs with symmetry. Here's how: Cut a design that has no lines of symmetry; unfold.
   a. Use the methods shown below to fold and cut newspaper. (Remember, the design you cut before unfolding cannot have lines of symmetry.) Tell how many lines of symmetry you get for each design.

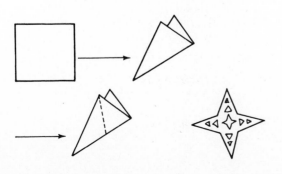

b. Write your name on a triangle-piece. The triangle-piece should have one vertex that measures $\frac{1}{6}$ of a turn. Place your triangle-piece on a sheet of paper. Trace it and your name. Flip along side AB and trace.

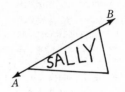 

after 1 flip

Repeat this procedure five times. Color your design. Does it have lines of symmetry? Does it have turning symmetry?
10. Select a cartoon figure from today's comics.
    a. Glue it on a piece of plain paper. Use a point p and a scale factor of 2 to enlarge the cartoon the way we did with the heart.
    b. Use squared paper (Material Sheet 17) to make a scale drawing of this cartoon.

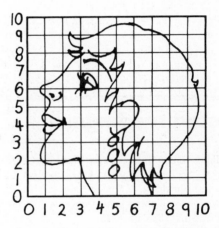

11. You can turn and slide plane figures through space to generate space figures. For example, if you turned the rectangle below about line $\overleftrightarrow{AB}$, the resulting "sweep" would be a cylinder.

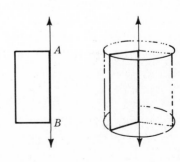

If you slide the polygon below along line $\overleftrightarrow{AB}$ through planes perpendicular to line $\overleftrightarrow{AB}$, you would generate a prism.

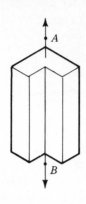

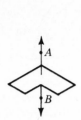

Tell what kind of space figures you would generate in each of the following cases.

a. Turn the triangle one full turn about line $\overleftrightarrow{AB}$.

b. 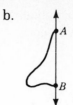 Turn the figure one full turn about line $\overleftrightarrow{AB}$.

c. 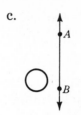 Turn the figure one full turn about line $\overleftrightarrow{AB}$. Be careful; this one is tricky!

12. Draw an inclusion diagram that illustrates the relationship among polygons.

→ Suggested readings

Charles, Randall I. "Some Guidelines for Teaching Geometry Concepts." *The Arithmetic Teacher*, April 1980, pp. 18-21.

Horak, V. M., and Horak, W. J. "Using Geometry Tiles as a Manipulative for Developing Basic Concepts." *The Arithmetic Teacher*, April 1983, pp. 8-15.

Jensen, Rosalie, and O'Neil, David R. "Let's Do It, Informal Geometry through Geometric Blocks." *The Arithmetic Teacher*, May 1982, pp. 4-8.

Mayberry, Joanne. "The Van Hiele Levels of Geometric Thought in Undergraduate Preservice Teachers." *Journal for Research in Mathematics Education*, January 1983, pp. 58-69.

O'Daffer, Phares, and Clemens, Stanley R. *Geometry: An Investigative Approach.* Reading, PA: Addison-Wesley, 1976.

Scheuer, Donald W., Jr., and Williams, David E. "Ideas." *The Arithmetic Teacher*, February 1980, pp. 28-31.

Van de Walle, John, and Thompson, Charles S. "Let's Do It, Cut and Paste for Geometric Thinking." *The Arithmetic Teacher*, September 1984, pp. 8-13.

Young, Jerry L. "Improving Spatial Abilities with Geometric Activities." *The Arithmetic Teacher*, September 1982, pp. 38-43.

Zurstadt, Betty K. "Tessellations and the Art of M. C. Escher." *The Arithmetic Teacher*, January 1984, pp. 54-55.

Seeing Is Believing
Children's Development of Geometric Ideas

■ *In this chapter, we describe how children between ages 2 and 12 develop geometric concepts, and we present many related activities.*

Knowledge in this area, although far from definitive, indicates that, as is true of number ideas, subtleties are involved in the development of geometric ideas, and that teachers must take into account special considerations when preparing activities for children. These considerations involve good timing, appropriate selection of materials and tasks, and the use of language that children can understand. ■

Objectives for the child

1. Draws models for familiar plane figures.
2. Identifies familiar plane figures.
3. Makes models for space figures.
4. Identifies familiar space figures.
5. Classifies familiar geometric figures.
6. Lists characteristics of familiar geometric figures.
7. Determines whether two plane figures are congruent by tracing, matching, and so on.
8. Determines by inspection whether two geometric figures are similar.
9. Checks by cutting and fitting whether two plane figures have the same size.
10. Identifies physical models for parallel segments or regions.
11. Identifies physical models for perpendicular segments or regions.
12. Locates points on a rectangular coordinate system.
13. Makes simple scale drawings.
14. Identifies projections of simple figures.
15. Uses a point and a scale factor to enlarge or reduce a simple picture.
16. Determines the outcome of a slide, flip, or turn.
17. Identifies figures having line symmetry or turning symmetry.

Teacher goals

1. Be able to identify and explain geometric concepts that children should develop in kindergarten through sixth grade.
2. Be able to diagnose the performance of children with respect to geometric concepts.
3. Be able to identify and carry out informal activities that will aid children in their development of appropriate geometric ideas.

At what age can children draw square figures?
At what age do children distinguish left from right?
At what age do children begin to use horizontal and vertical references to relate objects in their environment?

The purpose of teaching geometry in the elementary school is to help children acquire abilities to be used in describing, comparing, representing, and relating objects in the environment. The development of such abilities relies heavily on the kinds of experiences children have with real objects and on the ways in which they respond to these experiences. School experiences should therefore begin with *real things*—not pictures of real things or names for them, but real things! Objects, their attributes, the ways in which they are related to one another, and the kinds of actions that can be performed on them are the topics that should be studied.

Many geometry experiences should involve unstructured play activities in which children are encouraged to experiment, to find out, and to tell "why and what." These activities should not be confined to the early grades but should be a part of the total elementary program. Other experiences should involve more structured activities in which children are directed to collect data, make records, classify and organize information, and find patterns and generalizations. These activities should focus on the following skills:

1. Listing the attributes of geometric figures. For example, a square is "flat," has four sides, is closed, its sides have the same length, and so on.
2. Comparing geometric figures to see how they are alike and how they are different. For example, both circles and squares are flat, but a square has line segments for sides and a circle doesn't.
3. Identifying the results when geometric figures undergo change. For example, if this square piece of paper is cut into two parts and the pieces are refitted to make a triangle, then the shape of the paper is changed but its size is not.

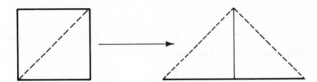

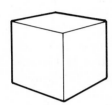

4. Identifying representations for geometric figures. For example, this is a picture of a cube, even though a cube isn't flat and the picture is.
5. Describing relationships among geometric figures. For example, if the length of each side of a square is doubled, then the area of the resulting square is four times as great.

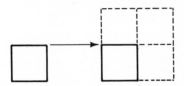

Children should also perform activities in which they learn to name figures, develop definitions, state generalizations, and practice skills. These activities should be very informal whenever new concepts are being introduced. Children should initially be allowed to use their own ideas, language patterns, and vocabulary. As they develop clearer, more refined concepts, they can be encouraged to make their language more precise.

Planning and sequencing experiences such as the ones we have just described requires knowledge of many geometric concepts, the way in which these concepts help us to understand the environment, and the order in which children acquire them. The available research does not describe precisely how and when children develop certain ideas, but it does indicate some unexpected patterns. Thus, adults cannot assume that ideas that seem evident to them are automatically evident to children.

In the next few sections, we outline children's development of some geometric ideas through the age 11. Descriptions are given for four overlapping age levels: before age 4, from 4 to 7, from 7 to 9, and from 9 to 11. The descriptions are followed by suggested classroom topics and related activities. Although some topics are sophisticated, you'll find that the related activities are very informal and rely heavily on the use of concrete materials.

Before age 4

Before age 4, children begin to acquire *topological* concepts, such as open/closed, inside/outside/boundary, separation/connectedness, order, and proximity. Children in this age group, however, generally do not develop ideas related to size, shape, direction, and angle, and there are some pretty good reasons why they don't.

Very young children are constantly exposed to changing views of familiar objects in their environment. For instance, family members are seen sitting and standing, seen from the front, back, and side, and seen close up and at a distance. Children thus perceive the shapes and sizes of objects changing as the objects move about in the environment. Consequently, they don't realize that the shapes and sizes of many objects in their environment are constant. Because an object may appear to be bigger or smaller from one point of view than from another, young children don't always attribute specific shapes to objects, and they sometimes can't identify specific shapes or correctly establish the relative sizes of the parts of an object. Finally, a child in this age group perceives the environment as a big stage with himself or herself front and center. How objects look from another point of view is not considered, horizontal and vertical references are ignored, and the child's representations of the environment are flat and randomly organized. To illustrate these claims, here is a 4-year-old boy's drawing of his mother.

Shane's picture of Mommy

The drawing indicates that the boy perceives his mother's head, in effect, as a *closed curve* and the features of her face as being *inside the curve*. Although these features are not placed appropriately *with respect to the horizon*, they *are in close proximity to each other*, and they are *not* in close proximity to the legs. In other words, the child has correctly decided which parts of the body are close together and which are not. The child also recognizes the *correct order* for the parts of the body and sees that the body is *connected*. The body and the hands enclose the arms. The body and the feet enclose the legs. The head, arms, and legs are *separated* by the body. (Notice the number of fingers on each hand. Does this surprise you?)

It's obvious from the picture that the boy has correctly conceptualized many geometric attributes of his mother's body but that others may have escaped him. Look at the size of the head in comparison to other body parts—notice that the picture is rendered with line segments and simple, closed curves that have approximately the same shape whether they represent eyes, nose, head, or body. The only reason that any part of the body is recognizable is that the parts are appropriately attached and organized; that is, the *topological* properties of the mother's body have been more or less observed.

Below is a drawing by a 3-year-old. Would you believe Santa Claus? Can you identify some topological attributes that the child has considered? Can you identify some that she hasn't? Does she make use of a horizontal reference?

"Santa Claus" by Shannon
Age 3

From ages 4 to 7

Between ages 4 and 7, children begin to develop some *Euclidean* concepts. Euclidean geometry is the branch of geometry that studies size, shape, direction, and angle. Perpendicularity and parallelism are important ideas in Euclidean geometry.

Children first become aware of direction and angularity. This awareness enables them to make broad distinctions among some simple closed curves; for example, they can distinguish circles from triangles and rectangles. But they may still confuse circles with noncircular ellipses and rectangles with other polygons. As their development progresses, they learn to make finer distinctions, discriminating between circles and ellipses and among triangles, rectangles, and parallelograms. They also learn to draw or copy models for these figures and even for composite figures.

Children in this age group also begin to see objects in relation to other objects; that is, they begin to *structure* space. At first, children view space as though they were in the center. Spatial relationships *among* external objects are not established, and the objects are perceived as a random collection. ("Here I am. There is the car; there is the sun; there is the ball; there is the butterfly.")

Gradually, children begin to recognize that there is structure and order in space. ("I am on the ground. The car is on this side of me; the airplane is not on the ground and is above me; the ball is on the ground and is on that side of me.") This conscious perception of objects in relation to each other leads to the development of ideas about left and right and about symmetry, as well as to the use of horizontal lines as references. Children thus begin to represent the world about them as something other than a random collection of objects. But they still do not consider the relative sizes of objects or their distances from one another, because they haven't developed the ability to conserve length, area, and volume. Furthermore, they don't consistently use horizontal references appropriately. The drawing below, for example, might have been done by a 6- or 7-year-old. Notice the chimney and the size of the child.

Another ability children in the 4-to-7 age group develop is the ability to place objects in appropriate positions by following a model and using two or more reference points. At first, the model and the copy must show the same positions with respect to the child, as illustrated below.

Teacher: "Put the truck in the right place in your copy."

Teacher Copy

Student Copy (The child's placing of the object anywhere in the shaded area indicates the use of reference points.)

Children gradually learn to place objects in appropriate positions even when the model and the copy do not show the same positions with respect to the child. Even so, children at this level may not be able to duplicate a model unless the model is extremely simple and the objects touch each other.

To sum up, children from 4 to 7 become aware that there is structure to space. They learn to express this awareness by sometimes using ideas of left and right and sometimes using horizontal references. Nevertheless, these devices are still used by the children to organize an environment they regard as surrounding *them.* At this age, they are unable to consider how an object might look from vantage points external to themselves.

■ Geometry concepts for kindergarten through second grade

How children develop geometric ideas has a great deal to do with teaching, but it's important to realize that our descriptions of this development are general. Children in both of the age groups already described differ tremendously as individuals, and their abilities are greatly affected by their previous experiences. At best, these descriptions are only a guide to be used in planning activities and in deciding on what can be realistically expected of children.

We are primarily concerned with the 4-to-7 age group here, because most children's formal schooling begins during this period. The activities we plan for children in this age group should be of three kinds:

1. activities designed to help children refine topological ideas they have begun to develop,
2. activities designed to help children make the transition from topological ideas to simple Euclidean ideas, and
3. activities designed to help children discover properties of space figures and their relationship to plane figures.

All three kinds of activities should be emphasized throughout this period of development, and they should be integrated so that geometric learnings begin to fit together.

Activities for the development of topological concepts

Teachers should plan activities that allow children to refine the topological ideas that they have begun to develop. Specifically, children should have experiences deciding what effects various changes will have on certain attributes of curves. In the following activity, the focus is on order and proximity. This activity illustrates how these attributes are affected by stretching, bending, or tying.

Example 11-1
An activity dealing with order and proximity

C-1

Materials: An elastic string with four beads of different colors—say, red, yellow, pink, and blue.

Teacher directions: Tell which beads are on the ends. Tell which beads are next to the yellow bead. Tell which beads aren't next to the yellow bead. Stretch the string. Now which beads are on the ends? Which beads are next to the yellow bead? (Etc.)

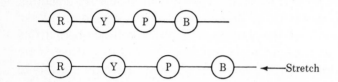

New teacher directions: If we're allowed to stretch the string of beads or wiggle it, but we aren't allowed to twist it, can we make the string look like this:

New teacher directions: If we twist the string like this, where will the different beads go? Try to make a design like this with your beads, and see whether you can find out. Will the red bead be close to the yellow bead? Was the red bead close to the yellow bead before?

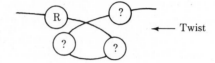

Here is an activity that illustrates the same concepts, except the model involves a plane region rather than a curve.

Example 11-2
An activity dealing with order, proximity, inside, outside, etc.

C-1

Materials: A square piece of white jersey with the face of a monkey, clown, bear, or the like drawn on it. Thumb tacks and a piece of corrugated cardboard. Pictures like the ones shown below.

Teacher directions: We can stretch this piece of material any way that we want as long as we do not tear or twist it. Can we make the monkey look like any of these pictures? Stretch and tack your material down to find out. Then tell why we can make a certain picture or why we can't.

Jersey Piece   Pictures

Children should also have experiences deciding when a common object has been represented appropriately. In Example 11-3, decide when certain attributes have been distorted or omitted.

Example 11-3
C-1
An activity that requires children to find irregularities in common figures

Materials: A set of teacher-made cards with funny-people pictures.

Teacher directions: Tell which pictures are right. Tell which pictures are wrong and why.

Example:

As well as learning topological attributes of objects and how these attributes are affected by certain changes, children should also learn to use vocabulary involving important prepositions. These prepositions include *in, on, under, inside, beside, between, upon, from, outside, above,* and so on. These words help children describe the relative positions of objects and enable them to place objects according to directions—for example, "Put the shoes in the box." The following activity will help children learn to use these prepositions to describe where objects are. Other activities should be planned in which children are to place objects when given oral directions.

Example 11-4
C-1
An activity that requires children to use appropriate vocabulary to describe position

Materials: A table with a box on it. Some familiar objects arranged in the box, to the side of the box, in front of the box, in back of the box, under the table.

Teacher directions: Tell where the comb is. Tell where the brush is. Tell where the shoe is. Tell where the bear is. Put the bear on the table. Put the car in the box. Put the shoe under the table.

on? in? under?

Activities for teaching about plane figures

As topological concepts are refined, Euclidean concepts should be developed. Activities should be planned that give children extensive experience with Euclidean shapes. These activities should require children to use other senses as well as sight for developing concepts. Abilities such as listing attributes, comparing shapes and sizes, and reproducing shapes should be emphasized. When children have learned to count to 10, they should be required to count the number of sides and the number of angles of some plane figures.

In the following three activities, Examples 11-5, 11-6, and 11-7, children identify and compare attributes of plane figures. Example 11-8 requires children to recognize plane figures that are congruent, regardless of how these figures are positioned. This activity can be used diagnostically as well as instructionally. In Example 11-9, children find congruent figures by matching boundaries. In Example 11-10, children are introduced to line symmetry. Although these kinds of activities are very informal, they provide the readiness needed for developing more formal concepts at later stages.

Example 11-5   C-1
An activity that requires children to find attributes of plane figures

Materials: Several sets of geo-pieces of different sizes and colors (Material Sheet 1).

Teacher directions: Look at this geo-piece. (Indicate a nonsquare rectangle.) Find all of your geo-pieces that look like this one. Tell me all you can about it. Does it have straight sides? Does it have corners? How many sides? How many corners?

This figure is called a *rectangle*. Trace one of your rectangles. Trace it several times to make a picture of something. Make a building. Make a tower.

New teacher directions: Put your rectangle-pieces away. Can you draw a rectangle without looking? (This activity should be repeated for other shapes.)

Example 11-6   C-1
An activity that helps children to find attributes of plane figures

Materials: Several sets of geo-pieces of different sizes and colors (Material Sheet 1). Geo-pieces should be cut from posterboard or other sturdy material. Blindfolds for children.

Teacher directions: I'm going to blindfold you and put a geo-piece in your hand. Feel it very carefully. Then see whether you can find another geo-piece with the same shape in the pile.

New teacher directions: Give me the geo-pieces back. Now take off the blindfold. Can you draw a figure like the one that was in your hand?

Example 11-7
An activity that requires children to classify plane figures

C-1

Materials: Geo-pieces (Material Sheet 1).

Teacher directions: Separate the geo-pieces into stacks so that all of the pieces in a stack have the same shape. How are the pieces in this stack (squares) like the pieces in that stack (circles)? How are they different? How are the pieces in this stack (squares) like the pieces in that stack (nonsquare rectangles)? How are they different?
(Repeat the activity for other pairs of figures.)

Example 11-8
An activity for evaluating children's ability to recognize shapes that are the same but positioned differently

C-1

Materials: Geo-pieces (Material Sheet 1). Index cards showing each geo-piece in various positions.

Teacher directions: These cards are all mixed up. Separate them into piles. Tell how you did it.

Example:

Example 11-9
An activity that introduces children to congruence ideas

C-1

Materials: Geo-pieces—two sets of 16 for each size— of different colors (see Material Sheet 1).

Teacher directions: Put the pieces into stacks so that the pieces in a stack match exactly.

Example:

Example 11-10
An activity that introduces line symmetry

C-1

Materials: Pattern-pieces with blue edges to indicate flip lines. Pieces of newspaper.
Teacher directions: Use a folded piece of newspaper to cut out each design. Make sure the blue edge is next to the fold. Unfold the newspaper to see what you've made.

 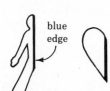   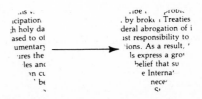

Activities for teaching about space figures

We have emphasized that the geometric concepts developed by young children are based on their experiences with objects in their environment. Even so, these early concepts generally do not include ideas that represent three-dimensional qualities. At least, such concepts are not formed in such a way that children can express them. What young children do learn are some attributes of plane figures and some topological attributes of objects. Children from 4 to 7 years of age may be able to separate models for squares from models for cubes, but if they are requested to "make one of these (a cube)," they'll likely draw a square. Even if they are given modeling clay to form three-dimensional models, they may find the task too difficult because of the number of corners, angles, edges, and so forth.

Nevertheless, it's important that children in the 4-to-7 age group have many experiences dealing with familiar objects and models for space figures. These should be built into exploratory activities in which children focus on simple concepts. The activities should be less structured than those presented for plane figures, since the concepts involved are much more complicated. These space-figure activities should emphasize discovering attributes of the space figures and finding relationships between space figures and plane figures. Here are three activities that illustrate these ideas.

Example 11-11   C-1
An activity that enables children to distinguish space figures that will fill space (stack without leaving "holes") from those that will not

Materials: Models for space figures: cubes, rectangular prisms, cones, triangular prisms, cylinders, irregularly shaped solid figures.

Teacher directions: Make a tower. Make it as tall as you can. Use any of the blocks that you want to use.

Question: Which blocks are best for making towers?

New teacher directions: Play a game. Make a tower. Each player places a block on the tower. The player to place the last block before the tower topples is the winner.

New teacher directions: Make a wall with the blocks. The wall cannot have holes. Tell which pieces you can use. Take a piece out of the wall. See whether you can turn it around and put it back into the wall so that it will still fit.

Example 11-12
An activity that requires children to classify space figures

Materials: Familiar objects—small boxes, cans, balls, cone-shaped water cups, paper-towel holders, and so on.

Teacher directions: Sort these objects into groups so that all of the objects in a group have the same shape. Tell what rule you used.

New teacher directions: For each object, make some tracings. Keep your tracings so that you can tell how you got them.

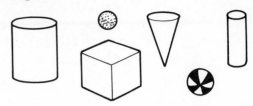

New teacher directions: Tell which of your figures will stack. Tell which will roll. (Etc.)

Example 11-13
An activity that enables children to see the relationships between space figures and plane figures

Materials: Commercial set of space figures. Clay rolled out in a tray.

Teacher directions: Make clay prints of your blocks. See how many different prints you can get for each block.

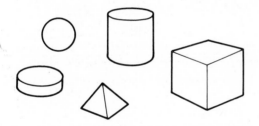

Can you use this ▢ to get this ☐ ?

Can you use this ▢ to get this ○ ?

Can you use this ▱ to get this _____ ?

Can you use this ▢ to get this • ?

What other solids can you use to get • ?

From ages 7 to 9

During the period from about ages 7 to 9, crucial developments take place in children's perceptions and abilities. These are based on much experimentation and form the basis for relating oneself realistically to the physical environment.

First, children in this age group develop the ability to conserve length and area. In other words, they begin to realize that the length and area of figures do not change when the figures are relocated in space. Understanding this idea allows children to investigate the sizes and shapes of objects and to decide how size and shape are related.

Second, children now become aware that horizontal and vertical references can be used simultaneously. This enables them to represent objects in space more realistically. Children's drawings of chimneys on housetops, for example, become vertical with respect to the horizon, whereas chimneys previously had been drawn perpendicular to the housetop.

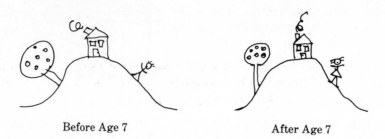

Before Age 7   After Age 7

The awareness of vertical and horizontal references also leads to the use of coordinate systems for structuring space. Children in this age group can recognize and use such a reference system for reproducing an arrangement of objects, whereas children in earlier stages of development ignore such references.

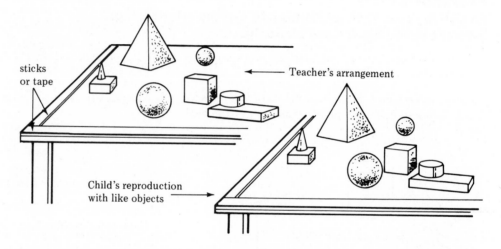

The ability to use such a reference system, however, is still in the early stages of development. Children will still have difficulty in simultaneously considering proximity, relative positions, and distances. Often, when duplicating patterns, children will observe proximity and relative positions but will represent distances inaccurately.

Children in this age group also begin to develop some idea of perspective; that is, they begin to consider how an object or set of objects looks from various points of view. For example, they can now learn to anticipate how the shadow of a stick will look when the source of light is located in different positions with respect to the stick. But this ability to consider objects from different vantage points is confined to very simple objects.

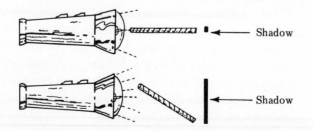

■ Geometry concepts for the second through fourth grades

The developments that begin to occur at about age 7 should be carefully considered in the preparation of student activities. In some cases, earlier activities should be repeated. These should also be extended and elaborated to account for the new developments.

At this level, children should engage in activities that will introduce them to ideas related to *perpendicularity, parallelism, similarity, size, congruence,* and *symmetry.* They should learn to name, describe, represent, and classify simple plane and space figures, as well as the parts of these figures. They should compare sizes and shapes of figures and begin to investigate simple shadow patterns and patterns obtained by sliding, flipping, or turning a simple shape. They should also have experiences using a simple coordinate system. The emphasis of these activities should be on refining concepts that have begun to develop at an earlier age and on developing intuitive ideas related to concepts that will be gradually refined during this and later age periods.

The following three sections present samples of the kinds of activities that should be included at this level. These examples highlight important ideas and methods, but all the topics that *could* be covered are far too numerous for us to consider here. As a teacher, you'll want to examine materials specially prepared for children so that you can acquire a more complete set of activities.

Activities for teaching about plane figures

The following five activities are designed to extend development of similar activities presented for an earlier level of development. In Example 11-14, children are to discover that size and shape are independent attributes. In Example 11-15, children learn to make a right-angle tester to find right angles of figures. Example 11-16 emphasizes congruence, Example 11-17 focuses on similarity, and Example 11-18 introduces line symmetry.

Examples 11-19 and 11-20 are of particular importance because these activities introduce concepts that are beginning to develop at this age level. In Example 11-19, children reproduce a scale drawing. In Example 11-20, children learn to identify and name perpendicular and parallel lines. Concepts used in Example 11-19 become more formal in Example 11-20.

Example 11-14

C-2

An activity to help children investigate the relationship between size and shape

Materials: Geo-pieces (Material Sheet 1).

Teacher directions: You can use your geo-pieces to make different shapes. For example, you can use the big and little triangles to make these designs.

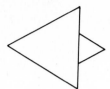

Questions: Do these designs have the same shape? Do they have the same size?

New teacher directions: Use the geo-pieces to make designs that have the same size but not the same shape.

Example 11-15

C-2

An activity in which children identify figures with right angles

Materials: A teacher-made ditto of a chart like the one below. The following geo-pieces and space figures. Material Sheet 23.

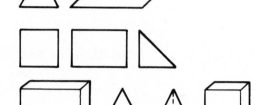

Geo-piece or space figure	Right angles?	
	Yes	*No*
□		
▭		
Etc.		
△		
▱		

Teacher directions: Make a right-angle tester by folding a sheet of paper two times. See if you can match its corner with the corners of the objects you have.
Fill in the chart to show the right angles you find.

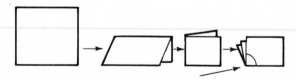

Example 11-16
C-2

An activity that introduces children to congruence of geometric figures

Materials: Teacher-made pattern-pieces and ditto.

Pattern-pieces   Ditto

Teacher directions: For each pattern-piece, do the following things. By matching with the pattern-piece, find every figure on the ditto sheet that has the same size and shape as the pattern-piece. Label each of these figures with the same letter as the matching pattern-piece.

Example 11-17
C-2

An activity that introduces similarity of geometric figures

Materials: Pattern-pieces and ditto.

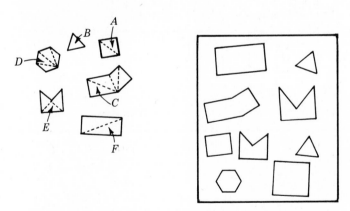

Pattern-pieces   Ditto

Teacher directions: For each pattern-piece, find every figure on the ditto sheet that has the same shape as the pattern-piece. They don't have to have the same size.

Example 11-18
An activity that introduces line symmetry

C-2

Materials: Mirror. Teacher-made ditto like the one below.

Teacher directions: Put a dotted line on each picture to show where the mirror should go if you want to get a familiar picture. Use the mirror to help you. (Teacher gives example.)

Example:

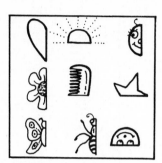

Example 11-19
An activity in which children make scale drawings

C-2

Materials: Pictures drawn on dot paper (15×15 arrays); blank pieces of dot paper in 15×15 arrays (see Material Sheet 16). Crayons.

Teacher directions: See whether you can use dot paper to copy the pictures *exactly*.

Example:

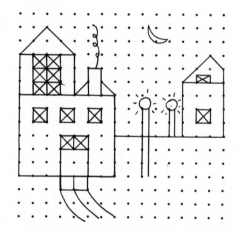

☞

Example 11-20
An activity that introduces parallelism and perpendicularity

Materials: Sheets of paper. Colored pencils. A teacher-made ditto like the one shown here.

Teacher directions: Fold a piece of paper two times like this. Now unfold it, and draw lines along the folds. Lines that cross like this are called *perpendicular* lines. How much of a turn is one of the angles made by perpendicular lines?

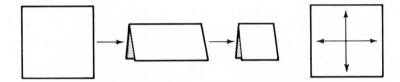

New teacher directions: Now fold another sheet of paper two times (in thirds, as shown below). Then unfold the paper, and draw lines along the folds.

Lines like these are *parallel* lines. Where have you seen parallel lines before?

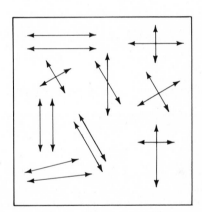

New teacher directions: Now, look at the ditto sheet. Put a ✓ next to the parallel lines. Put an x next to the perpendicular lines. Then get together with a friend. See whether you can find some places that show parallel and perpendicular lines.

Activities for teaching about space figures
The following three activities illustrate appropriate ways of teaching concepts about space figures. Example 11-21 helps children discover ways to make models of space figures; in this way, children should become aware of important attributes. In Example 11-22, children are to identify congruent space figures by touch alone. In Example 11-23, projections of space figures are identified.

Example 11-21   C-2
An activity that requires children to construct space figures

Materials: Models for space figures. The cylinder and the cone must each be marked with a line segment.

Teacher directions: Make covers for these figures by tracing their faces and surfaces.

For these, trace every face. Then cut the faces out and fit them together with masking tape.

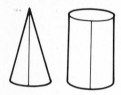

For these, draw a line segment on a sheet of paper.

Match the line segment of the space figure to the one on the paper. Roll the figure until the line segment goes around once and is in the starting position.

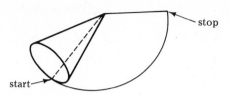

Then trace the "tops" and "bottoms." Your cover for the cone should look like this.
Cut out the pieces, and fit them together with masking tape.

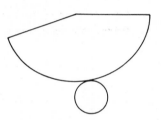

Example 11-22   C-2
An activity that introduces ideas related to congruence and similarity

Materials: A blindfold. A box with models for space figures. Models for space figures (two of each).

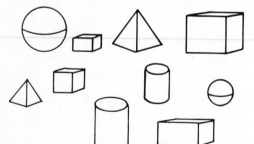

Models for space figures

Teacher directions: I'm going to put a block in your hand. See whether you can find a block in the box that has the same size and shape. See whether you can find a block that has the same shape but not the same size.

Example 11-23

An activity that requires children to use shadows to obtain certain projections

Materials: Flashlight. Large geo-pieces (Material Sheet 1). Pencil.

Teacher directions: Use the geo-pieces, the pencil, and the flashlight to make shadows. Sketch the shadows you get from each object. Then answer the questions.

Questions:

Can you get a shadow like this: ⬭ with this: ◯ ?

Can you get a shadow like this: ⬠ with this: ☐ ?

Can you get a shadow like this: ●

with this: ✏ ?

Etc.

From ages 9 to 11

The period from roughly ages 9 to 11 is one of the most dramatic stages of development for informal discovery of geometric ideas as children begin to be dissatisfied with flat interpretations of space. Horizontal and vertical references are not enough for interpreting the world about them. They become concerned with *filling* space and consequently discover the space-filling illusions that can be created with parallel planes. Projections for simple geometric figures can be predicted, and graphic representations can be discriminated and sometimes represented. A greater need to express perspective is felt. These advances are illustrated in the following picture, drawn by a 9-year-old boy.

By Shane
Age 9

Children at this age also have a great deal of curiosity, and their maturing mental capabilities enable them to become more conscious of details and relationships. Geometric ideas that might have been entirely missed earlier now become important to interpret. As a result, children's ability to find class relationships improves. Children who earlier would have insisted that "a square is a square and not a rectangle" begin to appreciate that "all squares are rectangles, but not all rectangles are squares." Children also become more able to consider a number of related concepts at one time. For example, their skill increases in simultaneously handling relative position, proximity, and distance. Consequently, their skill in using a rectangular coordinate system increases, and their scale drawings and maps become more accurate as well.

Notice that in the last few paragraphs we have used words such as "improves," "increases," and "begins" in reference to children's new abilities. These are key words. They should alert you to the fact that even though these abilities are growing, they are not yet "grown." For most children, the maturation of these abilities will take several more years. Furthermore, some ideas closely related to the concepts we have discussed cannot be appreciated by most children younger than 12—for example, "a point has no dimensions," "lines are an infinite set of points," and "between any two points there is another point." So, activities should continue to be intuitive, and children should manipulate materials in order to learn new concepts, rather than concentrate on abstract ideas.

■ Geometry concepts for grades four through six

Geometry activities for children in grades four through six should refine concepts that have been introduced in earlier grades. Attributes of plane figures should be more carefully studied, and these attributes should be used to slowly develop class relationships among figures.

In these grades, the study of space figures should become more structured. Parallelism and perpendicularity should be used to classify space figures and to make graphic representations of them. These ideas should then be used to develop coordinate systems, to reproduce arrangements of objects, and to make enlargements or scale drawings of figures.

Activities involving the measurement of length and area should become more precise, and the relationship between perimeter and area should be studied. Volume and angular measure should be reintroduced in greater depth. Finally, emphasis should be placed on considering when the following will occur:

1. two figures will have the same size
2. two figures will have the same shape
3. two figures will have the same size and shape
4. a figure will have line symmetry or turning symmetry

That's enough to scare the bravest teacher! But this list of ideas shouldn't overwhelm you. It's intended to guide you as you plan experiences for children. As you read through the activities that follow, you'll see that even sophisticated ideas can be handled very informally.

In Example 11-24, children investigate the relationship between size and shape. In 11-25, they are introduced to class-inclusion concepts for quadrilaterals. Such concepts as "a square is a rectangle, is a parallelogram, is a quadrilateral" should be learned. Being able to appreciate such subtleties will enable children to understand precise mathematical definitions that are dealt with in secondary school. Examples 11-26 and 11-27 are related; in these two activities children use rectangular coordinate concepts. In 11-28, children investigate some relationships between space figures and plane figures.

Example 11-24   C-3
An activity in which the sizes and shapes of figures are compared

Materials: Lots of geo-pieces—squares, triangles, rectangles, and so on.

Teacher directions: Do these things with your square-pieces:

1. Trace them to make figures that have the same size and shape.
2. Trace them to make figures that have the same shape but not the same size.
3. Trace them to make figures that have the same size but not the same shape.

Example:

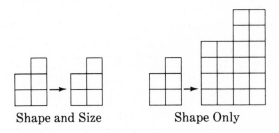

Shape and Size   Shape Only

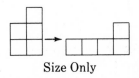

Size Only

Example 11-25   C-3
An activity that introduces class-inclusion ideas for quadrilaterals

Materials: Geo-pieces and geo-cards (see Material Sheets 23 and 24).

Teacher directions: Cut out all of the figures on this page (Material Sheet 23). Study the Quadrilateral card. How are the figures on this card alike? Go through all of the cut-outs and put a Q on the back of all cut-outs that are quadrilaterals.

Now study the Parallelogram card. Tell how the figures on it are alike and how they are different. Go through the cut-outs again. Put a P on the back of parallelogram figures.

Do the same thing for the Rectangle card and the Square card. Put R's and S's on the backs of the correct figures.

New teacher directions: Some figures have many names. Some names are better than others, however, because they tell us more. See whether you can answer these questions about the pictures of your figures and their names. Give reasons for your answers.

Questions: Can any square be called a rectangle? Can any rectangle be called a square? Can you find a rectangle that cannot be called a square? (Etc.)

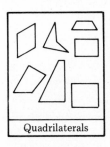

Quadrilaterals

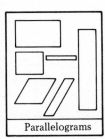

Parallelograms

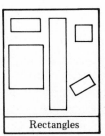

Rectangles

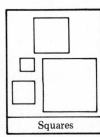

Squares

Example 11-26
An activity that introduces coordinate systems

C-3

Materials: Cut-outs such as those shown below. A grid with 16 squares. The cut-outs can be arranged on the squares of the grid.

Here are some rows and columns. Notice how the rows are numbered. Notice how the columns are numbered.

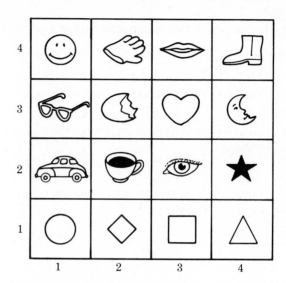

Teacher directions: This is a row of things:

This is a column of things:

Tell what's in the first row. Tell what's in the third row. Tell what's in the first column. Tell what's in the fourth column. (Etc.)

See whether you can figure out my pattern: The mouth is at (3, 4). What do you think the 4 stands for? What do you think the 3 stands for?

Tell where the car is. Give the number for the column first. Give the number for the row second.

I'm thinking of something at (3, 1). See whether you can read my mind.

Example 11-27

An activity that reinforces concepts related to flips, slides, turns, and symmetry

Materials: Many different geo-pieces (see Material Sheets 21 and 23). Paper for folding.

Teacher directions: You can trace geo-pieces to make designs that have line symmetry or turning symmetry. Make designs with the following kinds of symmetry. Examples are given to help you get ideas.

1. Turning symmetry.

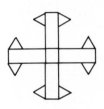

2. One line of symmetry.

3. Two lines of symmetry. (Does the figure you get have line symmetry?)

4. Turning symmetry but not line symmetry—$\frac{1}{4}$ turn.

5. Turning symmetry—$\frac{1}{6}$ turn.

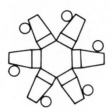

New teacher directions: You can also fold paper to make designs that have line symmetry. Do these things:

1. Fold once, and cut a design without symmetry. Unfold. How many lines of symmetry does your design have?
2. Fold twice, and cut a design without symmetry. How many lines of symmetry will the unfolded design have?
3. Fold three times, and cut a design without symmetry. How many lines of symmetry will the unfolded design have?

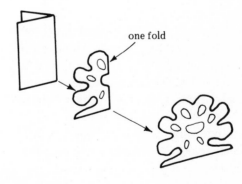

one fold

Example 11-28   C-3
An activity that shows relationships between space figures and plane figures

Materials: Clay models for these space figures.
Dull-edged knives.

Teacher directions: Try to make clay models for the space figures. Then guess what a certain cut will look like, and cut your model to find out whether you're right. Trace the cut to keep a record.

Cut this  to get this .

Cut this 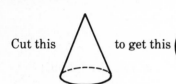 to get this ○ .

Cut this 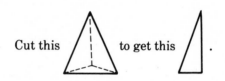 to get this .

Cut this 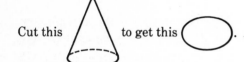 to get this .

Try some cuts of your own.

⭐ ——————————————————————————— ⭐

Ideas for enrichment: Constructions

Constructions that require only a ruler and compass are appropriate and motivational for many children in elementary school. The artistic aspect of this topic is exciting. It may appeal to children who are not necessarily whizzes at computation and give them a vehicle for excelling in mathematics. The whole subject of geometry at this level is enriching. Constructions provide more icing on the cake. We need to review a few basic constructions that you will show your children. Then let their imaginations loose. But first:

1. To construct the perpendicular bisector of line segment $\overline{AB}$, put your compass point on A and draw arcs above and below $\overline{AB}$. Then put the compass point on B and draw arcs above and below $\overline{AB}$. (Make sure your compass is spread apart far enough so that the arcs will intersect.) With a ruler, draw a line through the two points of intersection.

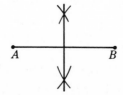

2. To construct the angle bisector of angle *ABC*, put your compass point on *B* and draw an arc that intersects both sides of the angle. Label these points of intersection *D* and *E*. Then move your compass point to point *D* and draw another arc in the interior of the angle. Now move the compass point to point *E* and draw another arc. These two arcs will intersect in a point. Label this point *F*. Draw a ray from *B* through *F* with a ruler.

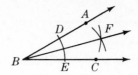

3. To draw beautiful pictures, use your imagination. Here's a sample:

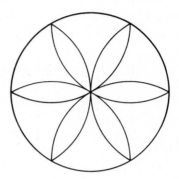

The computer corner

Here is a program entitled SYMMETRY, which allows children to learn by experiment. The program lets children draw figures on the screen; the program then produces mirror images. The results can be quite striking. Run the program a few times and create some interesting designs.

Example 11-29
Drawing designs with line symmetry

C-2, 3

Materials needed: SYMMETRY program on the MGB Diskette. Note: The teacher must load the program into computer memory for the student.

Teacher directions: This is a drawing program that will let you make designs that have line symmetry. To start, choose a color and a shape. After these choices are made, use the cursor keys to move to a location on the screen where you want to draw. When you press "D," the program draws shapes on the screen. You can change the shape by pressing "S." You can change color by pressing "C." Every time you draw, the computer copies what you draw on the opposite side of the screen, producing a mirror image or reflection.

Example: Use the program to produce some drawings. Try to draw a face, a butterfly, a heart, and so on.

Here is another activity that can be done using SYMMETRY. This activity improves motor coordination and spatial perception:

Example 11-30

C-2, 3

Materials: SYMMETRY program on the MGB Diskette. The left half of pictures that have been produced using the program.

Teacher directions: Here are some pictures. You are to draw pictures on the right side of the screen so that the computer is forced to draw pictures on the left side of the screen like these.

Example: Draw a picture on the right side of the screen that will make the computer draw this on the left side.

■ Computer tips

This program is open-ended. It allows for freedom of expression, discovery, and challenge. Programs like this can be very useful when planning experiences for special children.

☆ ————————————————— ☆

In closing . . .

In this chapter, we've outlined children's development of geometric ideas through age 11. Descriptions have been given for three age groups. Since abilities vary greatly at a given age, and since intellectual growth is continuous, the age groups overlap and are broadly organized.

Even though these age groupings are based on research into children's development of geometric ideas, a great deal of work remains to be done before such categories can be defined precisely. But even these fairly rough categories indicate the existence of certain patterns of development that we must consider if we want to be effective teachers.

Other considerations are equally important. First, children must interact with materials like those described in this chapter, and these interactions must involve more than visual discrimination. This means that manipulating materials, working in small groups, and expressing individual ideas are at least as important

as completing work from a textbook, taking part in large group instruction, and practicing skills. The teacher who settles for less, regardless of the teaching circumstances, will get considerably less.

Second, the fact that certain concepts are not expected to develop at a certain age doesn't mean that their instruction should be ignored. Play activities and appropriate teacher direction can still be useful. Who knows what insights may develop?

Third, in the final analysis, teachers are the most important researchers. Because of their day-to-day contact with children, they often become aware of children's capabilities long before these are studied by professional researchers. For this reason, it's important that you study behavior patterns of children and use what you find as a guide for constructing appropriate learning experiences in geometry.

It's think-tank time

1. Determine which of the activities in this chapter involve higher-order thinking skills and could be used to promote problem solving by elementary school children.

2. Select a learning objective and an activity appropriate for children in kindergarten through second grade. Make the necessary materials and carry out the activity. Give an account of what happens during the activity.

3. Select an activity that is appropriate for children in the third or fourth grade. Make the necessary materials and carry out the activity. Give an account of what happens.

4. Select an activity that is appropriate for children in the fifth or sixth grade. Make the necessary materials and carry out the activity. Give an account of what happens.

5. Find a 4-year-old, a 7-year-old, and a 9-year-old. Have each of them draw a picture of the same subject (a house, a boy, and so on). Analyze their pictures using concepts from this chapter. Bring the pictures to class for discussion.

6. Do the activities in any four of the example activities.

7. Using ideas developed in the Think-Tank section for Chapter 10, make up one activity for each of the four age groups discussed in this chapter.

8. Make up an activity requiring children to use Material Sheets 23 and 24. Write at least five good questions to ask children as they complete the activity. Carry out the activity with the children.

9. Make up an activity requiring children to use a point and a scale factor to produce a similar figure by enlargement.

10. Get a ruler and compass for these:
    a. Draw a line segment and construct its perpendicular bisector.
    b. Draw an angle and construct the angle bisector.
    c. Draw at least one beautiful design using only the ruler and compass.

11. Design three lessons that incorporate the use of the program SYMMETRY: one each for gifted children, one for average children, and one for children having special learning difficulties. Specify the objective and the intended level for each lesson.

➡

Suggested readings

Damarin, S. K. "What Makes a Triangle?" *The Arithmetic Teacher*, September 1981, pp. 39-41.

Dodwell, Peter. "Children's Perceptions and Their Understanding of Geometrical Ideas." In *Piagetian Cognitive-Development Research and Mathematical Education.* Reston, Va.: National Council of Teachers of Mathematics, 1971. (Pp. 178-188.)

Egsgard, John C. "Geometry All Around Us—K-12." *The Arithmetic Teacher*, October 1969, pp. 437-445.

Jackson, Stanley B. "Congruence and Measurement." *The Arithmetic Teacher*, February 1967, pp. 94-102.

Kidder, Francis Richard. "An Investigation of Nine-, Eleven-, and Thirteen-Year-Old Children's Comprehension of Euclidean Transformations." (Doctoral dissertation, University of Georgia, 1973.) *Dissertation Abstracts International, 34A*(1973):3238-3239.

Lindquist, Mary Montgomery (Ed.). *Selected Issues in Mathematics Education.* Reston, Va.: National Council of Teachers of Mathematics, 1981.

Lovell, Kenneth R. "Some Studies Involving Spatial Ideas." In *Piagetian Cognitive-Development Research and Mathematics Education.* Reston, Va.: National Council of Teachers of Mathematics, 1971. (Pp. 189-202.)

Mansfield, Helen. "Projective Geometry in the Elementary School." *The Arithmetic Teacher*, March 1985, pp. 15-19.

O'Daffer, Phares O. "Geometry: What Shape for a Comprehensive, Balanced Curriculum?" *Selected Issues in Mathematics Education*, May 1980, pp. 90-105.

Perciante, Terence Hugh. "The Influence of Visual Perception upon the Development of Geometrical Concepts." (Doctoral dissertation, State University of New York at Buffalo, 1973.) *Dissertation Abstracts International, 33A*(1973):5560.

Piaget, Jean, and Inhelder, Barbel. *The Child's Conception of Space.* London: Routledge & Kegan Paul, 1963.

Piaget, Jean, and Inhelder, Barbel. *The Child's Conception of Geometry.* New York: Harper & Row, 1964.

Robinson, G. Edith. "Geometry." In *Mathematics Learning in Early Childhood.* Reston, Va.: National Council of Teachers of Mathematics, 1975.

Shapiro, Bernard J., and O'Brien, Thomas C. "Logical Thinking in Children Ages Six through Thirteen." *Child Development*, 41(1970):823-829.

Suydam, Marilyn N., and Weaver, Fred J. "Geometry and Other Mathematical Topics." In *Using Research: A Key to Elementary School Mathematics.* Columbus, Oh: ERIC Science, Mathematics and Environmental Education Clearing House, 1975.

Wheatley, Grayson H., Mitchell, Robert, Frankland, Robert L., and Kraft, Rosemarie. "Hemisphere Specialization and Cognitive Development: Implications for Mathematics." *Journal for Research in Mathematics Education*, January 1979, pp. 76-79.

Williford, Harold. "What Does Research Say about Geometry in the Elementary School?" *The Arithmetic Teacher*, February 1972, pp. 97-104.

Wirszup, Izaak. "Breakthroughs in the Psychology of Learning and Teaching Geometry." In Larry J. Martin (Ed.), *Space and Geometry.* Columbus, Oh: ERIC Information Analysis Center for Science, Mathematics and Environmental Education, 1976.

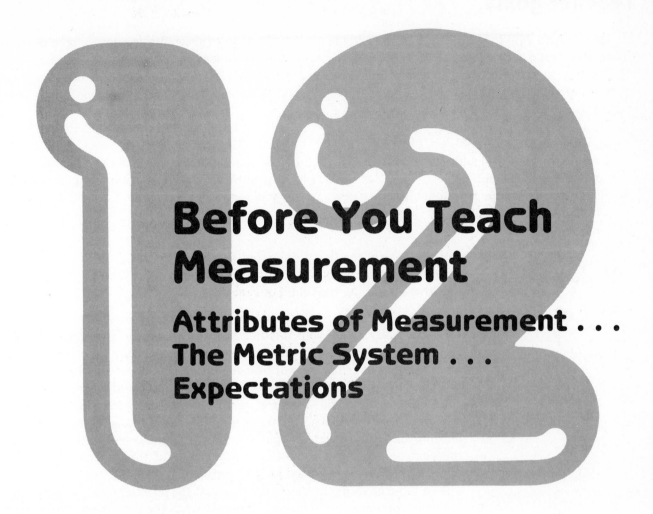

Before You Teach Measurement

Attributes of Measurement...
The Metric System...
Expectations

■ *This chapter presents ideas that the teacher should consider before beginning to design strategies for teaching measurement to children. We begin by discussing the importance of measurement as a means of mathematically describing our environment and then consider some important characteristics of measurement. Next, we introduce the metric system of measurement as the standard system of units to be used by elementary school children, and we suggest physical referents for these units. Finally, we summarize the concepts that are prerequisite to teaching formal procedures for measurement, emphasizing the need for readiness activities.*

The decision to use only metric units in the two chapters on measurement was a careful and deliberate one. One reason is that teachers of mathematics are assuming leadership roles in the conversion to the metric system in the United States, and future teachers must be willing and prepared to step into these positions. Another reason is simply that the metric system is easier to teach by itself. We realize, however, that people who first learned our customary units and then need to learn metric units must be treated gently.

In this chapter, also, we deliberately chose not to include activities for children. Again, this chapter is for you, the teacher. ■

Teacher goals

1. Be able to convince children of the importance of measurement.
2. Be able to provide many practical applications of measurement concepts.
3. Be able to identify attributes of measurement.
4. Be able to use metric units of measurement appropriately.
5. Be able to give physical referents for metric units of measurement.
6. Be able to provide relationships between English and metric units for informal comparison purposes.

7. Be able to describe the levels of children's development of measurement ideas.
8. Be able to associate the levels of concept development with approximate ages of children and with expectations for their behavior.
9. Be able to determine appropriate readiness activities.
10. Be able to diagnose the behavior of children by observing their participation in measurement experiences.

What's one of the first things that happened to you?
Try to imagine a day without measurement!
Is there anything that can't be measured?

Your weight and length were probably measured before you were 1 hour old. Ever since, you've either been measured or benefited from the results of previous measurements just about every day. If you think of all the measurements for yourself that you know, you'll probably be astonished at how long the list is. Besides your weight and height, you may know your pulse rate, blood pressure, temperature, and even your IQ, not to mention sizes for shoes, shirts, pants, sweaters, coats, hats, rings, and bathing suits. If you extend your list to include measurements of things in the environment, you'll see that the number of measurements and the diversity of things measured are truly amazing.

As a topic of study, measurement is vast. It's no exaggeration to say that the development of civilization is closely tied to the development of mathematics and measurement. Measurement provides a link between mathematics and the real world, imposing order on situations that otherwise would be chaotic. The use of formal measurement procedures goes back at least as far as the ancient Egyptians and Babylonians, whose accomplishments in this field still seem remarkable some 5000 years later. The ancient Greeks also used measurement to describe the physical environment, and the Greek mathematician Archimedes went so far as to suggest that anything that really exists can be measured. The search for ways to measure entities and the development and refinement of measuring procedures are milestones in the advancement of knowledge and technology.

→ Attributes of measurement

If you were asked what your height is, you would immediately recognize the question as one of measurement. Suppose you said that your height is 168 centimeters. Let's see what kind of information your answer provides.

First of all, we notice that you've used the metric system of measurement, because you're up to date and because it's sensible (more about this later). We can infer from your answer that you know what "height" refers to. That is, you know that there is a particular attribute that has been sorted out and called "height" and that for you, height can be described as the distance from the bottom of your feet to the top of your head. So, before we can measure, the *attribute* that we're interested in must be *isolated* and *described* in some realistic way.

Your response also has two parts to it—a number and a unit. It wouldn't have made sense if you had said just "168" or "centimeters." You had to tell both how many and how many of what. Thus, measures have both a *number* and a designated *unit*.

The unit that you used is an *appropriate* one: Since you were measuring a distance, you chose a *linear* unit of measure. It wouldn't have been reasonable to give your answer in square centimeters or in kilograms. The *relative size* of the unit is also appropriate. Kilometers would have been too large, and millimeters would have been too small. Also, the unit that you chose is a *standard unit*; that is, it has been precisely defined and is used by a large number of people. Therefore, measurement involves the use of an *appropriate, standard unit*.

Having selected your unit, you gave a number to tell how many of these units it takes to represent your height. In effect, you compared one centimeter to your height and decided that it takes 168 of these centimeters to represent it. Thus, measurement involves a *comparison* of the standard unit for measuring a certain attribute with the attribute possessed by a given object. The number completes the comparison by telling how many of the standard units are needed.

The fact that you gave a *whole number* to indicate the units implies that you rounded off your measurement. You're not *exactly* 168 cm tall, but that measurement is probably correct to the nearest centimeter. Measurement, then, is only *approximate*. No matter how small a standard unit you choose, you can't determine the exact number that represents a measurement, but you can choose how precise you want to be. You could have given your height to the nearest millimeter, for example, although this degree of precision probably isn't necessary in this situation.

Your height is an attribute that can be measured *directly*. That is, you can take a ruler or a tape that is scaled in centimeters and apply it to the distance from the top of your head to the bottom of your feet. Many other attributes are measured *indirectly*. For example, we can measure temperature by measuring the effect that a change in temperature has on mercury in a glass tube.

These attributes of measurement need to be considered in planning measurement experiences for children. Lessons should be designed in which children can actively discover measurement concepts, particularly those related to length, area, volume, weight, and capacity. In the later grades, study of the historical aspects of measurement should also be emphasized.

Using metric units of measure

The transition to the metric system of measurement in the United States will force us to use standard units that many of us did not grow up with. We feel at home with inches, quarts, pounds, and our 98.6°F body temperature, but we're going to have to get used to centimeters, liters, kilograms, and a normal body temperature of 37°C. The sooner we get started, the easier it will be for us. We'll see that the relationship among metric units for measuring an attribute is always based on a power of 10. Therefore, to convert from one unit to another within the system, we have only to multiply or divide by 10, 100, 1000, and so on—a relatively easy task. This conversion relationship is consistent with the place-value aspect of the base-ten numeration system, as well as with our monetary system. Thus, concepts that are a part of one of these systems will be applicable and useful in another. We'll also find that within the system, metric units for measuring one attribute are related to units for measuring another attribute, so not every unit has to be memorized independently. We'll be able to communicate better with the rest of the world, which already uses the metric system.

There are basic metric units for measuring attributes such as length and weight (mass). Smaller and larger units are designated by prefixes attached to the names for the basic units. The prefixes always mean the same thing, regardless of the basic unit. "Milli-" means one-thousandth, "centi-" means one-hundredth, "deci-" means one-tenth, and "kilo-" means one thousand. There are a number of other prefixes, but these are the ones that are used most frequently and that will be introduced to children first.

The basic unit for measuring length is the *meter* (the symbol for meter is m). For comparison purposes, we can say that a meter is a little longer than a yard. (Think of some things that measure about 1 meter.) Using our prefixes, we can determine that a millimeter is one-thousandth of a meter, a centimeter is one-hundredth of a meter, a decimeter is one-tenth of a meter, and a kilometer is 1000 meters. Let's look at some referents for these new units that we've made from the basic unit and the prefixes.

A *millimeter* (mm) is a small unit for measuring length. Since a millimeter is one-thousandth of a meter, there are 1000 millimeters in a meter. One millimeter is approximately the thickness of a paper match or of the wire in a standard paper clip. Try using a ruler calibrated in metric units to find something else that is about 1 millimeter long.

about 1 millimeter thick   about 1 millimeter thick

A *centimeter* (cm) is larger than a millimeter but smaller than a meter. Since a centimeter is one-hundredth of a meter, there are 100 centimeters in a meter. The width of an ice-cream stick is about 1 centimeter, and so is the width of one of your fingernails (probably). Think of some other things that measure approximately 1 centimeter. No doubt you've figured out that 1 centimeter is 10 millimeters, and one-tenth of a centimeter is 1 millimeter. (Stack ten paper matches and see whether they're about 1 centimeter high.)

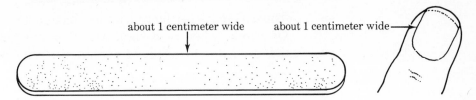

about 1 centimeter wide   about 1 centimeter wide

A *decimeter* (dm) is larger than a centimeter but still smaller than a meter. One decimeter is one-tenth of a meter, so there are 10 decimeters in a meter. We can also conclude that 1 decimeter is the same as 10 centimeters and 100 millimeters.

It's easy to find objects to represent 1 decimeter. Try to think of several. Find the length of a new piece of chalk, a bar of soap, your wallet, a box of crayons, or the width of a piece of toast. Even for adults, the best way to become familiar with metric units is to actually measure some objects.

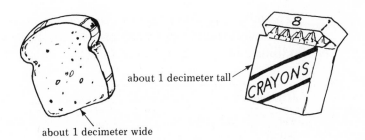

about 1 decimeter tall

about 1 decimeter wide

For measuring greater distance, we need a larger unit of measure. The most commonly used unit is the *kilometer* (km). For comparison purposes, we can note that a kilometer is approximately $\frac{5}{8}$ of a mile. A kilometer is 1000 meters, so a meter is one-thousandth of a kilometer. One kilometer is approximately 11 times as long as a football field. The distance from Chicago to Baltimore is approximately 1000 kilometers. How many kilometers is it from your town to Tucumcari?

Children's experiences with the kilometer as a unit of measure will eventually involve maps. No doubt you have a map that has a scale for kilometers on it. The procedure for determining distances is the same whether you're using miles or kilometers, so try finding some distances between familiar places to get the "feel" of this unit.

The fact that a kilometer is about $\frac{5}{8}$ of a mile is handy to know when we're considering such things as speed limits. Road signs usually show numbers of miles in multiples of 5; so, until the road signs change to kilometers, we can easily determine the number of kilometers for a certain number of miles in the following way.

The speed limit is 55 mph. What is the speed limit in kilometers per hour?

The ratio of miles to kilometers is about 5 to 8. But 55 is 5×11, so the number of kilometers is 8×11, or 88.

$$\frac{5}{8} = \frac{55}{?} = \frac{5 \times 11}{8 \times 11} \quad \text{or} \quad \frac{55}{88}$$

Thus, 55 miles per hour is approximately 88 kilometers per hour.

Of course, when road signs and speedometers are converted to metric units, even this little bit of arithmetic won't be necessary for using the metric system.

The following chart contains the most commonly used units of length in the metric system, their symbols, their definitions in terms of a meter, and familiar units that are approximately that length.

Millimeter	Centimeter	Decimeter	Meter	Kilometer

mm
.001 m
thickness of a
  paper clip
	cm
	.01 m
	width of a
	  fingernail
		dm
		.1 m
		width of a
		  piece of toast
			m
			1 m
			length of a
			  baseball
			  bat
				km
				1000 m
				width of Busch
				  Gardens (you
				  select a local
				  distance)

The prefix "deca" means 10, and the prefix "hecto" means 100. So, 1 decameter = 10 meters, and 1 hectometer = 100 meters. We deliberately left them out of the chart because we're predicting that common usage will skip from meters to kilometers, much as we now skip from yards to miles.

The units we've seen so far are used to measure length or distance and are therefore linear units. For measuring area and volume, the same pattern that is used in the English system is used in the metric. To measure area, we take the linear unit of measure and derive square units. Thus, for measuring area in the metric system, we can use square millimeters, square centimeters, square decimeters, square meters, and square kilometers. For measuring volume, we can use cubic millimeters, cubic centimeters, cubic decimeters, cubic meters, and even cubic kilometers.

The *cubic decimeter* is a very important unit of measure, because by definition it is a liter and because it is also used to define the kilogram, a unit of weight. A cubic decimeter can be represented by a box that is 1 decimeter long, 1 decimeter wide, and 1 decimeter high. The volume of the box is 1 cubic decimeter (or 1000 cubic centimeters, since its dimensions are 10 centimeters by 10 centimeters by 10 centimeters). When you're teaching nonlinear metric units to children, the cubic decimeter should receive a major emphasis.

The term *liter* (L) is not technically an official part of the metric system. It is simply the common name for a cubic decimeter, usually used when measuring liquid amounts. People also frequently use the term *capacity* rather than the term *volume* when speaking of containers for liquid. For example, we may say that the capacity of a gasoline tank is 80 liters. This is the same as saying that the volume is 80 cubic decimeters. For comparison purposes, we can say that a liter is a little more than a quart. So it isn't hard to imagine a liter of milk, a liter of cola, or a liter of wine.

The *volume* of this box is 1 cubic decimeter, or 1 L.

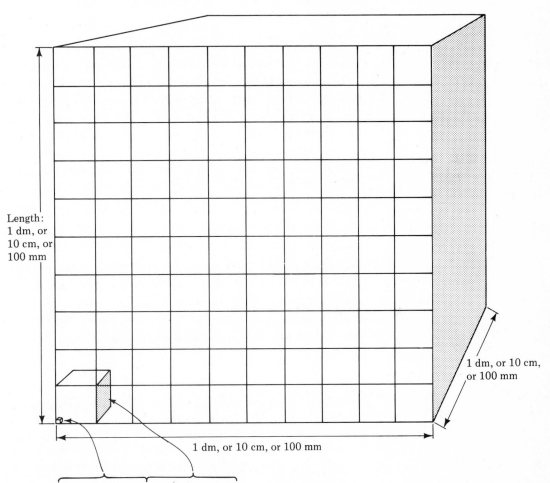

Length: 1 dm, or 10 cm, or 100 mm

1 dm, or 10 cm, or 100 mm

1 dm, or 10 cm, or 100 mm

Volume: 1 cubic millimeter, 1 cubic centimeter.

A *milliliter* (mL) is one-thousandth of a liter. Since we've just seen that a liter is 1000 cubic centimeters, we can conclude that a milliliter is 1 cubic centimeter. The milliliter is a handy unit of measure for small quantities and is used often in connection with medicine. One teaspoon holds about 5 milliliters. Machine-dispensed paper cups vary greatly, but many hold between 250 and 300 milliliters.

The cubic decimeter also provides a very close representation for metric units for weight. For all practical purposes, a cubic decimeter, or liter, of water at a certain temperature has a weight of one *kilogram* (kg). Since a kilogram is about 2.2 pounds, your weight in kilograms will be a nice, small number. It's easy to imagine a 1 kilogram package of hamburger or even a 10 kilogram turkey.

A small unit for measuring weight is the *gram* (g). There are 1000 grams in a kilogram, so a gram is one-thousandth of a kilogram. Since the cubic decimeter is 1000 cubic centimeters, we have 1 gram as the weight of 1 cubic centimeter of water. A nickel weighs about 5 grams. An individual-sized package of artificial sweetener weighs 1 gram.

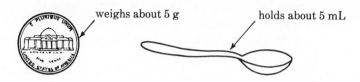

weighs about 5 g   holds about 5 mL

An even smaller unit commonly used in medicine dosages is the *milligram* (mg). A milligram is one-thousandth of a gram.

The technical distinction between weight and mass is not easily understood by today's elementary school children. We're suggesting that use of the term *weight* is fine for now. (The topic may be well-suited for an enrichment-type exploration for individual or group projects.)

To reemphasize the importance of the cubic decimeter and the cubic centimeter, we can demonstrate the relationships among some metric units with one model.

1 cm

This container holds 1 *cubic centimeter* or 1 *milliliter*. If it is filled with water, the water weighs 1 *gram*.

Thus, the following relationships exist: 1 cubic decimeter = 1 liter; 1 liter (of water) weighs 1 kilogram. So, 1000 cubic centimeters = 1000 milliliters; 1000 milliliters (of water) weigh 1000 grams. And, 1 cubic centimeter = 1 milliliter; 1 milliliter (of water) weighs 1 gram. The most commonly used metric units and their symbols are listed below. These symbols are prescribed by the U.S. Bureau of Standards.

Length		*Area*		*Volume*	
kilometer	km	square kilometer	km²	cubic kilometer	km³
meter	m	square meter	m²	cubic meter	m³
decimeter	dm	square decimeter	dm²	cubic decimeter	dm³
centimeter	cm	square centimeter	cm²	cubic centimeter	cm³
millimeter	mm	square millimeter	mm²	cubic millimeter	mm³

Capacity			*Weight*	
liter	L	(same as cubic decimeter)	kilogram	kg
milliliter	mL	(same as cubic centimeter)	gram	g
			milligram	mg

For measuring temperature in the metric system, the Celsius scale is used. This scale places the boiling point of water at 100°C and the freezing point of water at 0°C. On the Fahrenheit scale, water boils at 212°F and freezes at 32°F. So a temperature of 100°C is equivalent to 212°F, and a temperature of 0°C is equivalent to 32°F. An easy way to think of comparisons for other temperatures is to note the number of degrees between the freezing point and the boiling point on each scale. We have 100°

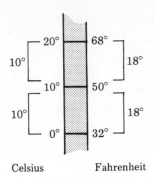

Celsius   Fahrenheit

between the freezing and boiling points on the Celsius scale, compared to 180° on the Fahrenheit scale. Thus, for every 10 Celsius degrees, there are 18 Fahrenheit degrees. An increase in temperature from 0°C to 10°C is equivalent to an increase on the Fahrenheit scale from 32°F to 50°F.

A ratio of 10 to 18 is equivalent to a ratio of 5 to 9. We can use this ratio to determine, for example, that 15 Celsius degrees are equivalent to 27 Fahrenheit degrees, so 15°C = 32°F + 27°F = 59°F. Of course, even simple conversions like these aren't necessary if you have a thermometer with a Celsius scale. Experiences with air and body temperatures will familiarize people with this scale. In a weather report, 0°C is pretty cold, 20°C to 25°C is nice, and 35°C is too hot. If the doctor says that your body temperature is 40°C, you're sick.

We've so far presented some of the metric units and some referents for them in an informal way. Our discussion is not meant to be a complete treatment of the metric system of measurement, nor are the comparisons with customary units meant to be exact. As you become familiar with metric units, you will undoubtedly explore their development and associated definitions more thoroughly. Activities for learning and for teaching concepts related to measurement and to the metric system will be presented in the next chapter.

For children who understand place value, learning to use the metric system of measurement is a natural extension of familiar ideas. In previous chapters we've stressed experiences with multiplication and division by powers of ten and maintained that these operations should not require computation. For example:

1. 12 × 10 = 120. 12 × 10 can be thought of as 12 tens, or 120.

    tens   ones

2. 8600 ÷ 100 = 86. 8600 can be thought of as 86 hundreds: 8600.

    hundreds

Then we extended these place-value ideas to decimals:

3. 1.68 × 100 = 168, or 1.68 hundreds is 168.
4. 54.3 ÷ 10 = 5.43, or there are 5.43 tens in 54.3.

With metric units of measurement, we readily find applications for these ideas.

1. How many millimeters in 12 centimeters? $12 \times 10 = 120$
2. How many meters in 8600 centimeters? $8600 \div 100 = 86$
3. How many centimeters in 1.68 meters? $1.68 \times 100 = 168$
4. How many centimeters in 54.3 millimeters? $54.3 \div 10 = 5.43$

Thus, the use of the metric system reinforces place-value ideas and provides practical reasons for continuing to emphasize them. Converting one unit to another unit appropriate for measuring a particular attribute always involves multiplying or dividing by a power of ten. So, the conversion is easy—for children who understand place value. It's also easy when compared to problems using the customary system, such as "How many inches in 1.68 miles?"

→ Getting ready to measure

Just as there are many concepts and skills that must be acquired before children are ready to begin operations on numbers, so there are many prerequisites to children's development of measurement ideas. Number concepts are developed earlier than measurement concepts. Children must have a sound understanding of the use of numbers to tell "how many" before they can begin to apply this understanding to a situation in which they decide "how many centimeters." They probably won't attain all the necessary concepts for measurement until they are ready to leave elementary school, but that is certainly *no* reason to exclude measurement experiences from the elementary mathematics curriculum. It *is* a reason to take a serious look at what children at various levels can be expected to do and to provide experiences that are consistent with these developmental stages. Piaget and other researchers who have concentrated on children's cognitive development have carefully examined these developmental stages. As we shall see, their conclusions with respect to measurement concepts can serve as a guide in selecting experiences for children.

Now, let's examine some concepts that are prerequisite to measuring length. Of these concepts, conservation of length or distance is one of the most important. For children, *distance* refers to the amount of separation between two objects, whereas *length* is a property of a single object. For example, two toy boats are a certain distance apart, and each of the boats is a certain length. Even though these ideas are different psychologically, both represent linear measurement, and conservation of length and distance is of primary importance in understanding them.

In the following experiment, a child who can conserve length will maintain that an object has the same length regardless of a change in its position. But while watching the same experimenter manipulate the sticks, a child who cannot conserve length may decide that:

▯ ▯
A   *B*   here *A* and *B* are the same length;

▯ ▯
A   *B*   here *A* is longer than *B*;

▯ ▭
A   *B*   here *A* is longer than *B*; and

▭ ▭
A   *B*   here *B* is longer than *A*.

And this child won't be at all concerned about the inconsistency of the responses!

A child who can conserve distance will maintain that two objects are the same distance apart regardless of the interposition of some third object. A child who cannot conserve distance may not be able to make a decision at all about the distance between two objects if some other object is placed between them, or the child may maintain that in the second case they are farther apart.

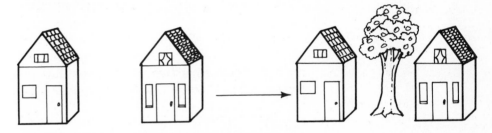

Children who do not understand the invariance, or constant nature, of length and distance are clearly not ready to measure. To measure, they must realize that changing the position of an object or of the measuring instrument does not alter the length of the object. Researchers believe that this ability to conserve length develops for most children at about age 7 and is operational, or can be used consistently, by age 8.

Another important concept that is prerequisite to an understanding of measurement is the separation of a unit into a number of subunits. To find the length of an object, the child must be able to imagine this length being separated into parts. This separation occurs through the selection of a smaller unit, and the smaller unit must be placed as many times as possible along the object to be measured. This procedure is called "iteration of the unit." For example, a child finds how many paper clips long a pencil is by moving the paper clip along the pencil as many times as possible.

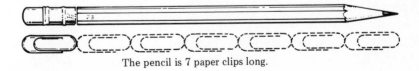

The pencil is 7 paper clips long.

Now consider this example.

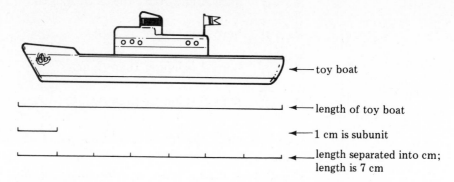

Thus, the length of the subunit must also be conserved. In addition, the child must accept this procedure as it applies to a measuring instrument. A ruler scaled in, say, centimeters (that is, separated into subunits) can then be placed on or beside the object to be measured and the object's length determined. According to Piaget, the generalized procedure for measurement of length can be intellectualized by most children at about age 8 or $8\frac{1}{2}$.

In connection with the development of number concepts in Chapter 1, we referred to the reflexive, symmetric, and transitive properties. Let's see how these properties are involved in the development of measurement concepts. The relation that we'll consider is *the same length as.*

1. This relation is reflexive. That is, if AB is a distance or length, then AB is the same length as AB. Children have to know that

moving $A \mathbin{\rlap{}} B$ to a position such as $\begin{array}{c} B \\ \square \\ A \end{array}$ or $B \mathbin{\rlap{}} A$ doesn't change its length

and that inserting an object between A and B, as in

doesn't alter the distance. In other words, children must *conserve* length or distance.

2. This relation is symmetric. Or, if AB is the same length as CD, then CD is the same length as AB. Children must be able to see that

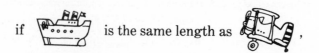

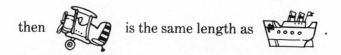

Or, if A●——————●B is the same distance as C●————●D, then C●————————●D is the same distance as A●——————●B.

3. This relation is transitive. That is, if *AB* is the same length as *CD* and *CD* is the same length as *EF*, then *AB* is the same length as *EF*. For children, this property can be illustrated as follows.

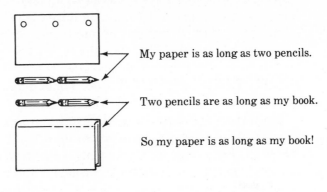

My paper is as long as two pencils.

Two pencils are as long as my book.

So my paper is as long as my book!

Or:

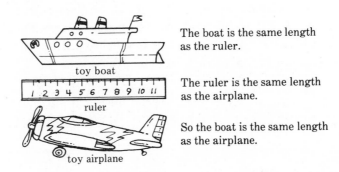

The boat is the same length as the ruler.

The ruler is the same length as the airplane.

So the boat is the same length as the airplane.

By using the transitive property, we can introduce a middle term—the measuring instrument—and go beyond a direct comparison of the two objects. So an understanding of this property is crucial to a complete understanding of measurement concepts.

The invariance, or constancy, of length or distance must be emphasized in connection with each of these three properties. Without these understandings, children are not ready for the development of concepts involved in systematic measurement.

■ Levels of development

We can distinguish three levels in the development of measurement concepts and assign age groupings to each of these levels, provided that we keep in mind that these groupings are broad and overlapping. The levels are suggested to provide some expectations about the performance of children. Of course, in a classroom situation you would use readiness activities and a great deal of observation as a basis for deciding on appropriate measurement experiences for your students.

We classify children as being at Level One if they have no measurement concepts. Level One is the category for most school children between ages 4 and 7. Children at this level, when faced with measurement tasks, use visual perceptions to justify their responses. They base their decisions on the end-points of paths and objects and do not conserve length or distance. If a measuring instrument is made available to them, it will probably be ignored, since they can see no need for this "middle term" for comparison purposes.

Level Two children are between no understanding and complete understanding of measurement. Many children between the ages 7 and 9 begin to see the value of the use of some sort of measuring instrument for comparison purposes, although they are likely to use it incorrectly. Conservation is not consistent at this level. Notice that a child of age 7 may be at either Level One or Level Two. Thus, diagnosis becomes particularly important at about the second grade.

Level Three children have achieved complete understanding of linear-measurement concepts. Children from 9 to 11 can be expected to be at this level. For them, measurement is operational or intellectual. Conservation of length and distance is consistent and unwavering. The reflexive, symmetric, and transitive properties are operational. The child can conceptualize an object as being separated into parts and then correctly use a measuring instrument to decide on the number of parts.

The fact that the *levels of understanding* do not overlap but that the *ages of children* at those levels do overlap is illustrated by the following diagram.

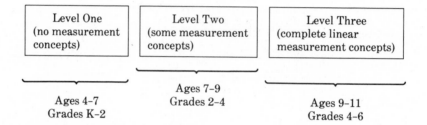

Many readiness experiences can be provided for children prior to the level of complete understanding of linear measurement. Children need the opportunity for exploration and experimentation, even if their behavior is initially trial-and-error. Readiness experiences involving manipulation of objects, along with questioning that directs children to think about measurement, are fundamental to their development of measurement ideas.

■ Measuring area

The units used for measuring area are derived from linear units. Thus, when attempting to establish concepts that are prerequisite to the measurement of area, teachers must include all of the concepts that are important to linear measurement. We build square units from line segments, for example, and arrive at units with which to measure the surface or area of two-dimensional figures.

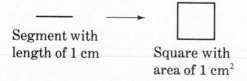

Segment with length of 1 cm

Square with area of 1 cm²

Children must realize that area units must be used instead of the linear unit and that the area of a figure doesn't change if the figure is moved about or rearranged. Conservation is as important in the measurement of area as it is in the measurement of length. Thus, the areas of *A*, *B*, *C*, *D*, and *E* are the same. Notice that the rearrangement of the figures in *C*, *D*, and *E* involves separating the figure into subparts. The fact that the area remains the same under this kind of rearrangement is a crucial one for children to understand.

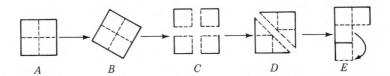

The areas of these shaded regions also are the same.

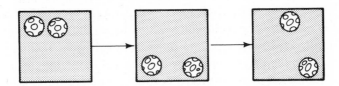

The reflexive, symmetric, and transitive properties discussed in connection with the measurement of length are also important to the measurement of area. Particular attention needs to be given to the transitive property, since it provides a justification for the use of units or a measuring instrument in comparing the areas of two figures. For example, consider the following diagrams.

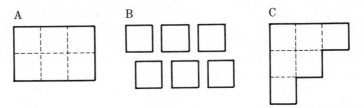

Figure *A* can be covered by the six small squares in Figure *B*. The six small squares in Figure *B* also cover Figure *C*. Therefore, the area of Figure *A* is the same as the area of Figure *C*.

The procedure we referred to as "iteration of the unit" is also important in the measurement of area. In early experiences like the one just described, children should have enough separate "units" to cover the figures they are comparing. Later, as they progress toward complete understanding of area measurement, they should learn to measure by repeating the unit as many times as possible.

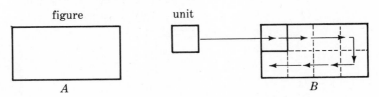

The unit will fit on Figure *A* 8 times. Or, Figure *C* below can be separated as in Figure *D*, showing that there are 11 small squares (or units) in the figure.

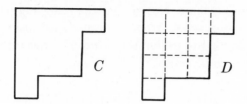

Conservation of area is also related to children's understanding of fractions. Being able to recognize each of these as $\frac{1}{2}$ of the square implies significant understanding.

The abilities involved in the measurement of area are probably developing at about the same time as those involved in linear measurement. Thus, by age 9, most children will be able to benefit from more systematic study. Experiences for younger children, in the form of readiness activities, have been demonstrated to be helpful in facilitating children's acquisition of concepts related to area measurement. At approximately age 11, formalization of the process can be expected; for instance, children can be asked to find the product of the length and width to determine the area of a rectangle. At each level, experiences for children should diagnose their abilities and needs. What the teacher learns can then guide the selection of other appropriate activities.

■ Measuring volume

Applying measurement procedures to three-dimensional figures involves all of the concepts previously discussed plus several more, so this development occurs later. Experiences for elementary school children should involve readiness activities focused on the exploration of physical situations. Extensive use can be made of blocks to build models of geometric figures and to experiment with measurement ideas.

The derivation of an appropriate unit for measuring volume and capacity builds on the units we already have for length and for area.

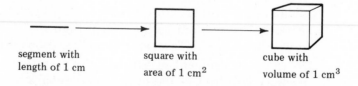

segment with
length of 1 cm

square with
area of 1 cm^2

cube with
volume of 1 cm^3

The idea of *displacement* should be considered in connection with conservation of volume. Thus, filling boxes to illustrate volume is appropriate, but so, too, is submerging models in water.

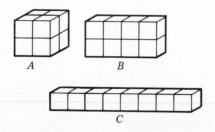

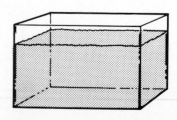

Here, *A*, *B*, and *C* can be built from the same number of little blocks. Therefore, *A*, *B*, and *C* have the same volume. Submerging *A*, *B*, and *C* will cause the water to rise to the same level; this also demonstrates that *A*, *B*, and *C* have the same volume.

■ Deciding on measurement experiences

Now that we've examined many of the concepts that must precede the introduction of formal procedures for measurement, it's apparent that decisions about appropriate experiences are pretty involved. But we have indications concerning what can be expected of children at different levels, and these must be considered. We know that many measurement activities should be exploratory. A wealth of commercial materials is available, and much of this is already in the classroom or easily accessible. Children should be involved in manipulating objects and in using trial-and-error methods to arrive at conjectures. Many of the experiences will be preparatory or readiness activities, and questioning is helpful in guiding children to important considerations. Finally, diagnosis must be a consistent and deliberate result of observing the experiences.

The computer corner

In this computer corner, we present a program called CHANGE. This program computes the number of ways that change can be made for a given amount of money less than or equal to $5.00. For example, if you enter the value of $.16, the program will tell you within a few seconds that there are six ways.

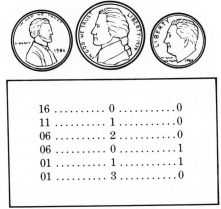

6 ways to make change for $.16

With this program, the teacher can design a number of activities that require the use of important cognitive or problem-solving skills. Here is an example activity that requires children to generate possible outcomes and organize their information in a chart.

Example 12-1
An activity that allows children to look for possibilities

Materials needed: The CHANGE program on the MGB Diskette.

Teacher directions: Run the CHANGE program to find out how many ways you can make change for the following values: $.05, $.06, $.10, $.12, $.15, $.18, and $4.37. Fill in a summary chart to show the number of ways for each value. Then fill in some charts to show the combinations for each value. Study your charts. Write some statements to tell the discoveries that you make. Is it realistic to list the combinations for $4.37?

Ideas for extention: Have students collect data for a number of dollar values. Have them look for patterns. See if they can come up with a systematic way to generate possible outcomes.

■ Computer tips

Computer activities can demonstrate ideas that might be difficult to demonstrate with traditional instructional tools. In this section, we have presented a program that can be used to construct such activities. If you use the program to collect data for a number of values, you will notice that the number of ways to make change increases rapidly as the dollar amount goes up. Coming to appreciate such a relationship would be difficult and the effects would not be as dramatic, even if one had a formula and a calculator.

In closing . . .

In this chapter, we've seen that measurement provides one very useful means of associating mathematics with the environment and that measures have certain limitations. We have given priority to the metric system of measurement and made suggestions about how to facilitate learning metric units for both children and adults. Informal comparisons have been emphasized, rather than converting from one system of measurement to another.

Research into children's development of measurement concepts has been briefly summarized,

with special attention to the implications of this research for the question of what kinds of experiences are appropriate in the elementary school.

The exercises in the Think-tank section for this chapter are designed to extend your knowledge of the topics we've discussed. Some of them will require reading from additional sources.

Now you should be ready to design some experiences for children!

It's think-tank time

1. Research and write a brief summary of the historical development of three of our customary units of measure (foot, pound, quart, and so on).
2. List 20 measuring instruments, and indicate what they are used to measure.
3. List four characteristics of measurement.
4. What are the basic metric units for length and weight?
5. What is a derived unit?
6. What do the prefixes "milli-," "centi-," "deci-," and "kilo-" mean?
7. Make a list of five things that are about this long:
   a. 1 meter b. 1 millimeter
   c. 1 centimeter d. 1 decimeter
8. Approximately how many kilometers are in 100 miles? 500 miles? 25 miles?
9. Name something that weighs about 1 kilogram.
10. Name something that will hold about 1 liter.
11. Why is the use of negative numbers more common with the Celsius scale for measuring temperature than with the Fahrenheit scale? Give an example.
12. Read and summarize an article concerning the acceptance of the metric system of measurement in the United States.

13. Read and summarize two articles that concern teaching measurement concepts to children.
14. a. What is meant by "conservation" as it applies to length, area, and volume?
    b. Give an example of the possible behavior of a child who cannot conserve length.
    c. Give an example of the possible behavior of a child who cannot conserve area.
    d. Give an example of the possible behavior of a child who cannot conserve volume.
15. Demonstrate that the reflexive, symmetric, and transitive properties apply to the relation *has the same area as.*
16. a. What is meant by "iteration of a unit"?
    b. How can iteration of a unit apply to the determination of volume?
17. Complete the following activity.

Selecting Proper Units

Here are some things that can be measured. Select the *best* unit for each. Then use the letter next to your selection to find your metric message.

Length of your shoe
t. centimeter   u. kilometer   v. meter
Volume of a thimble
o. cubic centimeter p. cubic meter q. square centimeter
Contents of a coffee cup
l. liter   m. milliliter   n. kiloliter

Weight of a paper clip
e. liter   a. gram   i. kilogram

Weight of a turkey
i. cubic centimeter   j. gram   k. kilogram

Area of a postage stamp
c. square kilometer d. square meter e. square centimeter

Area of your bedroom floor
e. meter   a. square meter o. cubic meter

Room inside a refrigerator
q. meter   r. square meter s. cubic meter

Thickness of a dime
w. millimeter   x. centimeter   y. milligram

Length of an ant
c. milligram   d. milliliter   e. millimeter

Weight of a flea
a. milligram   b. milliliter   c. millimeter

Distance from Boston to Atlanta
r. meter   s. decimeter   t. kilometer

Distance around your front yard
d. centimeter   e. meter   f. liter

Your weight
p. milligram   q. gram   r. kilogram

Diameter of your wrist watch
u. millimeter   o. meter   e. milligram

Weight of a rubber band
n. milligram   m. kilogram   o. liter

Weight of a dog
h. milligram   i. kilogram   j. liter

Volume of your classroom
r. cubic centimeter s. square meter t. cubic meter

MESSAGE:

_ _ _ _ _ _ _ _ _ _ _ _ _ _ _ _ _ _!

18. Fill in the blanks.
   a.   1.68 m = _____ cm
   b.   38 cm = _____ mm
   c.   480 g = _____ kg
   d.   5 km = _____ m
   e.   17 L = _____ mL
   f.   258 mL = _____ L
   g.   1 mg = _____ g
   h. 4 m^2 = _____ cm^2
   i. 7 m^3 = _____ cm^3
   j. 1.78 cm^2 = _____ mm^2
   k. 1 m^3 = _____ mm^3
   l.   60 kg = _____ mg
   m.   3 mm = _____ m
   n. 25 km^2 = _____ m^2
   o. 86 cm^3 = _____ mm^3
   p. 17.9 dm = _____ cm
   q.   35°C = _____ °F
   r. 1 dm^2 = _____ mm^2
   s.   41°F = _____ °C
   t. 1 dm^3 = _____ cm^3

19. Make a chart that shows both Celsius and Fahrenheit, with the Celsius temperature in multiples of five from −40°C to 100°C.

20. Make charts for area and volume similar to the one for length (page 266) in this chapter, showing the symbol, the definition in terms of a square meter and cubic meter, and a familiar unit for each.

Suggested readings

Bruni, James V., and Silverman, Helene J. "Organizing a Metric Center in Your Classroom." *The Arithmetic Teacher*, February 1976, pp. 80–87.

Burton, Grace M. "Metrification of Elementary Mathematics Textbooks in the Seventies—the 1870's, That Is." *The Arithmetic Teacher*, 1979, pp. 28–31.

Cathcart, W. George. "Metric Measurement: Important Curricular Consideration." *The Arithmetic Teacher*, February 1977, pp. 158–160, 400–401.

Clason, Robert G. "1866—When the United States Accepted the Metric System." *The Arithmetic Teacher*, January 1977, pp. 56–62.

Copeland, Richard W. *How Children Learn Mathematics: Teaching Implications of Piaget's Research.* New York: Macmillan, 1974.

Engel, C. William, and Lichtenberg, Donovan R. *The Metric System of Measurement* (a series of 12 filmstrips). Glenview, Ill.: Educational Projections Corp., 1975.

Fowler, Mary Ann. "Let's Take Another Look at Teaching the Metric System." *The Arithmetic Teacher*, December 1978, pp. 15–16.

Hirstein, James J., Lamb, Charles E., and Osborne, Alan. "Student Misconceptions about Area Measure." *The Arithmetic Teacher*, March 1978, pp. 10–16.

Hollis, Loye Y. ("Mickey"). "Mathematical Concepts for Very Young Children." *The Arithmetic Teacher*, October 1981, pp. 24–27.

Hovey, Larry, and Hovey, Kathi. "The Metric System—An Overview." *School Science and Mathematics*, February 1983, pp. 112–121.

Johnson, Martin L. "The Effects of Instruction on Length Relations on the Classification, Seriation, and Transitivity Performances of First- and Second-Grade Children." *Journal for Research in Mathematics Education*, May 1974, pp. 115–125.

Labinowicz, Ed. *The Piaget Primer.* Menlo Park, Calif.: Addison-Wesley, 1980.

Leutzinger, Larry P., and Nelson, Glenn. "Let's Do It with Powers of Ten." *The Arithmetic Teacher*, February 1980, pp. 8–13.

Lindquist, Mary Montgomery, and Dana, Marcia E. "The Neglected Decimeter." *The Arithmetic Teacher*, October 1977, pp. 11–17.

Lovell, Kenneth. "Summary and Implications." In *Research on Mathematical Thinking of Young Children.* Reston, Va.: National Council of Teachers of Mathematics, 1975.

McGinty, Robert. "One Point of View, Current Status of the Metric System." *The Arithmetic Teacher*, October 1984, pp. 3–4.

Moredock, H. Stewart. "Geometry and Measurement." In *Sixty-ninth Yearbook of the National Society for the Study of Education.* Chicago: University of Chicago Press, 1970. (Pp. 167–235.)

Peavler, Cathy Seeley. "Metricating—Painlessly, Cheaply, Cooperatively." *The Arithmetic Teacher*, October 1974, pp. 533–536.

Piaget, Jean, Inhelder, Barbel, and Szeminska, Alina. *The Child's Conception of Geometry.* New York: Basic Books, 1960.

Pottinger, Barbara. "Measuring, Discovering, and Estimating the Metric Way." *The Arithmetic Teacher*, May 1975, pp. 372–377.

Schussheim, Joan Yares. "Metric Week Celebration All Year Long." *The Arithmetic Teacher*, April 1980, pp. 34–36.

Smart, James R. *Metric Math: The Modernized Metric System.* Monterey, Calif.: Brooks/Cole, 1974.

Tabler, M. Bernadine, and Jacobson, Marilyn Hall. "Ideas: Fill It Up Please." *The Arithmetic Teacher*, January 1981, pp. 27–32.

Taloumis, Thalia. "The Relationship of Area Conservation to Area Measurement as Affected by Sequence of Presentations of Piagetian Area Tasks to Boys and Girls in Grades One through Three." *Journal for Research in Mathematics Education*, November 1975, pp. 232–242.

Threadgill, Judith. "Let's Metricate Parents Too!" *The Arithmetic Teacher*, December 1978, pp. 18–19.

Trueblood, Cecil R., and Szago, Michael. "Procedures for Designing Your Own Metric Games for Pupil Involvement." *The Arithmetic Teacher*, May 1974, pp. 404–408.

Viets, Lottie. "Experiences for Metric Missionaries." *The Arithmetic Teacher*, April 1973, pp. 269–273.

Wagman, Harriet G. "A Study of the Child's Conception of Area Measure." (Doctoral dissertation, Columbia University, 1968.) *Dissertation Abstracts International*, *30A* (1969):1350.

Superstitious? Not Us . . .

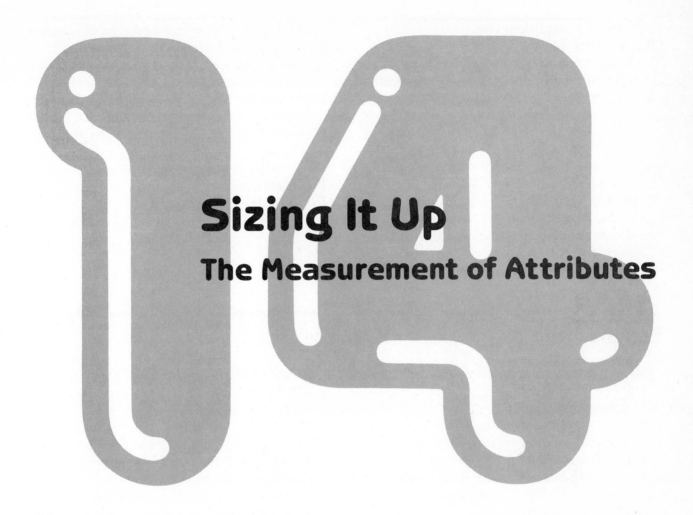

Sizing It Up
The Measurement of Attributes

■ This chapter concentrates on methods of presenting measurement ideas to children in the elementary school. The topics discussed include measurement of length, area, volume, weight, angles, time, and temperature. The methods for each of these topics are intended to help children become familiar with measuring units, to provide exploratory experiences, and to develop concepts informally through active participation in measurement situations. The teacher is urged to stress relationships among topics and to encourage children to use estimation throughout all measurement experiences. A short section on concepts related to money is also included.

Again, we're using only metric units. Although children will be experiencing customary units at home and in various parts of their environment, measurement instruction in school can be consistently metric. The schools must lead in the voluntary conversion to the metric system in the United States. ■

Objectives for the child

1. Conserves length and area.
2. Conserves volume.
3. Identifies situations that call for the measurement of length, area, volume, and weight.
4. Compares and orders on the basis of length, area, volume, capacity, and weight.
5. Recognizes the need for standard units of measurement.
6. Estimates the number of standard units for a specific measurement.
7. Selects appropriate units for measuring length, area, volume, and weight.
8. Measures length, area, and volume by iteration of a unit.
9. Determines perimeter and area of polygons.
10. Recognizes the relationship among units for measuring length, area, and volume of geometric figures.
11. Recognizes the relationship between metric units for measurement of volume and weight.
12. Compares and orders angles on the basis of size.
13. Uses protractor to measure angles.
14. Reads clocks.
15. Reads thermometers.

Teacher goals

1. Be able to plan instructional strategies that will enable children to learn concepts related to measuring length.
2. Be able to plan instructional strategies that will enable children to learn concepts related to measuring area.
3. Be able to plan instructional strategies that will enable children to learn concepts related to measuring volume.
4. Be able to plan instructional strategies that will enable children to learn concepts related to measuring weight.
5. Be able to plan instructional strategies that will enable children to learn concepts related to measuring angles.
6. Be able to plan instructional strategies that will enable children to learn concepts related to measuring time and temperature and to reading clocks and thermometers.
7. Be able to diagnose the performance of children with respect to each of these topics and to prescribe remedial experiences for children who have not achieved expected levels of performance.

How thick is a sheet of notebook paper?
Why is a clock harder to read than a thermometer?
How can *you* teach children to measure?

In Chapter 12, we considered the ideas prerequisite to teaching and learning about measurement. Now we're ready to concentrate on methods for presenting measurement concepts to children. There is a wealth of commercial material for the topic, and these materials are probably nearby. So, check out your supplies and get set.

Measuring length

For children in the early grades, experiences for introducing linear measurement should focus on the idea that there really is an attribute, determined by the distance between two points, that can be compared, ordered, and measured. Sometimes this attribute is called *height*. In this case, questions such as the following are appropriate: "Who is taller?" "Who is shorter?" "Which is higher?" "Which is lower?" "How tall are you?" "How high is the ceiling?" Sometimes the attribute is called *length*, and sensible questions would include "Which is longer?" "Which is shorter?" and "How long is the pencil?" Sometimes the attribute is called *width*, as in "Which is wider, the door or the window?" "Which is narrower, the sidewalk or the driveway?" and "How wide is your desk?" And sometimes the attribute is called *distance*, and we ask "Which is farther?" "Which is closer?" or "How far is it from your home to your school?" The question "How deep is it?" is also appropriate for length. Even a question such as "How big around is your waist?" must be answered by measuring length. Is it any wonder that young children are confused by the number of possible situations to be interpreted?

Teachers, then, must provide examples of these situations, beginning with objects that children can touch, move, compare, and order according to length. The following concept-development activities are samples of such experiences.

Example 14-1
A concept-development activity for order

C-1

Materials: Box of familiar objects—pencil, chalk, toothbrush, safety pin, lipstick, fork, comb, emery board, wooden spoon. One object in the box should clearly be the shortest and one clearly the longest.

Teacher directions: We could sort these out in a lot of ways. Today we're going to look at how long these things are.

Questions: Which is the longest?

Which is the shortest? Pick out some that are about the same length.

It's harder to tell which is longer when the objects are almost the same length. Can you do it? How? Be sure the ends are together. Now line them up in order.

Example 14-2   C-1
A concept-development activity for order

Materials: Children in your class.

Teacher directions: Let's see who is taller.

Questions: Who is taller, Vicki or Lynn? Who is shorter, Brad or Willie? Where would Tom fit in with Vicki and Lynn? Where do I fit in? Let's have five people stand in line from shortest to tallest.

Example 14-3   C-1
A concept-development activity for measurement

Materials: Objects in the classroom.

Teacher directions: Find something that's wider than your desk top.

Questions: Is your book wider than your desk? Is our table wider than your desk? Is my desk wider than your desk? What's the widest thing you see?

Example 14-4   C-1
A concept-development activity for distance

Materials: None.

Teacher directions: Let's talk about things that are far away.

Questions: Which is farther from your desk, the lunchroom or the library? Which is farther from our school, the shopping center or downtown? How can you tell? Name a place that you've been to that is far away.

Examples 14-1 through 14-4 involve comparing and ordering length without the use of a measuring unit or numbers. Soon after experiences such as these, children can begin to measure with a nonstandard unit. This experience will enable them to practice a measuring procedure and at the same time lead them to see the need for standardizing units. The following examples illustrate this kind of concept-development activity.

Example 14-5

C-1

A concept-development activity that allows children to compare measures obtained by using different units

Materials: Strips of pink construction paper. Scissors.

Teacher directions: Cut your strip of paper so that you have a piece that is about as long as your little finger. Call this a "pinky."

Questions: How many pinkies long is your desk? How many pinkies long is your pencil? How many pinkies long is your nose?

New teacher directions: Get a longer strip of construction paper and make a pinky ruler. Find measurements for things on this list (width of a desk, length of the eraser, width of the door, height of the desk, width of our math book).

Questions: Since we're all measuring the same things, why aren't our numbers the same? This question presents a good possibility for further discussion.)

Example 14-6

C-1

A concept-development activity that allows children to compare measures obtained by using the same unit

Materials: Jumbo paper clips. Familiar objects.

Teacher directions: Use your paper clip to measure with.

Questions: How many paper clips long is the pencil? How many paper clips tall is this cup? How many paper clips wide is your notebook paper? Why are we getting the same numbers now? Can you tell how many paper clips tall you are? How? How many paper clips does it take to go around your neck?

At the first- and second-grade levels, activities such as these should be emphasized. Eventually, we want children to realize that using units of the same length is more reasonable than using individually selected ones. Third-graders may profit from lessons concerning the historical development of standard units. They can also be convinced that the metric system of measurement allows us to communicate with more people and provides some nice patterns that are consistent with what they are learning about place value.

Now we're ready to introduce the centimeter as a unit for measuring length. Here are several sample concept-development activities. Hundreds more are possible, so make up some of your own, too!

Example 14-7

C-2

A concept-development activity for using centimeters

Materials: Rulers and sharpened pencils.

Teacher directions: This line segment is 1 centimeter long: _____. Your ruler is marked off in centimeters. Draw a line segment that is 1 centimeter long. My little fingernail is about 1 centimeter long. Find something in this room that's about 1 centimeter long.

Questions: Not many. This should be an active measurement lesson for children. Make a list of all the things they found.

Example 14-8

C-2

A concept-development activity for measurement with estimation

Materials: Rulers. Worksheet with various line segments from 1 cm to 15 cm in length.

Teacher directions: First estimate how many centimeters long these line segments are. Then measure each line segment to the nearest centimeter and see how close you were.

Example 14-9

C-2

A concept-development activity for measurement with estimation

Materials: Rulers.

Teacher directions: First, using the edge of your book, draw line segments that you think are about 5 cm long, 10 cm long, 12 cm long, 15 cm long, 1 cm long. Measure the segments and write down their lengths. Now, under each of your segments, use your ruler to draw a segment that is as long as your original segment was supposed to be.

Questions: Were most of your segments too long or too short? Can you do it better next time?

After children have had a good deal of practice using rulers to measure lengths in centimeters, it's a logical extension to measure lengths for which a straight-edge measuring instrument is not the most appropriate tool. The next three activities are samples.

Example 14-10
A concept-development activity for comparison

C-2

Materials: String for each child.

Teacher directions: Measure these objects, using your piece of string. You'll be surprised at some of the results.

Questions: Which is longer, the height or the distance around my cup?

List some other things you can compare by using your string.

... wastebasket ...   ... box of oats ...

Example 14-11
A concept-development activity for measurement

C-2

Materials: String. Rulers or tape measures. A teacher-made ditto like this one.

Teacher directions: Let's measure some things that aren't straight. Use your string to measure, and then find out how many centimeters long your measurement is with the ruler.

Questions: Find the distance around your ankle (knee, head). Find the distance around the light switch (desk top). Keep good records of your measurements. Now find the distances along the paths shown on the ditto sheet. Which path is longest? Which path is shortest? Why?

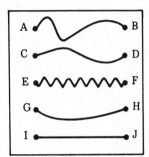

Example 14-12
A concept-development activity for fun

C-3

Materials: Pieces of string for each child.

Teacher directions: Let's measure some things that aren't straight.
1. Put the string around your thumb.
2. Now double that length by folding your string. This much string should fit around your wrist.
3. Now double the wrist length. The new length should fit around your neck.
4. Now double the neck length. This much string should fit around your middle!!

Make up some equations for these:

1 middle = 8 thumbs   1 neck = 2 wrists   etc.

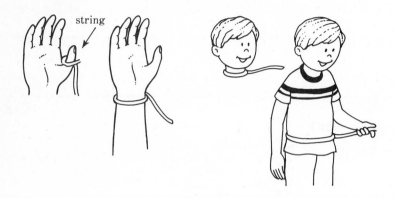

After children become reasonably familiar with the centimeter as a unit for measuring length, they are ready to look at some longer units. Experiences with other units should be similar to those already described; that is, there should be many opportunities for comparing, ordering, and then actually measuring the lengths of objects or of distances. At this point, the teacher should stress relationships between familiar and new units. Example 14-13 is a sample activity for introducing the meter; it can be carried out with children in the third grade or above.

Example 14-13
A concept-development activity using meter sticks

C-2

Materials: Meter sticks, one for each pair of students.

Teacher directions: This stick is 1 meter long, so it's called a "meter stick." I'm going to give each of you a partner. One partner uses the meter stick to find some things that measure about 1 meter. The other partner makes a list of the things you find. After five things are listed, switch jobs, Find out how many centimeters long the things on your list are and record that also.

The next two examples are concept-development activities for introducing the kilometer.

Example 14-14   C-3
A concept-development activity for distance

Materials: Meter sticks. Handout with scale drawing.

Teacher directions: Kilometers are used to measure long distances. One kilometer is 1000 meters. Mark off 10 meters in the hall. How many times this distance is a kilometer?

How fast do you walk in kilometers per hour? The distance you've marked in the hall is 10 meters. Walk up and down that distance for 5 minutes, and keep track of the number of times. How many meters did you walk? Multiply by 12 to find the number of meters per hour. How many kilometers per hour is this?

Can we mark off 100 meters on the playground? How many times this far is a kilometer? Eight kilometers is about the same as 5 miles. How many kilometers would 10 miles be?

Draw a picture showing two towns that are 40 miles apart. How many kilometers is this? Make a scale similar to the one we just did.

New teacher directions: For a class project, let's find something that's 1 kilometer from our school.

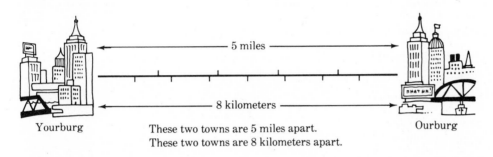

Yourburg — 5 miles — Ourburg

8 kilometers

These two towns are 5 miles apart.
These two towns are 8 kilometers apart.

Example 14-15   C-3
A concept-development activity with maps

Materials: Rulers, maps (see Material Sheet 25).

Teacher directions: There are 1000 meters in 1 kilometer. Maps often have a scale that tells how many kilometers are represented by 1 centimeter. Find the scale on your map.

Questions: How many centimeters are between Chicago and Dallas on your map? How many kilometers does that represent?

The millimeter is a useful unit for measuring small distances. The following examples are designed to give children experience with this unit.

Example 14-16   C-3
A concept-development activity using smaller units

Materials: Ruler.

Teacher directions: When we separate a centimeter into ten equal parts, each of these parts is a millimeter. The thickness of a paper clip is about 1 millimeter. Find some things that measure about 1 millimeter. List them.

Example 14-17   C-3
A concept-development activity using millimeters

Materials: None.

Teacher directions: Draw the smallest girl you can. Name her "Milly Meter."

Milly Meter

Questions: How long is Milly's leg? How long is Milly's arm? How long is Milly's hair?

The following activities extend concepts related to perimeter. Children in elementary school need many preparatory experiences like these; there is plenty of time for formulas in junior high school.

Example 14-18   C-3
A concept-development activity for perimeter

Materials: Rulers and geo-pieces (see Material Sheet 1).

Teacher directions: Pick out one of the big squares.

Questions: How long is one of the sides? What is the distance around it? Instead of measuring all four sides and then adding, what else could you have done? Answer these same questions for one of the little squares.

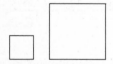

New teacher directions: Pick out one of the big triangles.

Questions: How long is one of the sides? What is the distance around it? Since all three sides are the same length, what could you have done instead of measur-

ing all three sides and then adding? Answer these same questions for one of the little triangles.

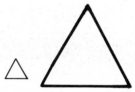

New teacher directions: Pick out a rectangle that isn't square.

Questions: How long are the sides? What is the distance around it? Is there any other way you could have figured this out besides measuring all sides and then adding? Answer these same questions for one of the little rectangles.

Example 14-19
A concept-development activity for perimeter

C-3

Materials: Dot paper (see Material Sheet 16).

Teacher directions: Perimeter means the distance around a figure. Draw rectangles on your paper with the dimensions listed in the table (the distance between the dots on your paper is 1 unit). Find the perimeter of each rectangle. Find another way to determine the perimeter besides adding.

Width	Length
3 units	2 units
1 unit	4 units
1 unit	1 unit
2 units	3 units
.	
.	
.	
(Etc.)	

Example 14-20
A concept-development activity for perimeter

C-3

Materials: Light switch, pencil and paper, rulers.

Teacher directions: Our light switch is 7 cm wide and 11 cm long. We know that the distance around it is $7 + 7 + 11 + 11$, or $(2 \times 7) + (2 \times 11)$, or 36 cm. What other dimensions could it have and still have this same perimeter? Draw figures to show them.

Examples:

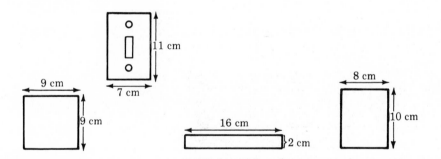

Example 14-21
A concept-development activity for perimeter

C-3

Materials: Rulers.

Teacher directions: Draw any polygon. Trade papers with someone. Find the perimeter of the figure you received.

Questions: Who has the largest perimeter? Who has the smallest perimeter? Who drew the figures?

Example 14-22
A concept-development activity for perimeter

C-3

Materials: Meter stick.

Teacher directions: Find the perimeter of this room. Find the perimeter of the hall. (Etc.)

Example 14-23
A concept-development activity that enables students to name and define important parts of the circle

C-3

Materials: Large circular sheets of paper (see Material Sheet 22).

Teacher directions: Fold one of your circle-pieces in half. Then unfold, and draw a line segment along the fold.

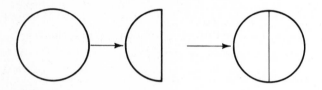

Now fold it in half again; draw a line segment along the second fold.

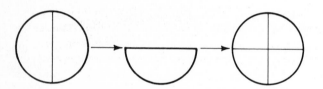

Questions: Where do you think the center of the circle is? Label the center of the circle with an *O*. Each one of these line segments is called a *diameter* of the circle. Can you tell what a diameter is?

New teacher directions: Make some new folds. Draw line segments along the new folds. Use a piece of string to measure some of the diameter segments. Tell what you find out.

New teacher directions: Now separate one of the diameters into two parts by darkening half of it. One of these parts is called the *radius* of the circle. Where are the end-points of the radius? Label the radius and one of the diameters with their names.

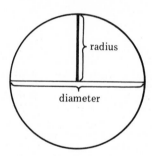

The plural of "radius" is "radii." Measure some of the radii of your circle. Tell what you find out.

Example 14-24

A concept-development activity for comparing the diameter and circumference of a circle

Materials: For each child, circle-pieces cut from construction paper (Material Sheet 22), two sheets of lined notebook paper taped together so that lines meet (as shown below), crayons.

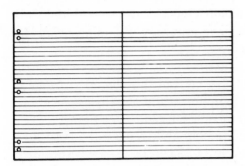

Teacher directions: We're going to find the distance around some circles. This distance is called the *circumference* of the circle. Here's how we'll do it. Mark a letter at the edge of each circle.

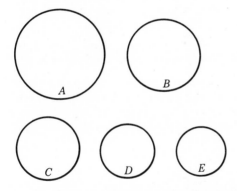

Then take a circle and roll it like a wheel along one of the lines on your notebook paper. Start at the letter and roll the circle until you get back to the letter. Use your crayon to make a line segment that shows how far you rolled. Label the segment with the letter of the circle. Do this for all of your circles.

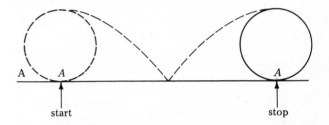

start   stop

New teacher directions: Now you have a chart that shows circumference for your circles. Let's fix the chart so that it shows diameters. Fold circle *A* in half to find a diameter. Then use the diameter to mark off line segments on line *A* of the chart.

diameter

circumference

Questions: Study your chart. About how many diameters does it take to make a circumference? (A little more than 3.) If you know the diameter of any circle, you can multiply by 3.14 to find the approximate circumference. This is the formula $C = \pi D$, and π is approximately 3.14.

As we've seen, concepts related to the measurement of length should be developed through active participation by children in a variety of experiences. These should be designed in such a way that children can examine and explore different types of situations and select appropriate measuring units and procedures. Estimation should be a part of each activity; children should be encouraged to make guesses and later to revise them. Activities for evaluation exercises similar to items 17 and 18 in the Thinktank section for Chapter 12 can also be used with children.

Measuring area

Experiences for introducing the measurement of area should be similar to those for introducing the measurement of length; that is, children should focus on the attribute of area, the meaning of concepts related to area measurement, and appropriate units for measuring.

Introductory lessons (through grade four) should involve covering regions and determining the number of units needed to cover them. Children should be led to discover that linear units are not appropriate for measuring area but that square units are.

Eventually, in fifth or sixth grade, you may want to discuss some of the formulas for determining the area of polygons, but this should grow out of many experiences in which children estimate and informally explore relationships by manipulating objects and models for geometric figures.

The next two activities are designed to help children see the need for a nonlinear unit of measurement. Example 14-25 can be used with children at the first- or second-grade level. Many activities like it should be presented before the introduction of grid paper or dot paper.

Example 14-25   C-1
A concept-development activity for area

Materials: Tiles. Prepared pictures.

Teacher directions: Use tiles to cover Leo's door. Now use some other tiles to cover Lily's door.

Questions: Who has the bigger door, Leo or Lily?

New directions: Draw a picture of a house. Find how many tiles it takes to cover the front of the house.

Questions: Can you think of some problems in which you would need to know how much flat space something takes up?

Example 14-26

C-3

A concept-development activity for area

Materials: Material Sheet 17.

Teacher directions: Remember our light switch that is 7 cm wide and 11 cm long?

Suppose we want to tell someone how much space it takes up on the wall. How could we do it? Draw a picture of the light switch on your squared paper. How many little squares are in your picture? We could tell someone that the light switch takes up as much space on the wall as 77 little squares.

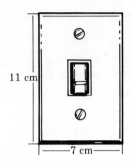

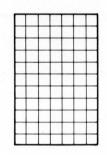

The order in which the teacher introduces standard units will undoubtedly depend in part on the materials used. Since we considered the centimeter first in measuring length, we'll begin with the square centimeter as a unit for measuring area.

Example 14-27

C-3

A concept-development activity for area

Materials: Worksheet with rectangular figures like those shown below. Transparency made from Material Sheet 17.

Teacher directions:

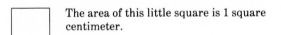

The area of this little square is 1 square centimeter.

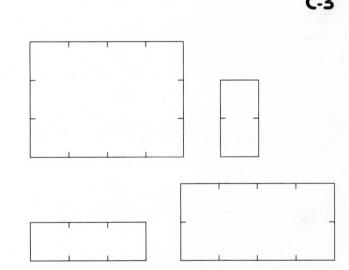

Questions: How many of these little squares can you fit on each of the figures on your worksheet? Use the transparency to find out.

Each little square is a square centimeter. What is the area of each of these figures in square centimeters?

Example 14-28

C-3

A concept-development activity for finding area

Materials: Dot paper (see Material Sheet 16).

Teacher directions: If you connect four dots, you can make a square that is 1 square centimeter. Draw some squares of different sizes. You can find the area of each square by counting the little squares contained in it.

Questions: What is the area of each square? Can you find another way to determine the area of a square besides counting?

Example:

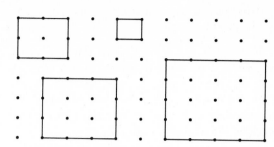

Example 14-29

C-3

A concept-development activity for finding area

Materials: Dot paper (see Material Sheet 16).

Teacher directions: Draw some rectangles on your paper.

Questions: What is the area of each rectangle? Can you find another way to determine the area of a rectangle without counting all of the little squares?

Example 14-30

C-3

A concept-development activity for finding area

Materials: Dot paper (see Material Sheet 16).

Teacher directions: Look at this figure.

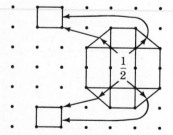

Question: What is the area of the figure? (10 cm²)

New teacher directions: Draw five figures on your dot paper. See whether you can find their areas.

The line segments connecting these 4 dots enclose an area of 1 square centimeter.

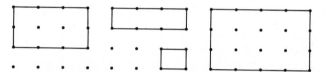

Units for measuring area can be derived from each of the units used for measuring length. Thus, once children have a good understanding of the square centimeter, they can readily apply the concepts they've acquired to the square decimeter, square meter, square kilometer, and square millimeter. The following activity uses the square decimeter as a vehicle for emphasizing the conservation of area, a concept that should be stressed in connection with each of the units for measuring area.

☞

Example 14-31
A concept-development activity for conservation of area

Materials: Scissors. Construction paper. Rulers.

Teacher directions: Cut out four squares that each have sides 1 decimeter in length.

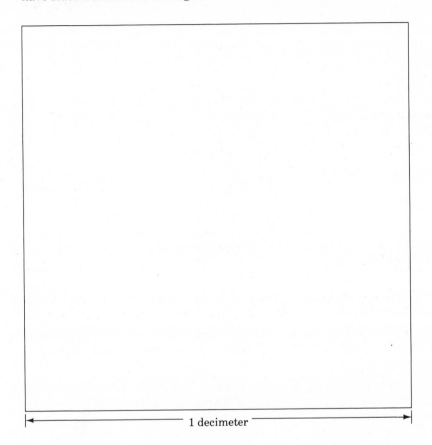

This is a square that is 1 decimeter on a side. Its area is 1 square decimeter.

├──────────── 1 decimeter ────────────┤

Take one of your squares and fold it like this. Then cut along the fold, and use your 2 pieces to make any figure that is not a square. What is the area of your new figure?

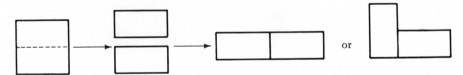 or

Take one of your squares and fold it like this. Then cut along the fold, and use your 2 pieces to make any figure that is not a square. What is the area of your new figure?

 or

Take a square and cut it any way you want. Make a figure from your new pieces. What is the area of your new figure?

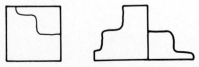

The next activity is a sample for introducing children to the square meter.

Example 14-32
A concept-development activity for area

C-3

Materials: Chalk and masking tape. Meter stick.

Teacher directions: If I draw a square on the chalkboard that is 1 meter on a side, the area will be 1 square meter. Let's try to find the area of our classroom floor in square meters.

Questions: Can we mark off the length in meters? Use the masking tape. How many? Now mark off the width in meters. How many? How many square meters is the area of our floor?

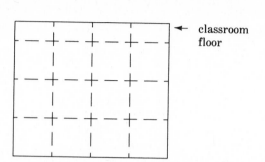

← classroom floor

We have 12 squares, each having an area of 1 square meter. We have 4 parts that will just about fit together to make 2 more squares that each have an area of 1 square meter. So, altogether the area is about 14 square meters.

The preceding activities have been primarily concerned with the area of squares and rectangles, because these are the figures that should be emphasized in the elementary grades. But elementary school children can benefit from considering the area of triangles, nonrectangular parallelograms, and other figures.

The following activities illustrate some ideas related to these other kinds of figures. With many children in the fifth and sixth grades, formulas can be developed for determining areas, but they are not a necessary outgrowth of these experiences.

☞

Example 14-33

C-3

A concept-development activity for areas of triangles

Materials: Construction paper. Scissors.

Teacher directions: Cut out a right triangle like this. Now cut out another right triangle that's the same size, and let's see whether we can find the area of our right triangle.

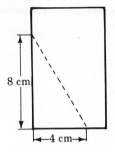
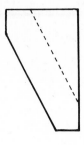

Questions: Can you fit your two triangles together to form a familiar figure? What kind of figure is this? What is its area? Since the two triangles are the same size, the area of one of them will be one-half as much as the area of the whole rectangle. What is the area of one triangle? (16 cm²)

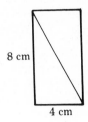

☞

Example 14-34

C-3

A concept-development activity for area of triangles

Materials: Construction paper. Scissors.

Teacher directions: Cut out any triangle. Then cut out another triangle that's the same size.

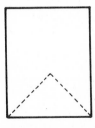
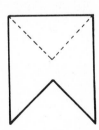

Questions: Can you fit your two triangles together to form a rectangle? (Probably not.) Let's look at one of them. Can you make two right triangles out of this triangle?

Now you have three triangles. Can you fit them together to form a rectangle? (Yes.)

What dimensions would you have to know to find the area of the rectangle you've made? We can call the length and width of the rectangle the *height* and the *base*. Then the area is the height times the base. Since your two triangles were used to form the rectangle, the area of one triangle would be one-half the area of the rectangle.

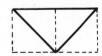

What is the area of your triangle? ($\frac{1}{2}$ the height times the base.)

Example 14-35

A concept-development activity for area of parellelograms

Materials: Construction paper. Scissors.

Teacher directions: Cut out a parallelogram like this one.

Since we know how to find the areas of rectangles, see whether you can decide how we might cut the parallelogram to form a rectangle. Right! Cut off a right triangle, and fit it on the other end.

C-3

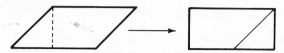

Questions: Is the base of the rectangle the same as the base of the parallelogram that we started with? How do you know? Is the height of the rectangle the same as the height of the parallelogram? How do you know? What is the area of the rectangle? (Base × height.) Is the area of the parallelogram the same as the area of the rectangle? (Yes.)

Example 14-36

A concept-development activity for area

C-3

Materials: Dot paper (see Material Sheet 16).

Teacher directions: Let's find another way to determine the area of a figure. Connect your dots like this. Color the corners.

We could count to determine the number of square centimeters in this figure by putting the 4 corner pieces together to get 2 more squares. Or, we could separate the figures into some rectangles and right triangles and find the area of these parts.

Questions: Can you think of another way to find the area?

The area of the big square is 16 square centimeters. What is the area of the shaded part?

What does that leave for the area of our original figure? 14 square centimeters. Now try this one.

At the elementary level, children's experiences with measurement of area should be informal and should emphasize meaning. There will be opportunities at the junior high-school level to formalize and extend concepts. The sample activities that we've provided, therefore, are not intended to be a complete treatment of the measurement of area. Rather, they are intended to provide children with some informal exposure to ideas that will be covered again in later grades.

Measuring volume

Standard units for measuring volume can be derived from the units used for measuring length. For example, a cube that is one centimeter on an edge has a volume of one cubic centimeter (1 cm³). Children should be led to see the need for cubic units, and—again—early experiences should be informal and provide opportunities for manipulating objects. The cubic decimeter should also be stressed because of its relationship to units for measuring weight and capacity. The two following activities reflect these emphases.

Example 14-37   C-3
A concept-development activity for volume

Materials: Centimeter cubes. Square centimeter paper (see Material Sheet 17). Cellophane tape. Scissors.

Teacher directions: Make a box that is 8 centimeters by 6 centimeters by 2 centimeters.

Questions: How many centimeter cubes would it take to fill the box?

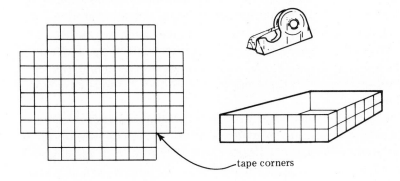

tape corners

New teacher directions: Now make a box that is any size you choose, and we'll make a chart.

Suppose that I made a box that was 20 centimeters by 2 centimeters by 4 centimeters. Can you guess how many cubes it would hold? How do you know?

Who?	Dimensions	Number of centimeter cubes to fill box
Charlene	2 cm, 3 cm, 1 cm	6
Larry	3 cm, 5 cm, 2 cm	30
.	.	.
.	.	.
.	.	.

☞ _____

Example 14-38
A concept-development activity for volume

C-3

Materials: Material Sheet 17. Scissors and tape (or glue).

Teacher directions: Make a decimeter cube. Since your box doesn't need a top, you can tape or glue five square decimeters together. The volume of your cube is 1 cubic decimeter.

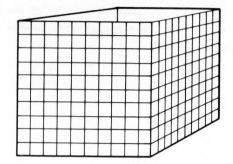

Questions:

  How many centimeter cubes could you put on the first layer?
  How many layers are there?
  Then how many centimeter cubes will fill the box?
  So the volume of the cube is 1000 cubic centimeters.

_____

For most children, lessons on volume should be confined to these types of exploratory activities and should involve using cubes to fill boxes they've made.

Nonstandard containers can be used effectively to discuss concepts related to volume. Children should be led to see, for example, that it is imperative for purposes of communication to have standard units. They can see advantages to using small cereal boxes or match boxes rather than boxes they've made or unusually shaped, irregular containers. Then they can discuss the cubic centimeter and cubic decimeter as being the best unit—for communicating with the most people and with the least chance of being misunderstood.

The model for the cubic decimeter that children should make in connection with their study of volume provides a physical representation for a liter. The volume of the cube is 1 liter. Thus, if the cube is constructed from plastic or heavy waxed paper, it will actually hold a liter of water. The volume of other containers can then be compared directly to the liter.

Since the liter is 1000 milliliters, comparisons for 750 mL, 500 mL, and 250 mL can easily be made. At this point, the teacher should emphasize estimation, and children should be actively involved in filling containers and in pouring from one container to another.

Commercial or teacher-made glass or plastic containers that are marked off in milliliters and liters should also be made available. Children should fill these containers with water and pour from nonstandard to standard containers and vice versa.

Thus, teaching strategies for the measurement of volume are similar to those for other types of measurement. Exploratory experiences should be designed that lead to familiarity with the units, and estimation should be encouraged. But, most of all, children should be actively involved. Here's a sample activity.

Example 14-39
A concept-development activity for capacity

Materials: Decimeter cube made from plastic or from milk-carton material. Containers that children have collected and brought in. Possibly, a funnel for pouring into small-mouthed bottles. Water.

Teacher directions: This box holds a liter. Let's fill it with water.

Questions: Do you think that any of our other things hold a liter? Which ones? How many Santa Claus mugs would it take to fill the box with water?

→ Measuring weight

The model for the cubic decimeter also provides a representation for the kilogram. A cubic decimeter, or liter, of water weighs very close to 1 kilogram. Thus, a milliliter of water weighs very close to 1 gram. Children need experiences designed to familiarize them with these units. Again, physical comparison is the suggested procedure, and estimation and direct involvement are encouraged.

The next three activities illustrate ways of introducing the kilogram and the gram.

Example 14-40
A concept-development activity for weight

Materials: Balance. Clay ball that weighs 1 kilogram. Familiar objects.

Teacher directions: This clay ball weighs 1 kilogram. Find something that will balance the ball.

Questions: Which weighs more, the clay or your book? Which weighs less, the clay or my billfold? Can you find something that weighs about the same as the ball?

 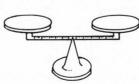

Example 14-41
A concept-development activity for weight

C-2

Materials: Balance. Familiar objects.

Teacher directions: The side of the balance with the heavier object will go down.

Questions: What will the balance look like if I put a pencil on this tray and a cup on the other tray? Which is heavier? Guess which of the things I mention is heavier, and we'll check your guess with the balance: the chalk or the eraser? the scissors or the ruler? the penny or the nickel?

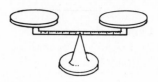

New directions: This lead piece weighs 10 grams. Try to find something that will balance it. (Etc.)

Example 14-42
A concept-development activity for weight

C-3

Materials: Bathroom scale calibrated in kilograms.

Teacher directions: Guess Greg's weight in kilograms. Greg weighs 40 kg. Now guess Gary's weight. Gary weighs 42 kg. Now guess Lori's weight.

Questions: Do we have some volunteers? What do you think our principal weighs? (Etc.)

Measuring angles

As you recall from Chapter 10, or from high school geometry, an angle can be defined as the union of two different rays that have a common endpoint. Rays extend infinitely, so we can't measure their length, and it wouldn't make sense to try to measure the endpoint. So what is there about an angle to measure? Why do we say that angle *A* is larger than angle *B* or that angle *C* is smaller than angle *B?* In talking about the sizes of these angles, we're referring to the "spread" between the rays.

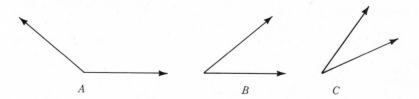

The protractor is a measuring instrument used for measuring angles in units called *degrees.* Any angle has a measure between 0° and 180° associated with it. These figures illustrate how a protractor is used to measure angles.

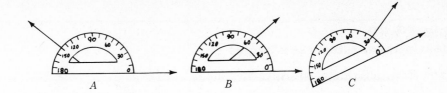

Children's experiences with measurement of angles should emphasize estimation and should allow for actual measuring experiences, as illustrated in the next two activities.

Example 14-43
A concept-development activity with angles

C-2

Materials: Scrap paper for each child. (Newspaper works well.)

Teacher directions: Fold your paper like this.

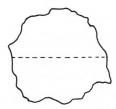

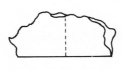

Find some angles that have the same measure as the angle that you've just made. The measure of this angle is 90°. Now fold again, like this.

Questions: What is the measure of the new angle? (45°)

New teacher directions: Use and

 to estimate the measure of these angles:

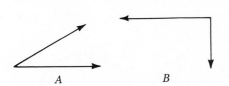

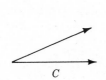

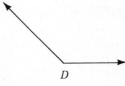

☞

Example 14-44
A skill-development activity with angles

Materials: Protractors. Rulers. A teacher-made ditto.

Teacher directions:
Measure these angles.
Record your measurements.
Draw some angles and measure them. Draw an angle that measures 85°. (Etc.)

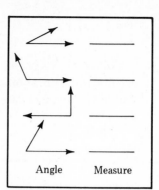

Angle   Measure

Measuring time and temperature

Most people probably don't often use the term *measure* in connection with time and temperature. But a clock measures the passage of time, and a thermometer measures the effect of a change in heat energy on a certain substance in a sealed glass tube. Children will be expected to be familiar with devices that are in common use for measuring time and temperature. Therefore, we must provide them with some appropriate experiences to help them "read" clocks and thermometers.

■ Time

In learning to "read" clocks, children should learn that both the movement of the clock's hands at a certain rate and the position of the hands at a given point in time are significant. It's a good idea to start with 12 o'clock. Children can first be helped to discover that the minute hand moves all the way around in one hour. Then they should see that the hour hand moves from 12 to 1 to indicate one hour, and that, in another hour, the hour hand will have moved from 1 to 2.

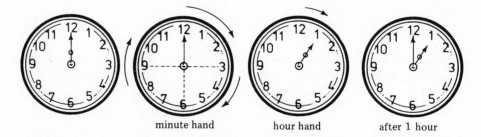

minute hand   hour hand   after 1 hour

Children should watch as clocks are turned ahead to emphasize the full revolution of the minute hand for each unit moved by the hour hand. Then they can experiment individually with a real clock or with clocks made from paper plates.

The movement of the minute hand must be separated into intervals—30-minute, 15-minute, and then 5-minute segments of time. Counting by fives is a helpful skill that children should have developed before this experience. They will have already established that in one hour the minute hand makes a full turn. The teacher should convince children that 1 hour is the same as 60 minutes and then have them decide that one-half hour is 30 minutes, represented by one-half of a revolution by the minute hand. Later they will need to learn that in 5 minutes the minute hand moves a certain distance—always the same part of a revolution—and that the clock can be separated into these smaller segments for telling time more precisely.

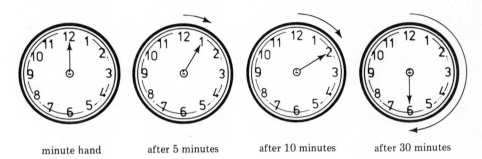

minute hand   after 5 minutes   after 10 minutes   after 30 minutes

Of course, it's difficult for young children to put all of these concepts together. A good case can be made for waiting to develop some of these ideas until children have more of the mathematical skills that are necessary for full understanding. In the first and second grades, most children should concentrate on hour and half-hour intervals. Not until later grades can the assimilation of concepts involving hours and minutes be expected. Clock-reading skills should also be included in other subjects besides arithmetic. Whenever they're taught these skills, children should have their own clocks for developing and practicing them.

The following activities are examples of appropriate lessons for children.

Example 14-45
A skill-development activity with clocks

S-2

Materials: Clocks—at least one for every two or three children (they can be cast-off, donated, scavenged, or even broken clocks). Cards labeled with events; e.g., I go to school. I eat lunch.

Teacher directions: Set your clock at 12 o'clock. Now show me 12:30. Show me 12:15. Show me 12:05, or 5 minutes after 12.

Show the time you go to bed. Show the time school is out. Show the time when your favorite TV show comes on.

Put the cards in order from what you do first to what you do last. Now show us what time you do those things. Let's write the time on the back of the card.

Example 14-46
A concept-development activity with time

C-2

Materials: 0–9 cards. Clocks. Time board.

Teacher directions: I'll set the clock. You pick the cards to fill in the time board.

Questions: Look at the little hour hand. It's just past 1. Are hours left or right of the two dots? [Left.] So put a 1 in that slot. Look at the longer minute hand. How many minutes past the hour is it? [25] So put a 2 and 5 to the right of the two dots. What does the time board say? [1:25] Who wants to set the clock for us? [I do! I do!]

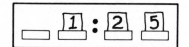

Of course, as the digital clock becomes more popular, children will have less difficulty in learning to tell time.

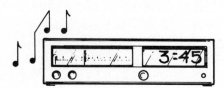

■ Temperature

Reading a thermometer is simple in comparison with reading a clock. (But with the Celsius scale for temperature, negative numbers will be used more frequently than with the Fahrenheit scale, and this deserves special classroom attention.) Some rather elaborate devices are available for teaching children to read thermometers, and if you have such devices, use them. If not, just let children practice associating a point on a number line with a number that represents the temperature, or let them practice reading the scale on the thermometer in the room. Use your imagination to make up activities. Here is one example.

Example 14-47

A concept-development activity for temperature

C-3

Materials: Two thermometers for measuring air temperature, one placed on a wall in the classroom and one on a wall outside the building.

Teacher directions: Let's keep a record of the temperature inside and outside for the month. We'll check it every day at 11:30.

Let's keep a record of the daily high temperature and the daily low temperature for this month. We can get these from television or from the newspaper.

Watch or listen to the news tonight. Write down the high and low temperatures in the United States, and we'll compare them with our own high and low temperatures.

I'll bring the newspaper. Let's pick a partner-city for this week and compare its temperature with ours.

Estimation

Estimation should be a deliberate part of activities involving measurement. Children should be encouraged to guess, revise their guesses, and then guess again. Their guesses will become quite accurate as they're exposed to familiar referents for the attribute they're measuring. For example, after a few days of paying attention to the temperature during freezing weather, children will be likely to estimate that the temperature of a similar day is 0°C or below. Or, given some books that weigh close to 1 kilogram, they'll be able to estimate quickly the weight of other books or similar objects. Their improved accuracy is both rewarding and reinforcing, and the ability to estimate various measurements is a particularly practical skill.

An estimation activity that's especially effective is having children estimate one another's height. The next example shows one way in which you could direct this experience.

☞

Example 14-48

C-2

A concept-development activity involving estimation of height

Materials: A device for measuring heights, probably between 100 cm and 200 cm. (It's easy to make with a meter stick and tape you can stick to the wall.)

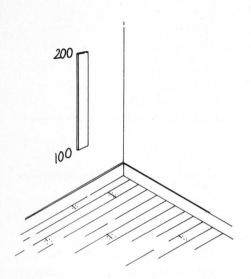

Teacher directions: Write down how tall you think Julie is.

Questions: Did anyone write 50 centimeters? How about 200 centimeters? What's the smallest number that we have? The largest? All right, my height is 168 centimeters. Does anyone want to change his or her guess for Julie's height? Now what are the largest and the smallest numbers that we have? Let's measure Julie. She's 143 centimeters tall. Guess Gordie's height. Now guess Mike's height. (This procedure can be carried out for each child in the class.)

New teacher directions: Let's make a chart to show our results. The total shows how good a guesser you are. Let's see who has the lowest total.

Who?	*My guess*	*Actual height*	*How far off I was*
Julie	150 cm	143 cm	7 cm
Gordie	152 cm	150 cm	2 cm
.	.	.	.
.	.	.	.
.	.	.	.
			Total 36 cm

The next activity also involves estimation, but it can be used for evaluation. Children are asked to select numbers that fit a measurement situation and a prescribed standard unit. Their ability to estimate will depend on their familiarity with the standard units, so an observant teacher can detect if there is a need for further preparatory or exploratory experiences either for the entire class or for small groups of children.

Example 14-49
A skill-development activity for metric awareness

Materials: Prepared handout.

Teacher directions: Use 1 or 10 or 100 to fill in the blanks so that the sentences make sense.

Note: This handout should be followed by a similar one that has been modified as shown below.

Directions: Place a decimal point in each of these sentences so that the sentence makes sense.

1. My pencil is _____ centimeter(s) long.
2. Our holiday turkey weighed _____ kilogram(s).
3. Grace drinks _____ liter(s) of milk a day.
4. My fingernail is _____ millimeter(s) wide.
5. The playground is _____ meter(s) long.
6. No one in our class weighs as much as _____ kilogram(s).
7. My bicycle is _____ meter(s) tall.
8. The chalkboard eraser is _____ decimeter(s) long.
9. Brad's mother gave him _____ milliliter(s) of cough syrup.
10. When you cook a hard-boiled egg, the temperature of the water is _____ °C.
11. On the interstate highway, it's safe to drive at a speed of _____ kilometer(s) per hour.
12. It's getting ready to snow outside; the temperature is _____ °C.
13. Debby's puppy weighs _____ kilogram(s).
14. Two nickels weigh _____ gram(s).
15. Our classroom is _____ meter(s) wide.

---

1. My pencil is 126 centimeters long.
2. Our holiday turkey weighed 104 kilograms.
   . . . (Etc.)

Money

Concepts related to money have been presented in other sections of this book. There, the ideas were always related to place value. Our monetary system—with pennies and dimes and with bills for one dollar, ten dollars, hundred dollars, and thousand dollars—provides a natural and familiar place-value model. We not only used this to reinforce the meaning of numerals but also to give meaning to the computational procedures that children must learn. For example, in regrouping situations, exchanging one ten for ten ones, and vice versa, is sensible and fair, especially at a bank.

But young children need instruction that focuses on all of the coins. They need to be taught the value of each coin, relationships between coins, and techniques for determining the value of several coins.

A crucial idea to this development is the fact that any amount (and we'll use amounts less than one dollar) can be directly related to pennies. $0.27 is the same as 27¢. When showing 27¢ with as few coins as possible, we can use one quarter and two pennies, because we know 1 quarter = 25 pennies and 25 + 1 + 1 = 27. "With as few coins as possible" is a convenient and logical restriction to make. "With as many coins as possible" would force us to use all pennies. Children should be led to conclude that there are usually quite a few possible combinations that will have the same value. In representing 27 as 25 + 1 + 1, we're relying on counting knowledge that children have before they can handle certain sums. If you watch most adults handling coins, you'll see them counting at least as frequently as adding.

The two activities that follow provide opportunities for finding relationships between coins and for finding values of several coins. Other activities that involve simulated purchases are excellent for children.

Example 14-50
A concept-development activity for showing relationships among coins

C-1, 2

Materials: Charts. Coins, or cards with coin pictures stamped on them (see Material Sheet 6). Coin cards can be made like this:

Teacher directions: Pick out 1 nickel. How many pennies are in one nickel? Fill in the chart. Pick out 2 nickels. How many pennies? Fill in the chart. [Etc.] (Repeat for dimes and quarters.)

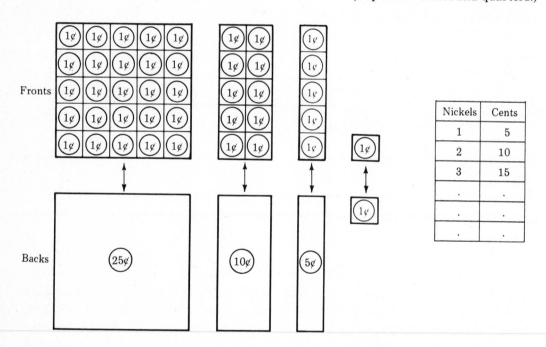

Nickels	Cents
1	5
2	10
3	15
.	.
.	.
.	.

Questions: Many, many questions relating these ideas.

☞

Example 14-51
A concept-development activity for determining values of coins

C-2, 3

Materials: Coin cards from Example 14-50 or real coins (see Material Sheet 6).

Teacher directions: Let's see if we can find out how much this is. Put the one that's worth the most first. Turn them over. The first one's worth 25¢. Now, the first two together are worth 35¢. Finally, all of them together are worth 48¢.

New directions: You must pay for this 37¢ toy boat. What coins will you use? (Etc.)

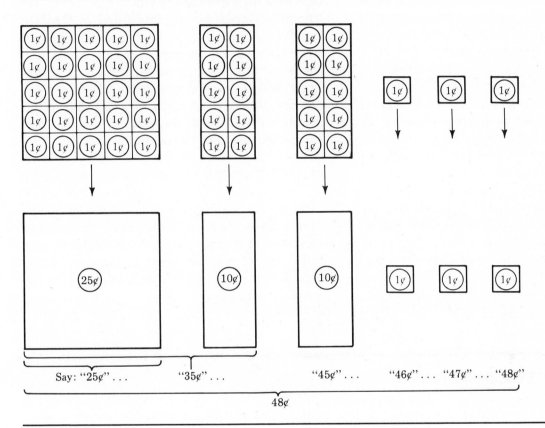

A common mistake made by adults is to use a decimal point *and* a ¢ sign, so you find ".49¢" listed as a price. This is actually less than one-half a penny! You may have even seen something like "$.49¢." Who knows what that means!

Ideas for enrichment: π

The number system that we have discussed in this book is the rational number system. There is one irrational number that elementary school children should know something about. That number is π. It appeared in this chapter when we were dealing with circles. This is the ratio of the circumference of a circle to its diameter. $\pi = C/D$ and is a little more than 3. This ratio was known very early in history and is mentioned by Chinese and Greek mathematicians. A biblical reference to a circular "molten sea" in II Chronicles 4:2 gives a distance around it as 30 cubits and the distance across as 10 cubits $(30/10 = 3)$. Archimedes calculated the value of π to be between 3-10/71 and 3-1/7.

When we use 3-1/7 or 3.14 in our calculations, we are only using a rational approximation for π. Since it is an irrational number, it cannot be represented in the form a/b where a and b are integers, and it cannot be represented by a terminating or repeating decimal.

Interest in π can be generated by having children measure the distance around a lot of different size circles with string, and then measuring the distance across those circles. When they divide their two numbers, they will all get a number a little greater than 3.

Some children will be interested in the fact that the decimal never ends or repeats. It can be carried out as far as anyone wants to take it. There are tables listing at least the first 10,000 digits, and the person who holds the record for memorizing the most digits of π is listed in the *Guiness Book of World Records*.

Here are just a few of the digits:

3.14159265358979323846264333 . . .

Of course, memorizing a large group of unrelated digits is not very "enriching." Encourage interested students to do a little research and find out what they can about this unusual number. Then they can share their results with the rest of the class.

The computer corner

Money is probably one of the most important topics in measurement because related ideas are applied every day. Such everyday situations require children to manage such situations as the following:

2 dimes is the same as 20 pennies,
8 can be made using 1 nickel and 3 pennies,
2 quarters is 1 dime and 4 pennies more than 36.

These requirements are generally made long before children are comfortable with multiplication and division concepts. Unfortunately, in most school curriculums there are too few opportunities for children to practice these ideas. So, in this corner we provide a program called TOYSTORE. This program presents simulations where children pretend to be clerks in a toystore making change for customers who buy toys (for a more thorough description of the program see Chapter 16).

Example 14-52
This is an activity where children must make change as efficiently as possible in terms of pennies, nickels, dimes, and quarters

C-2, 3

Materials: The program TOYSTORE on the MGB Diskette.

Teacher directions: In the program you will have to make change as efficiently as possible. You can never use more than 4 pennies, 1 nickel, 2 dimes, or 3 quarters. Sometimes, to make efficient change you will not use a certain coin. In situations such as this, you will enter "0" for the coin that is not used.

Example: Ryan wants to buy a red wagon that costs $1.64. He pays you with $2.00. What change will you give him?

Number of pennies? ____
Number of nickels? ____
Number of dimes? ____
Number of quarters? ____

■ Computer tip

There should be a close relationship among computer activities that children complete, class discussions, and seat work. Too often, the computer is used as a toy, for the presentation of "cute" activities that might not be germane to the goals in an instructional unit of study. Occasionally this is okay, but in general this should not be the case. It's the teacher's task to make connections. In the case of the program TOYSTORE, the teacher should provide readiness activities in which children act out problems with real or play money; afterward, children should practice at the computer with TOYSTORE. When all of the children have had a chance at the computer, there should be class discussion. The teacher should lead the discussion extending concepts presented in the program. Here's an example:

Today, the cash register contains only dimes, pennies, and quarters with which to make change. Let's do some problems and make change using only these coins. Our change must be made as efficiently as possible.

A toy puppy costs $3.63.
You're paid $4.00.
What change do you give?

In closing . . .

In this chapter, we've provided concepts and sample activities related to measuring length, area, volume, weight, angles, time, and temperature, and concepts related to money. The relationships among units for length, area, and volume have been carefully pointed out, as have the relationships between the metric units for volume and weight.

Our approach has been determined by our belief that children's experiences with these measurement concepts should be informal and should emphasize meaning. Measurement concepts will be reviewed, extended, and formalized at the junior high-school level. Thus, teachers in the elementary school can allow children the opportunity to manipulate objects and to actively participate in the measurement process. Estimation should be an integral part of the development of these concepts.

Once again, we encourage teachers to have fun with these mathematical concepts!

It's think-tank time

1. Determine which of the activities in this chapter involve higher-order thinking skills and could be used to promote problem solving by elementary school children.
2. Give formulas for each of the following, and explain how they were determined.
   a. perimeter of an equilateral triangle
   b. area of a square
   c. perimeter of a regular polygon with n sides
3. Find answers to the following questions, and describe your procedure.
   a. How thick is a sheet of notebook paper?
   b. How much money would you have if you had a stack of dimes 1 kilometer high?
4. a. Determine the area of triangle ABC by using the indirect procedure described in Example 14-36.

   b. Use the formula for the area of a triangle to determine the area of triangle ABC.
   c. Draw four more triangles in this figure that have $\overline{AB}$ as a base and the same area as triangle ABC.
5. The formula for determining the area of a circle is $A = \pi r^2$, where r is the radius of the circle and π is approximately 3.14. Give a rationale for delaying the development of this formula and its related concepts until the junior high years.

6. a. How many cubic millimeters are in a cubic meter?
   b. If you could lay these little millimeter-cubes out in a line, side by side, how far would they reach?
7. How much does a grain of rice weigh? Describe your procedure.
8. Fold your newspaper as indicated in Example 14-43 to get a straight edge. Now fold the straight edge to get three approximately congruent angles. Now you can represent angles of 30° and 60° as well as 90° and 45°.
   a. Use this "angle measurer" to draw:
      1) a right triangle with one 30° angle.
      2) a right triangle with two 45° angles.
      3) a triangle with three 60° angles.
      4) a triangle with two 30° angles.
   b. Now measure the lengths of the sides of your four triangles. What conclusions can you reach from each of your four cases?
9. Outline an instructional strategy or activity that could be used to teach children concepts related to each of the following topics. Be sure to specify the concept or concepts that you are considering.
   a. measurement of length
   b. measurement of area
   c. measurement of volume
   d. measurement of weight
10. Describe an activity that will help children determine an approximate value for π.

→ Suggested readings

Aman, George. "Discovery on a Geoboard." *The Arithmetic Teacher*, April 1974, pp. 267–272.

Bachrach, Beatrice. "No Time on Their Hands." *The Arithmetic Teacher*, February, 1973, pp. 102–108.

Bork, Ron. "Speed Trap." *The Arithmetic Teacher*, April 1979, p. 18.

Brougher, Janet Jane. "Discovery Activities with Area and Perimeter." *The Arithmetic Teacher*, May 1973, pp. 382–385.

Bruni, James V., and Silverman, Helene. "Developing the Concept of Linear Measurement." *The Arithmetic Teacher*, November 1974, pp. 570–577.

Carpenter, Thomas P. "The Performance of First-Grade Students on a Non-Standard Set of Measurement Tasks." In *Technical Report No. 211*. Madison, Wis.: Wisconsin Research and Development Center for Cognitive Learning, 1971.

Hecht, Anne T. "Environmental Problem Solving." *The Arithmetic Teacher*, December 1974, p. 42.

Hiebert, James. "Units of Measure: Results and Implications from National Assessment." *The Arithmetic Teacher*, February 1981, pp. 38–43.

Hildreth, David J. "The Use of Strategies in Estimating Measurements." *The Arithmetic Teacher*, January 1983, pp. 50–53.

Hunt, John D. "How High Is a Flagpole?" *The Arithmetic Teacher*, February 1978, pp. 42–43.

Jeffers, Verne. "Using the Digital Clock to Teach the Telling of Time." *The Arithmetic Teacher*, March 1979, p. 53.

Jensen, Rosalie. "Multilevel Metric Games." *The Arithmetic Teacher*, October 1984, pp. 36–39.

Jensen, Rosalie, and O'Neil, D. R. "Meaningful Linear Measurement." *The Arithmetic Teacher*, September 1981, pp. 6–12.

Lappan, G., and Winter, M. J. "Sticks and Stones." *The Arithmetic Teacher*, March 1982, pp. 38–41.

Leutzinger, Larry P., and Nelson, Glenn. "Meaningful Measurements." *The Arithmetic Teacher*, March 1980, pp. 6–11.

Lichty, Shirley Lindinger. "Dinner Time." *The Arithmetic Teacher*, March 1980, p. 16.

Lichty, Shirley Lindinger. "How Would You Measure Me?" *The Arithmetic Teacher*, November 1980, p. 28.

Lindquist, Mary Montgomery, and Dana, Marcia E. "Gad Zucchs! Zucchini in Your Math Class?" *The Arithmetic Teacher*, December 1978, pp. 6–10.

Lindquist, Mary Montgomery, and Dana, Marcia E. "Independent Mathematics Centers for Everyone." *The Arithmetic Teacher*, April 1979, pp. 4–9.

McClintic, Joan. "Capacity Comparisons by Children." *The Arithmetic Teacher*, January 1979, pp. 19–25.

Ott, Jack M., Sommers, Dean D., and Creamer, Kay. "But Why Does C = πd?" *The Arithmetic Teacher*, November 1983, pp. 38–40.

Rees, Jocelyn Marie. "A Measurement Fantasy." *The Arithmetic Teacher*, January 1978, pp. 16–17.

Riley, James E. "It's About Time." *The Arithmetic Teacher*, October 1980, pp. 12–14.

Shaw, Jean M. "Meaning Metrics: Measure, Mix, Manipulate, and Mold." *The Arithmetic Teacher*, March 1981, pp. 49–50.

Thiessen, Diane. "Measurement Activities Using the Metric System." *The Arithmetic Teacher*, October 1979, pp. 36–37.

Thompson, Charles S., and Van de Walle, John. "A Single Handed Approach to Telling Time." *The Arithmetic Teacher*, April 1981, pp. 4–9.

Thompson, Charles S., and Van de Walle, John. "Let's Do It, Learning About Rulers and Measuring." *The Arithmetic Teacher*, April 1985, pp. 8–12.

Toher, Gertrude C. "Measuring Up and Up." *The Arithmetic Teacher*, December 1978, p. 41.

Weir, Merlene J. "A Money Center." *The Arithmetic Teacher*, April 1981, pp. 18–20.

Zirkle, Ronald Elmer. "The Role of Manipulatives in Schematic Approaches to Area Measurement for Middle School Students." (Doctoral dissertation, Virginia Polytechnic Institute and State University, 1980.) *Dissertation Abstracts International*, *41A*(1981):3933A.

15

Making Numbers Count

Organizing, Representing, and Interpreting Data

■ This chapter presents statistics and probability ideas that are appropriate for elementary school children. Because applications of these ideas should be explicitly identified, teachers are encouraged to design experiences that make use of familiar language and that draw from real-life or interdisciplinary situations. We suggest that children participate in the selection of problems and that they collect and organize data. Description and interpretation of the data are emphasized, along with the formulation of tenable hypotheses.

Many of the ideas discussed in the chapter should be reserved for the upper elementary grades, although intuitive notions from statistics and probability can be developed earlier. In all elementary level grades, the situations should be simple enough that the data do not overwhelm students or obscure the problem-solving procedures. And the situations should always be interesting! ■

Objectives for the child

1. Participates in the selection of interesting problems to be analyzed.
2. Collects and records data accurately and systematically.
3. Organizes data efficiently.
4. Represents data by using graphs and charts.
5. Describes data accurately.
6. Makes valid interpretations of data.
7. Formulates hypotheses to be tested.
8. Tests hypotheses through analysis of data.
9. Determines all of the possible outcomes in simple probability experiments.
10. Determines the probability of an event occurring and represents this probability by a fraction.
11. Uses numbers appropriately to analyze simple situations involving statistics or probability.

Teacher goals

1. Be able to relate ideas from statistics and probability to children's experiences and to use familiar language to express these ideas.
2. Be able to assist children in the selection of appropriate problems to be analyzed.
3. Be able to assist children in the collection of data.
4. Be able to assist children in efficiently organizing their data.
5. Be able to help children with accurate description and valid interpretation of data.
6. Be able to present interesting and appropriate problems for children to explore.

Water consumption in the United States comes to about 1876 gallons (more than 7000 liters) per person per day. Do you know anyone who drinks that much water?

There are 1.2 television sets per family in the United States. Do you know any family that has 1.2 television sets?

Can you tell who's going to be elected president from the results of the first precinct that reports?

What are the chances of rain tomorrow?

The use of concepts from *statistics* and *probability* is a part of daily life. All of us are exposed to descriptions and predictions that use numbers in a variety of contexts. The more we understand about how numbers are used, the more we will be able to make appropriate decisions—for instance, whether to take an umbrella to work if the weather report says there's a 10 percent chance of rain.

Children also encounter concepts from statistics and probability in and out of school. Their own conversations include many statements of a statistical nature, such as:

Ty Cobb holds the record lifetime batting average of .367.

My family drinks about two cartons of soft drinks a week.

I go swimming about once a week in the summer.

I spend part of my allowance on candy and part of it on potato chips, and I save some of it.

The dentist said I had only one cavity, and that's not too bad for a kid my age.

Statements that imply various degrees of probability also are often used by children—for example:

I might go to Mary Kay's after school.
Grandma will probably come to see us this weekend.
If you eat a lot of candy, you'll probably get cavities.
If you stay out in the sun, you'll almost certainly get sunburned.
There's no way it's going to snow in July.

Through magazines, newspapers, and especially television, children are exposed to descriptions of data and to predictions based on those data. Such statements occur frequently, from news reports to commercials. Thus, children are familiar with them, although they need help in integrating statements that involve statistics or probability if they are not to be misled by "good-sounding" but flawed arguments.

In school, ideas from statistics and probability are useful because of their applicability to various subjects. Graphs, charts, and predictions based on those ideas are used in many school subjects. Ideas from mathematics can therefore be used to help children interpret and understand situations that they encounter throughout the curriculum.

Statistics

The introduction of statistical procedures at the elementary level should include discussions of particular problems, of the types of data that it would be helpful to have, methods for collecting and tabulating the data, and interpretation or description of the data. Problems should be kept simple, and the necessary calculations shouldn't be overwhelming. The calculator, however, does make it possible to attack problems easily now that would have previously required tedious calculations.

After children have become acquainted with probability as a mathematical means of making predictions, they can learn to draw inferences from data they collect. The initial emphasis should be on description of the data, however, with prediction coming later.

Teachers can take advantage of the statistical expressions that children are already being exposed to by calling their attention to them and by discussing them in class. The following activity suggests a follow-up assignment to an introductory lesson at the fourth-grade level or above.

Example 15-1
A reinforcement activity for statistical ideas

Materials: None.

Teacher directions: Your homework assignment is to watch TV tonight. Make a list of everything you hear that has to do with statistics.

Questions: What are some key words to listen for? (Average, annual, monthly, daily, per person) What are some good programs to watch if you want to hear these words? (News, weather, sports, documentaries, commericals)

New teacher directions: Find a newspaper article that uses statistics. Bring it to school tomorrow.

New teacher directions: Make a list of some sentences you hear people say that have to do with statistics. For example, "Lisa said that her average in spelling is 84."

■ Collecting and organizing data

Children also should be encouraged to find statistical information in other textbooks and to discuss what they find there in mathematics class. Directions such as "Get your Social Studies book out; we're going to study math" are amusing to children, and they point out that mathematics is used everywhere.

There is much material available in newspapers, magazines, encyclopedias, and other textbooks in which data have already been collected and organized. This is useful for discussion and provides information pertaining to statistical procedures. Children will also benefit from collecting and organizing their own data, and the activity means more to them if it's actually about them.

The following activities illustrate the collection and organization of data that concern the children themselves. They are designed to develop concepts related to statistics, and they're appropriate for children at any grade level. Let them use squared paper to record their favorite pets, dessert, birthday, birth month, etc. (You can even use this in special education classes.)

Example 15-2
A concept-development activity with graphs

C-1, 2, 3

Materials: Blocks. Material Sheet 15. Calendar.

Teacher directions: Let's draw a graph to show something about our class. You use graph paper. I get to use the blocks.

Here's the problem. We want to show how many of us have birthdays on Sunday this year, how many on Monday, and so for each day of the week.

Questions: How can we do this? We could just list everybody's name and put the day of his or her birthday next to it. What's wrong with that? Kathy says that it would be better to list the days and then put a block for each person. Can you show this with your graph paper? Be neat!

Which day has the most? Which day has the least? How long will this chart be useful? If you move away, we'll have to knock your block off!

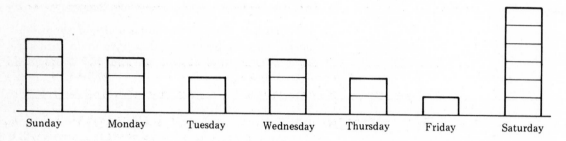

New teacher directions: We can find the *average number* of birthdays per day by moving some blocks around.

The average number of birthdays per day is 3. Does that mean we have 3 people with birthdays on Friday? Why not?

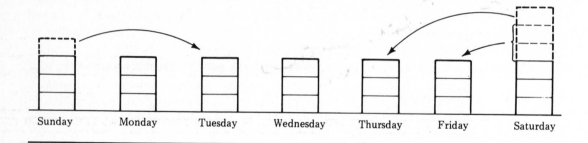

Example 15-3
A concept-development activity for finding an average

Materials: Strips of construction paper for each child (approximately 3 cm wide and longer than anyone's foot). Cellophane tape. Scissors.

Teacher directions: Let's find the average foot length in our class. Cut your paper strip so that it's the same length as your foot. Now let's tape all of our strips together.

Questions: We've put everyone's foot size together. Now, how can we find the average foot size? How many centimeters long is our big, long strip? How many people's feet do we have? How many centimeters is that for the average foot?

Let's cut our strip into equal parts that stand for the average foot in our class. Now, who has a foot that is about that long? Is your foot shorter or longer than average? Suppose we all bought a pair of shoes in this size (indicating the average size). What would happen?

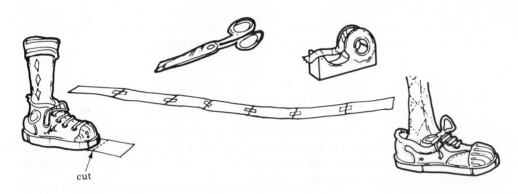

cut

Example 15-4
A concept-development activity for finding an average

Materials: None.

Teacher directions: Let's find the average number of brothers and sisters of our class.

Questions: How can we find the average number? Let's make a chart first, before we find the total. Look at the chart. What can we use besides addition to find the total?

Since there are 20 people in class, how many brothers and sisters is this per person?

Does that mean that each person has 2 brothers and sisters? What does it mean? If the total had been 41, the average would be 2.05. In other words, the average is between 2 and 3 and close to 2. (No one had 2.05 brothers and sisters. . . .)

Number of brothers and sisters	8	7	6	5	4	3	2	1	0
Class members			I	I	I	III	ⅢⅡ	ⅢⅡ	III

$$T = (1 \times 6) + (1 \times 5) + (1 \times 4) + (3 \times 3) + (5 \times 2) + (6 \times 1) + (3 \times 0) = 40$$

Example 15-5
A concept-development activity for statistics

C-1, 2, 3

Materials: Material Sheet 15.

Teacher directions: You select a topic. Decide how we can select the data, and then decide the best way to organize them.

Questions: Should we work in groups to collect the data? How can we organize our data? Should we use a bar graph, line graph, or circle graph?

Possible topics: Color of pencils, shoes, shirts, socks, hair, eyes, bicycles, cars, bedspreads.
Number of pets, buttons on clothes, things in pockets or purses, aunts or uncles, cousins, pieces of jewelry, pennies.
Birth months, ways the children get to school, what they ate for breakfast, what they did last summer, etc.

Almanacs and books of world records are other sources of interesting data for use in discussions of statistics. Whatever source you use, the emphasis should be on interpreting or describing the information so that children understand where the numbers come from and what they tell us.

■ Drawing inferences

Many of the experiences just described involve collecting data from students in class. Suppose that we wanted to find some information about the whole school. One way to do so would be to predict on the basis of a small sample that would apply to a larger group—that is, to draw *inferences.* Inferential statistics is widely used. For example, pollsters carefully select a group of people to interview and then infer from their answers what the opinions are for the general population. Examples of this kind of procedure, drawing inferences from the examination of sample data, should be included in the presentation of statistical ideas, as illustrated by the following activity.

Example 15-6

C-3

A concept-development activity for statistics

Materials: A big sack of blue and green marbles.

Teacher directions: Let's do an experiment. We'll try to determine—without counting—whether there are more blue or more green marbles in the sack. Draw five marbles without looking. Record your numbers and colors. Put them back.

> Kurt got 3 green and 2 blue ones.
> Dan got 1 green and 4 blue ones.
> Paulann got 5 blue ones.
> Wade got 1 green and 4 blue ones.

Questions: On the basis of these four draws, what do you think?

Let's let everyone draw and then find our totals for blue and for green. We drew 82 blue marbles and 18 green ones. Do you think there were more blue ones in the sack? What do you think the ratio of blue to green might be? Could you have made a good guess after Kurt's draw? Why or why not?

Open the sack and count. There are 100 blue and 25 green. So the ratio of blue to green is 4 to 1.

Other activities for children that involve statistical ideas and graphing, in particular, would include making pictographs or line graphs and interpreting circle graphs. Pictographs are particularly effective with elementary school children.

Who caught more fish?
Who caught fewer fish?
How many fish for the three of them?
How many more for Sally than Jane?

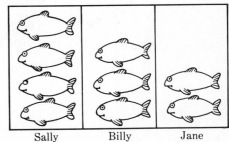

An appropriate sequence of graphing activities for elementary school children is the use of concrete representations—such as stacking blocks or books—then picture graphs, followed by bar graphs and line graphs. Circle graphs can be interpreted but are difficult to construct in many cases.

The next activity provides an application of statistical data that is of direct interest to elementary school children.

Example 15-7

C-3

A concept-development activity for using charts

Materials: A table like the one below. Calculators.

Predicting your adult height		
Age in years	*Boys*	*Girls*
5	1.62	1.51
6	1.53	1.42
7	1.45	1.35
8	1.39	1.29
9	1.33	1.24
10	1.28	1.18
11	1.23	1.13
12	1.19	1.08
13	1.15	1.04

These numbers are rounded to the nearest hundredth. They are based on information about people and their heights.

Teacher directions: Find your age in the table. The number beside your age tells what you must multiply your height by to find out about how tall you'll be when you're grown up.

Questions: How many centimeters tall are you, Phyllis? You're 150 centimeters tall, 11 years old, right?

For an estimate of Phyllis' adult height, look at the table. What number would Phyllis multiply by 150? Right, 1.13. Use the calculator if you want to. $150 \times 1.13 = 169.5$, or about 170 centimeters.

Now find your own number and your predicted height.

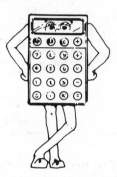

The following activity combines ideas about graphing to represent statistical data and the use of calculators and percents. With the 10×10 square, it also emphasizes how easy it is to handle large samples.

☞
Example 15-8
A concept-development activity for statistics

C-3

Materials: Data that children have collected. Calculator(s). Graph paper.

Teacher directions: Let's draw a graph to show what we found out about the favorite ice cream for our school. Our data are in this table.

Favorite ice cream	
Chocolate	126
Vanilla	81
Strawberry	45
Chocolate Chip	27
Maple Nut	9
Total	288

Questions:

What percent of the children interviewed like chocolate? $\frac{126}{288} = .4375$, or about 44%.
What percent like vanilla? $\frac{81}{288} = .28125$, or about 28%.
What percent like strawberry? $\frac{45}{288} = .15625$, or about 16%.

What percent like chocolate chip? $\frac{27}{288} = .09375$, or about 9%.
What percent like maple nut? $\frac{9}{288} = .03125$, or about 3%.

Note: Because of rounding, some adjustment may have to be made sometimes to be sure the percents total 100.

If your grid is separated into 100 parts, how can we show 44%? Right—just color in 44 of the 100 parts.

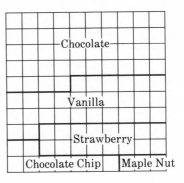

The preceding activities, as well as activities performed in other contexts, should teach children that graphing is just one more of a number of efficient ways to represent mathematical ideas. Both the graphing of data and the interpretation of graphs are important for children to learn. The preceding discussion contains only a few of the possible instructional activities; others are discussed elsewhere in the text. Development of graphing techniques can begin as early as the first grade and should be extended throughout the elementary school years.

→ Probability

We've seen that by using statistical procedures, we can look at a small sample and draw inferences, or make predictions, about the larger population from which the sample was selected. In probability, sometimes called the "science of chance," we can look at the *total* population and make some predictions about the small sample we will select. For instance, if Example 15-6 were altered so that we started with the knowledge that 100 of the marbles were blue and 25 were green, we could predict that someone who drew 5 marbles would be likely to get more blue ones than green ones.

Statistics and probability are usually studied together. Children should certainly be provided with experiences from both topics at the same grade level, and the experiences should have the same characteristics. That is, probability experiences should draw from real-life situations; they should make good use of the language patterns of the children; and they should cut across subjects of study in the elementary school. The problems should be kept simple, and children should collect, organize, and interpret their own data.

Children also can suggest problems to be studied. The study of statistics and probability should be mathematically and psychologically rewarding for elementary school children. In short, children can learn a lot and have fun at the same time.

■ Using familiar language

We've already said that probability statements are part of a child's language. Children should be made aware that many everyday sentences have an element of chance involved and that even sentences about certainties or impossibilities can be classified according to probability.

A slight modification of Example 15-1 provides a good awareness activity for children. The following activity is another useful one for introductory purposes. Remember that in the elementary grades the language used should remain rather informal, even though the words the teacher uses have precise definitions in the formal study of probability.

Example 15-9   **C-3**
A concept-development activity involving probability

Materials: Three shoe boxes, labeled "certain," "uncertain," and "impossible." 3 × 5 cards with sentences on them.

New teacher directions: Now draw a card from the "uncertain" box. Classify your sentences as "likely" or "unlikely."

 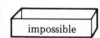

Teacher directions: Draw a card and read it to us. Decide which box to put it in, and tell us why.

The "likely" sentences are more probable than the "unlikely" ones, even though they're still "uncertain."

Our school will be struck by lightning.	Our teacher is older than any of us.
The cafeteria will serve milk today.	Logan will have oysters for breakfast tomorrow.
My dog will be a mathematician.	Trisha will make an A on the spelling test.

■ Determining possible outcomes

Experiences in which children actually determine the probability of an event are appropriate at the elementary level if they are not complex. The possible outcomes, however, should be easy to determine and to list. The children should keep good records for their experiments and should be assisted in analyzing their results.

The following are examples of activities that lead up to the determination of the probability of an event and that require children to list all possible outcomes.

Example 15-10
A concept-development activity for probability

C-3

Materials: Children working in pairs (one flipper, one recorder). One coin.

Teacher directions: Toss the coin 20 times. Keep a record of what you get.

H	⌐⌐⌐⌐	12
T		8
Total		20

← data for one pair of students

Questions: What results could you get for each toss? How many tosses for the whole class? How many heads? How many tails?

H	12	10	9	8	7	11	15	7	11	8	98
T	8	10	11	12	13	9	5	13	9	12	102
Totals	20	20	20	20	20	20	20	20	20	20	200

← class data

☞

Example 15-11

A concept-development activity for probability

Materials: Children working in pairs (one spinner-spinner, one recorder). One spinner per pair.

Teacher directions: Spin the spinner 20 times (spin it again if it lands on a line). Keep a record of how many reds and how many blues.

Questions: What colors could you get? How many spins for the whole class? How many reds? How many blues? Did anybody get green?

R	ⱶⱵⱵ IIII		9
B	ⱶⱵⱵ ⱶⱵⱵ I		11
Total			20

← data for one pair of students

R	9	8	8	7	13	11	10	8	14	9	97
B	11	12	12	13	7	9	10	12	6	11	103
Totals	20	20	20	20	20	20	20	20	20	20	200

← class data

red   blue

There are many possible activities similar to these, in which there are only two possible outcomes, each being as likely as the other. Children's familiarity with fractions will lead them to decide rather quickly that the number of actual occurrences of each outcome is approximately one-half of the total. Experiments with situations with more than two possible outcomes can then be conducted, such as in the example that follows.

☞

Example 15-12

A concept-development activity for probability

Materials: Children working in pairs (one tosser, one recorder). One die per pair (have children bring dice from one of their games at home if you don't have any).

Teacher directions: Toss the die 30 times. Be sure to shake it well before each toss. Keep a record of what you get.

Questions: What could you get? How many tosses for the whole class? How many 1s? 2s? 3s? 4s? 5s? 6s? Did anybody get a 15? A 0?

1	IIII	4
2	ⱶⱵⱵ II	7
3	III	3
4	ⱶⱵⱵ I	6
5	IIII	4
6	ⱶⱵⱵ I	6
	Total	30

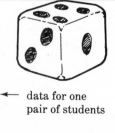

← data for one pair of students

In the three previous examples, we assumed that the coin, spinner, and die were *fair*. That is, we assumed that we were just as likely to get a head as a tail, or just as likely to get a red as a blue, or just as likely to get a 1 as a 2, 3, 4, 5, or 6. Children can be convinced of the significance of this assumption by considering an example such as the following.

Example 15-13
A concept-development activity for probability

Materials: Spinner.

Teacher directions: We're going to have the girls play the boys. If the spinner lands on blue, the girls get a point. If the spinner lands on red, the boys get a point.

Questions: Is that OK with everybody? What's the matter, Leroy?

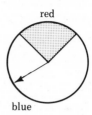

■ The probability of an event

Children should discuss the question "What would you expect?" in performing the kinds of activities we've seen so far. Doing so will help them to see the need for the number that is associated with an event which is called the *probability of an event.* For example, the probability of getting a head when you flip a coin is $\frac{1}{2}$; the probability of getting a 5 when you toss a die is $\frac{1}{6}$; and the probability of the spinner landing on blue in Example 15-13 is $\frac{3}{4}$. The probability of something happening that is *impossible* is 0. The probability of something happening that is *certain* to occur is 1. Thus, the probability of an event is a number that is less than or equal to 1 and greater than or equal to 0. Elementary school children do not have to acquire this general concept, however. It's enough that they examine situations that have a small number of possible outcomes and become intuitively convinced that fractions can be used to represent probabilities in these specific cases. The following concept-development activities illustrate these kinds of situations.

Example 15-14
A concept-development activity for probability

Materials: A penny for each child.

Teacher directions: Flip your coin.

Questions: What did you get, Shawn? A head. What did you get, Pam? A tail. What did you get, Joey? A tail. Did anyone get something besides a head or a tail? How many possible outcomes are there?

Conclusion: Since it seems that we are just as likely to get a head as a tail and that there is 1 chance out of 2 of getting a head, we say: "When a coin is flipped, the probability of getting a head is $\frac{1}{2}$."

Example 15-15

A concept-development activity for probability

C-3

Materials: Four cards, like these.

Teacher directions: Mix up the cards and turn them over. Now pick one.

Questions: What did you get? Your chances of getting a 2 are 1 out of 4. What are your chances of getting a 3? The chances of getting an even number are $\frac{2}{4}$. What are the chances of getting an odd number?

Example 15-16

A concept-development activity for probability

C-3

Materials: One die for each child.

Teacher directions: Toss your die.

Questions: What did you get? What are the possible outcomes? How many possible outcomes are there? What is the probability of getting a 2?

Conclusion: Since it seems that we are just as likely to get any one of the six numbers of dots as any other, there is 1 chance out of 6 of getting a 2. So the probability of getting a 2 is $\frac{1}{6}$.

New questions: What is the probability of getting 8 dots? (0) What is the probability of getting less than 4 dots? What are the possible numbers? (1, 2, 3) How many of these numbers are there? (3) So the probability of getting less than 4 is $\frac{3}{6}$, or $\frac{1}{2}$.

Use this picture to represent what we've just found out. Shade in the columns for numbers less than 4. You've shaded $\frac{3}{6}$ of the big rectangle.

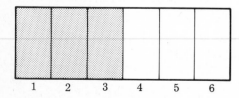

New questions: What is the probability of getting an odd number of dots? What are the possible numbers? How many are there? Shade in the columns for odd numbers. You've shaded in $\frac{3}{6}$ of the big rectangle.

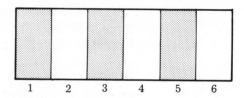

Conclusion: There are 3 chances out of 6 to get an odd number of dots, so the probability is $\frac{3}{6}$, or $\frac{1}{2}$.

After many experiences like the ones in the preceding examples, children can move on to more complicated situations. They must clearly understand the questions that are being asked, and they should always list the possible outcomes. Here are some examples.

Example 15-17

A concept-development activity for probability

Materials: Cards with the numbers 4, 5, 6, and 7.

Teacher directions: Draw two cards. Find the sum of the numbers on your cards.

Questions: What's the probability that our sum will be odd? First, list all of the possibilities.

Conclusion: There are 4 chances out of 6 to get an odd sum, so the probability of getting an odd sum is $\frac{4}{6}$, or $\frac{2}{3}$.

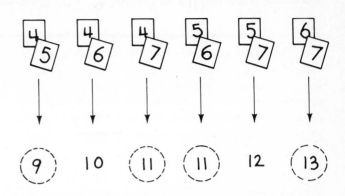

Experiments involving probability are also good topics for class projects. Collecting data from flipping two coins a large number of times will yield a close approximation to $\frac{1}{4}$ for the probability of two heads occurring. Children should first analyze the possible outcomes when two coins are flipped and make predictions concerning results.

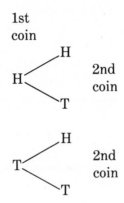

Example 15-18

A concept-development activity for probability

Materials: Cards with the numbers 1 to 9.

Teacher directions: Select two cards.

Questions: What's the probability that the two numbers will have no common factors other than 1? First, list all of the possibilities.

*1, 2						
*1, 3	*2, 3					
*1, 4	2, 4	*3, 4				
*1, 5	*2, 5	*3, 5	*4, 5			
*1, 6	2, 6	3, 6	4, 6	*5, 6		
*1, 7	*2, 7	*3, 7	*4, 7	*5, 7	*6, 7	
*1, 8	2, 8	*3, 8	4, 8	*5, 8	6, 8	*7, 8
*1, 9	*2, 9	3, 9	*4, 9	*5, 9	6, 9	*7, 9	*8, 9

The pairs that are asterisked have no common factors other than 1.

New questions: How many pairs in all? 36. How many asterisked pairs? 27. So the probability is $\frac{27}{36}$, or $\frac{3}{4}$.

Ideas for enrichment: Interesting data

There is a wealth of available resources that relate to these topics. Every newspaper and practically every magazine we get has interesting statistical information presented in tables, charts, and graphs. Keep your file of interesting uses (and misuses) of data current. Be a clipper and saver and have a recent almanac handy! These pieces of information are nice to use in word problems you construct or they can start class discussions. The statistical approach to collecting the data is certainly a worthwhile discussion topic as well as an appropriate analysis of the data.

The information below concerns the 1985 behavior of Americans *daily*. To answer the question, "How much is this per person?" use a population of 240 million for the United States.

1. Americans spend $700 million on recreation and entertainment.
2. Americans drink 15.7 million gallons of beer and ale.
3. Americans use 450 billion gallons of water in their homes, factories, and farms.
4. Americans buy 4 million eraser-tipped pencils.
5. Americans eat 170 million eggs.
6. Americans eat more than 400 000 bushels of bananas.
7. Americans print 62.5 million newspapers on 23 000 tons of newsprint.
8. Americans spend $14.3 million on lottery tickets.
9. Americans put 70 million quarters into arcade game machines.
10. Americans take 30 million sleeping pills.
11. Americans eat 12.5 million pounds of cheese.
12. Americans use 2.5 million tubes of toothpaste.
13. Americans send 9 million greeting cards.
14. Americans eat 3 million gallons of ice cream and ice milk.
15. Americans eat 815 billion calories of food. This is roughly 200 billion more than they need to maintain a moderate level of activity. That's enough extra calories to feed everyone in Mexico, a country of 80 million people!

The computer corner

In this corner, we present a program called HIGHLOW, which lets students play a game of chance. The program keeps an ongoing record of correct and incorrect responses and indicates whether or not each response is the most probable one. After several turns, the player is given a report that analyzes strategies used to make guesses. Run the program several times using guesses that are the most probable. Then play the game using guesses that are not probable. Think of ways you could use the program to build probability concepts. The following is an example.

Example 15-19
Using probability to make choices

Materials needed: The program HIGHLOW on the MGB Diskette.

Notes: The teacher must load the program into the computer's memory.

Teacher directions: This is a game of chance, called "High, Low, and In-Between." The computer will show you three cards, each with a number between 1 and 99, including 1 and 99. First, the computer will display two of the cards. Your job is to guess whether the number on the third card will be between, less than, or greater than both numbers on the first two cards.

Example:

First two cards

| 17 | | 42 | Your guess, "higher."

Last card

| 82 | You win! 82 is greater than either 17 or 42.

Points are awarded for correct answers. The player with the highest score after ten turns is the winner.

■ Computer tips

A game can often do more to get a concept across to children than other kinds of instructional material. Computer games can be especially useful because they can give directions, monitor responses of players, keep score, and prevent conflicts that can arise with traditional game formats. This saves the teacher lots of time that might otherwise be spent arranging materials, giving directions, and settling disputes. The computer can also be a *player* when a child wants to play and there's no one else available.

In closing . . .

In this chapter, we've presented some simple ideas from statistics and probability that can be developed with children in the elementary school. Emphasis has been on the collection of data and its subsequent interpretation. Children should be encouraged to formulate hypotheses and to devise procedures to test their hypotheses. Teachers should thus be able to present interesting and appropriate problems for children to explore.

It's think-tank time

1. Determine which of the activities in this chapter involves higher-order thinking skills and could be used to promote problem solving by elementary school children.
2. Modify Example 15-1 so that it fits discussions of probability.
3. Change the experiment in Example 15-6 as follows. You're given the information that the ratio of blue marbles to green marbles is 3 to 2. The marbles are to be drawn out of the bag two at a time. Predict how many times out of 100 draws a child will draw two green marbles.
4. Describe a problem, appropriate for fifth- or sixth-grade students, that (a) makes use of a circle graph to represent the data and (b) makes use of a line graph to represent the data.
5. List ten more sentences that could be used in Example 15-9.
6. Suppose that you flip three coins.
   a. List all of the possible outcomes.
   b. What is the probability of getting three heads.
   c. What is the probability of getting two heads and one tail?
7. Suppose that you toss two dice.
   a. What is the probability of getting a 7 for your sum?
   b. What is the probability of getting less than 4 for your sum?
   c. What is the probability of getting a prime number for your sum?
   d. What is the probability of getting an even number for your sum?

8. Suppose that you were to spin a spinner like the one shown below.

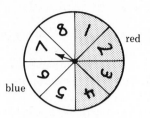

   a. What is the probability of getting an even number?
   b. What is the probability of getting blue?
   c. What is the probability of getting an even number on blue?
   d. If you spin twice, what is the probability that the sum of your two spins will be 10?
9. Often, we cannot determine the *exact* probability for an event. For example, the probability of the sun "coming up" tomorrow is *close to* (but not exactly) 1, and the probability of an elephant walking down the hall is *close to* 0.
   a. Make up five more examples in which the probability of an event is *very close to* 1.
   b. Make up five more examples in which the probability is *very close to* 0.
10. Have students play the game in Example 15-19 and direct them to make both probable as well as improbable choices. Have them keep a record of their results. After collecting a sufficient amount of data, have them compare the ratios of correct responses for probable choices to that of correct responses for improbable choices. Have them discuss their findings.
11. Use the statistical information in Ideas for enrichment and write ten word problems.

→ Suggested readings

Baretta-Lorton, Mary. *Mathematics Their Way.* Menlo Park, Calif.: Addison-Wesley, 1976. (Chapter 6.)

Billstein, Rick. "A Fun Way to Introduce Probability." *The Arithmetic Teacher*, January 1977, pp. 39-42.

Bright, George W., Harvey, John G., and Wheeler, Margariete Montague. "Fair Games, Unfair Games." In *Statistics and Probability for Elementary Teachers.* 1981 Yearbook of the National Council of Teachers of Mathematics. Reston, Va., 1981.

Choate, Stuart A. "Activities in Applying Probability Ideas." *The Arithmetic Teacher*, February 1979, pp. 40-42.

Christopher, Leonora. "Graphs Can Jazz Up the Mathematics Curriculum." *The Arithmetic Teacher*, September 1982, pp. 28-30.

Curcio, Frances Rena. "The Effect of Prior Knowledge, Reading and Mathematics Achievement, and Sex on Comprehending Mathematical Relationships Expressed in Graphs." (Doctoral dissertation, New York University, 1981.) *Dissertation Abstracts International, 42A*(1982): 3047-3048A.

Davies, Pauline. "Creative Mathematics: Two Activities." *The Arithmetic Teacher*, October 1976, pp. 418-424.

Enman, Virginia. "Probability in the Intermediate Grades." *The Arithmetic Teacher*, February 1979, pp. 38-39.

Gawronski, Jane Donnelly, and Mcleod, Douglas B. "Probability and Statistics: Today's Ciphering." In *Selected Issues in Mathematics Education.* National Society for the Study of Education and National Council of Teachers of Mathematics, 1980. (Pp. 82-90.)

Gilbert, Robert K. "Hey Mister! It's Upside Down." *The Arithmetic Teacher*, December 1977, pp. 18-19.

Hoemann, Harry W., and Ross, Bruce M. "Children's Understanding of Probability Concepts." *Child Development*, March 1971, pp. 221-236.

Horak, Virginia M., and Horak, Willis J. "Let's Do It, Collecting and Displaying the Data Around You." *The Arithmetic Teacher*, September 1982, pp. 16-20.

Horak, Virginia M., and Horak, Willis J. "Let's Do It, Take a Chance." *The Arithmetic Teacher*, May 1983, pp. 8-15.

Jacobson, Marilyn Hall, and Tabler, M. Bernadine. "Ideas, Lift-Off Gameboard." *The Arithmetic Teacher*, February 1981, pp. 31-36.

Jones, Graham. "A Case for Probability." *The Arithmetic Teacher*, February 1979, pp. 37, 57.

Juraschek, William A., and Angle, Nancy S. "Experiential Statistics and Probability for Elementary Teachers." In *Teaching Statistics and Probability.* 1981 Yearbook of the National Council of Teachers of Mathematics. Reston, Va., 1981.

Klitz, Ralph H., Jr., and Hofmeister, Joseph F. "Statistics in the Middle School." *The Arithmetic Teacher*, February 1970, pp. 35-36.

Lai, Theodore. "Bingo and the Law of Equal Ignorance." *The Arithmetic Teacher*, January 1977, pp. 83-84.

Leffin, Walter W. "A Study of Three Concepts of Probability Possessed by Children in the Fourth, Fifth, Sixth and Seventh Grades." *Technical Report No. 170.* Madison, Wis.: Wisconsin Research and Development Center for Cognitive Learning, 1971.

O'Neil, David R., and Jensen, Rosalie. "Let's Do It, Looking at Facts." *The Arithmetic Teacher*, April 1982, pp. 12-15.

Pincus, Morris, and Morgenstern, Frances. "Graphs in the Primary Grades." *The Arithmetic Teacher*, October 1970, pp. 499-501.

Romberg, Thomas A., and Shepler, J.L. "Retention of Probability Concepts: A Pilot Study into the Effects of Mastery Learning with Sixth Grade Students." *Journal for Research in Mathematics Education*, January 1973, pp. 26-32.

Shaw, Jean M. "Let's Do It, Dealing with Data." *The Arithmetic Teacher*, May 1984, pp. 9-15.

Shulte, Albert P. "A Case for Statistics." *The Arithmetic Teacher*, February 1979, p. 24.

Shulte, Albert P. "Teaching Statistics and Probability." 1981 Yearbook of the National Council of Teachers of Mathematics. Reston, Va., 1981.

Siegel, Murray H. "The Statistical Survey: A Class Project." In *The Agenda in Action*, 1983 Yearbook of the National Council of Teachers of Mathematics. Reston, Va., 1983.

Souviney, Randall J. "Quantifying Chance." *The Arithmetic Teacher*, December 1977, pp. 24-26.

Stone, Janine S. "Place Value and Probability (with Promptings from Pascal)." *The Arithmetic Teacher*, March 1980, pp. 47-49.

Wall, Curtiss E., and White, Ann M. "Reading, Interpreting, and Constructing Tables, Charts, and Graphs." *Monograph of Virginia Council of Teachers of Mathematics*, 1978.

Webb, Leland D., and McKay, Jamie D. "Making Inferences from Marbles and Coffee Cans." *The Arithmetic Teacher*, September 1979, pp. 33-35.

16

Computers and Mathematics Instruction
Status and Direction

■ *In this chapter, we discuss the implications for instruction of microcomputer technology. Since changes in technology are occurring so rapidly, an accurate picture of "current practice" would be out-of-date before the ink on this page dries. So we feel an obligation to discuss specific uses in relationship to general categories that will continue to have potential. We will define broad categories for instructional computing and demonstrate programs for each. Our programs are small and limited and not intended to compete with commercial programs. Each, however, has its own important message. The categories, the programs, and all the messages should help you build a frame of reference for viewing instructional computing that will not erode with time and the change of technology.*

You will notice that we don't talk much about programming. Why? Because this is not a chapter on programming. This is a chapter concerned with improving mathematics instruction. We admit that there are important programming concepts that children should learn, and we believe that the act of writing and running your own program can be magic for most students whatever the subject matter. But we cannot allow the teaching of programming to crowd out important subject matter development.

Before we leave the chapter, we discuss important issues related to instructional computing. These issues are not mathematics specific. They are issues concerning equity, appropriate uses of computers at different age levels, and so on. All educators should be concerned with and willing to deal with these issues when necessary. This chapter is a long one, but there's a lot to say about this topic. ■

Teacher goals

1. Be able to describe the major parts of a microcomputer using appropriate computer terminology.
2. Be able to explain the difference between primary and secondary memory.
3. Be able to explain the difference between software and hardware.
4. Be able to define three categories for instructional computing.
5. Be able to describe programs for Computer Directed Instruction, giving examples of desirable features.
6. Be able to describe programs for Computer Enhanced Instruction, giving examples of desirable features.
7. Be able to describe programs for Computer Managed Instruction, giving examples of desirable features.
8. Be able to describe a general model for instruction that integrates Computer Directed, Enhanced, and Managed Instruction.
9. Be able to describe the role of the teacher in a computer-managed instructional environment.
10. Be able to discuss issues related to instructional computing.

Almost every adult in our country has had some experience with a microcomputer. This experience may be limited to toying with one at a department store counter or watching one in use in an office or home or on television. On the other hand, this experience may have involved hours "online" completing intricate computer tasks. Even those who have never touched a microcomputer or actually seen the real thing can probably identify one and probably have notions of how they're used.

When considering how microcomputers might be used to improve instruction, your concept or view depends on your previous experiences with microcomputers. Other personal factors also affect your view, such as your academic background, interests, and willingness to accept change. Even the type and power of the machine you have used or seen will affect your view.

Many people think of a microcomputer as a tool that can be programmed to teach—that is, the computer provides the learner with instructions, asks questions, evaluates answers, and provides feedback just as does a live teacher. Other people see the computer as a tool that manages the paper work of instruction, calculating grades, preparing reports, and keeping inventories. Still others think of the computer as the object of study. They picture classes full of students sitting behind micros learning to write computer programs, learning to identify parts of the computer, or learning historical facts related to computer development.

All of these views are valid, but each alone is far too limited. Microcomputer technology has vast potential for improving the quality and extending the limits of instruction. To realize this potential, educators must build a framework in which to study and design instructional practices. This framework must be comprehensive, flexible, and built to accommodate constant change.

In this chapter, we will explore this potential. We certainly can't build an entire framework, but we can help you select and arrange a few building blocks. We will begin by discussing some of the general features of microcomputers and describing the tasks they can perform well. Then we will define general categories of instructional use, giving examples of computer programs that belong in each category. Of course, the examples are related to mathematics teaching. We will end the chapter by listing critical issues related to using computers for instruction that we feel all educators should consider.

What is a microcomputer?

What is a microcomputer? What can it do that humans can't?

A microcomputer is a machine that operates using electricity. Because of this, it can operate at high speed. Ideally, electrical current flows at the speed of light—187,000 miles per second—so the microcomputer's operation approaches this speed. Such variables as heat resistance slow things down. Nonetheless, micros can do their work more than just quickly! They can't do anything humans can't; they simply perform some tasks faster and more accurately.

A microcomputer has four main parts—the *microprocessor*, the *primary memory*, and the *input* and *output devices*. For any serious use, a micro must also be able to access *secondary memory devices*. The input and output devices are just what you would expect: input devices allow you to enter information, and output devices allow the computer to send information to the outside world. The *keyboard* is a common input device, and a *monitor* is a common output device. A *disk drive* is a device that can input information stored on a *floppy diskette* or a *hard disk* into the computer's memory. A disk drive can also accept information outputted from computer memory and store this output on a floppy diskette or hard drive.

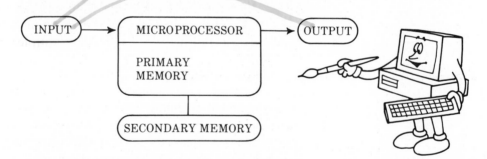

A microcomputer can't do anything by itself. A *program*, a set of well-organized instructions that the computer can interpret, must be entered into the computer's primary memory. Once a program is in memory, the instructions in that program can be executed. There are several ways a program can be inputted into the computer's memory. You can *key in* a program, using the keyboard. On the other hand, a program may be stored on a *secondary storage device*, such as a diskette. The disk drive, much like a record or tape player, finds and reads the program on the diskette and enters it into the computer's memory. This sure beats typing a program every time you want to use it!

The microprocessor is the most important part of the microcomputer. You might say that it is the computer's brain. When information is entered or inputted, it goes to the microprocessor, which interprets the information and routes it to primary memory or to other devices installed in or connected to the computer. The processor is like a traffic policeman in that it regulates, directs, and controls the flow of information.

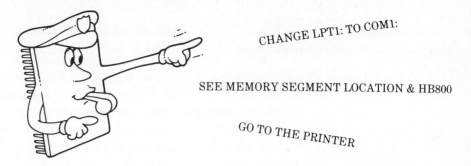

CHANGE LPT1: TO COM1:

SEE MEMORY SEGMENT LOCATION & HB800

GO TO THE PRINTER

Years ago, processors were very large. But the microprocessors in today's machines are tiny, and thus the term *micro*processor. This development has made it possible to manufacture small, powerful machines that are inexpensive but which deliver the potential of this technology to the ordinary classroom.

What makes the potential so extraordinary? To appreciate the answer, we must consider two things. First, we must consider the difference between *software* and *hardware*. The input and output devices, the microprocessor, and primary memory are all computer parts referred to as hardware. Programs stored on floppy disks, tapes, or hard drives are called software. The ability to store programs externally as "software" makes it possible to develop and constantly refine an unlimited number of programs.

Earlier we said that microcomputers operate very quickly and are capable of discriminating between current flowing at high or low voltages. These capabilities make it possible for a microcomputer to answer simple true or false questions quickly. For example, if you drew a card from a regular deck of playing cards and indicated "true or false" to the statement "The card is a spade," you would be making the kind of decision that computers can make. So why do computers appear to be so intelligent? Because of their speed, they can make millions of simple true or false decisions in very little time. Taken together, these decisions can lead to the completion of large complicated tasks that might take humans years to complete.

Finally, because of its ability to make decisions and because it operates on electricity, the microcomputer can be connected to any other electrical device. Once connected, the microcomputer has the potential to operate and regulate the other device. Think about it! This renders the microcomputer as the ultimate medium, one that will have greater effect on education than any medium before it!

Using computers to improve instruction

Before microprocessors and microcomputers were developed, computers were gigantic, expensive to operate, and difficult to program. Nevertheless, people with vision saw the potential for instruction. Excited by this, they began to experiment, and from this, a category of instruction known as Computer Assisted Instruction (CAI) grew. As the years passed and technology improved, CAI gained wide acceptance among educators. But unfortunately, as technology improved, the category was expanded to include almost any application related to instruction. Because of the wide variety of constantly changing applications, the term has never been clearly defined. To discuss instructional computing intelligently, however, we need clearly defined terms with which to tag things. Thus, we will replace Computer Assisted Instruction with three categories: Computer Directed Instruction (CDI), Computer Enhanced Instruction (CEI), and Computer Managed Instruction (CMI).

In the next three sections, we will explore CDI, CEI, and CMI. For each category, we will demonstrate programs that meet the category's criteria and have instructional significance. The programs we will share are designed to demonstrate a few important instructional ideas. Although all the programs could serve useful purposes in the classroom, they are not designed as complete utilities. At the end of the chapter, in Think-tank time, you will be given activities and exercises that will reinforce the concepts that have been developed in these three sections.

Computer Directed Instruction

The definition of this category is implied by its name; a program in this category literally directs instruction. The program may present an entire lesson, a set of lessons, or part of a lesson. The assumption is that learners, by pressing the right keys, are able to move through lessons without outside help. This does not assume anything about the learner or the quality of the lesson. The definition simply states that once a CDI program has been installed in the memory of a computer, the necessary conditions for completing the lesson involve only the computer and the learner.

CDI programs can be used to accomplish many goals in mathematics, including the following: Developing concepts and skills, practicing concepts and skills, applying concepts and skills, solving problems, and simulating experiences.

A CDI program may deliver a simple practice lesson requiring short answers. On the other hand, the program may deliver a lesson that demands considerable time, requiring learners to apply both concepts and skills to solve problems. When choosing a CDI program for use in the classroom, we must evaluate it in two ways.

First, we must decide what educational purposes the program achieves and whether these purposes accommodate our instructional plan. If the program meets our instructional needs, then we must determine *what the computer program does that other traditional instructional materials can't.*

After all, if the instructional goal can be achieved using pencil and paper, then how can using a microcomputer be justified?

Second, we must decide whether the program is technically sound. Is the program well-formatted? Is the reading level appropriate? Is the program user-friendly? Will the program adequately handle mistakes the user might make? Are graphics and sound used appropriately? In this section, we are concerned with the first kind of evaluation, so we will demonstrate several CDI programs and list the characteristics of each.

■ ARRAY, ARRAY: A program for concept development

The first CDI program that we demonstrate is one we used in Chapter 4. This program is called ARRAY. Its purpose is to help students learn the basic multiplication facts while associating each fact with a rectangular array.

Here's what the program does. First, it asks for the date and the learner's name. Then the program provides a sequence of 20 exercises for the learner to complete. Each exercise uses questions to relate a multiplication fact with a multiplication model such as the following.

How many rows?

How many in each row?

How many in all?

Fill in the blanks.

$$\underline{\hspace{2cm}} \times \underline{\hspace{2cm}} = \underline{\hspace{2cm}}$$

Students must examine the model, then input the number of rows, the number in each row, and the number in all. The result is a multiplication number sentence. The number of rows and the number in each row are randomly generated by the program, although the number of rows and the number in each row cannot exceed 9. After each exercise, learners are told whether the response is correct. If correct, positive reinforcement is provided. If incorrect, the learner has another opportunity to do the exercise. When 20 exercises have been completed, a printed progress report can be obtained. A progress report looks like this:

Manard Muftit   Dec. 2, 1988

Manard, you missed 3 exercises out of 20. Your score is 85%. Here are the exercises that you missed:

Your answers	The correct answers
$6 \times 7 = 48$	$6 \times 7 = 42$
$9 \times 6 = 56$	$9 \times 6 = 54$
$7 \times 5 = 30$	$7 \times 5 = 35$

Correct your mistakes, try to improve. Then try the program again.

Notice that the report gives the student's name, the date, the number correct, the percentage correct, a message, and a list of errors in addition to the correct answers. If desired, the learner can get a printed copy of the report. Then the learner can elect to do another exercise set or exit the program. If you have the MGB Diskette, you should run this program a couple of times to see how the goals are accomplished. The file name for the program is ARRAY.

The following are features of the ARRAY program.

- The program certainly meets CDI conditions. After the program is installed in memory, it is possible for learners to take a lesson without outside help.
- Concept development and practice are provided through the use of models and appropriate questions. Each exercise demonstrates a model related to numbers in a multiplication sentence, thus constantly reinforcing the concept to be acquired.
- The program tells the learner whether or not responses are correct and allows incorrect answers to be corrected. If the learner gets the exercise wrong a second time, the program provides the correct answers and gives the learner a new exercise.
- The program provides a progress report and an option to get a printed copy. The report lists incorrect responses and gives corrections. This information may be used for remediation purposes.
- If the learner gets four consecutive exercises wrong, the program automatically exits the learner with a message for him or her to see the teacher. The program does not allow the learner to sit and practice "getting wrong answers."

Now let's determine features the program doesn't have.

- The program does not require the learner to apply concepts or solve problems.
- The program does not allow teachers to modify the types of exercises that are presented without actually rewriting substantial parts of the program. Nor does it allow teachers to select the level or number of exercises to be presented.

Assume that this program meets all the following technical requirements: It is formatted properly, the reading level is appropriate, the graphics and sound are used effectively, and so on. Instead of asking, *Is the program good?* we should ask, *Under what circumstances does the program have instructional value?*

The program builds and reinforces a simple concept, and will be effective in situations where learning basic multiplication facts is a goal. Once the teacher has introduced the basic facts, the program can be used to provide limitless practice with models that are accurate and neat. Herein lies the value of the program:

It can provide sound practice on a learning objective that most educators agree is important.

Precious teacher time is conserved that might otherwise be exhausted in providing drill in probably a less systematic fashion.

It provides records that keep teachers informed of learners' progress and difficulties.

What are the dangers in using such a program? First, such programs are often overused. Students become bored, and negative learning begins to take place. Even worse, such programs are used in place of other learning activities for developing higher cognitive abilities—problem solving and so on. Like most programs, this one should only be used when it meets specific instructional needs.

■ AUTHOR: A program that can be customized

The next program we demonstrate is one we prepared for Chapter 9. The learner is presented with a stimulus for which a response is to be entered. Along with the stimulus, the program provides a line of text that somehow clarifies the stimulus or gives a *tip* as to what answer is expected. The program compares the learner's response with three possible responses that are built into the program. In the event that the learner does not enter one of the built-in responses, the program simply tells the learner that he or she has entered an answer that the program is "not looking for." For example, the program might ask students to translate word problems or arithmetic information to number sentences or expressions:

MATH TRANSLATION
The computer is thinking of number expressions. It will give you hints that will help you guess each one. Read the hints carefully, then try to guess the expressions. You will get 3 tries for each expression. Good luck!

MATH TRANSLATION
2 bowls, each with 3 pears. A box has 5 more pears. How many in all?

Use a small "x" for multiplication. Don't forget parentheses.

Acceptable answers:

$(2 \times 3) + 5$
$(3 \times 2) + 5$
$5 + (2 \times 3)$
$5 + (3 \times 2)$

Run the AUTHOR program on the MGB Diskette to get a feel for how it works. Then return to this place in the text so that we can explain how you can modify the program to design your own lessons.

Designing your own lessons

A teacher using this program can develop custom lessons. To do so, the following materials must be prepared.

1. Prepare a set of directions. The directions you prepare should indicate what students are to do to complete the lesson. You can use up to 10 lines with no more than 35 characters in each line. These directions should be planned in advance and then entered into the BASIC program using DATA STATEMENTS starting with line 2100. (See Exercise 8 at the end of the chapter for a blank grid that can be used for planning directions.)

Here is how the directions for our lesson were prepared.

```
              MATH TRANSLATION
- - - - - - - - - - - - - - - - - - - - -
The computer is thinking of number
- - - - - - - - - - - - - - - - - - - - -
expressions. It will give you hints
- - - - - - - - - - - - - - - - - - - - -
that will help you guess each one.
- - - - - - - - - - - - - - - - - - - - -
Read the hints carefully, then try
- - - - - - - - - - - - - - - - - - - - -
to guess the expressions. You will
- - - - - - - - - - - - - - - - - - - - -
get 3 tries for each expression.
- - - - - - - - - - - - - - - - - - - - -
Good luck!
- - - - - - - - - - - - - - - - - - - - -
```

The tiny blanks are not typed into the DATA STATEMENTS. They are intended only as a guide to help you plan the directions.

2. Prepare the exercises.

For each, you must design (1) a stimulus, (2) a hint or tip that gives information about the stimulus, and (3) up to four acceptable responses. There are limitations. The stimulus and the tip are limited to two lines of 35 characters, and each of the responses is limited to 20 characters. The characters for acceptable responses must be entered exactly as they are to be used in the program. You cannot be careless using upper and lower cases of letters. Students using the program must also enter the appropriate cases of letters. If we design a program so that case is ignored, then JANUARY and january will be accepted as correct when we really want January. To aid you in planning, we have provided a form in Exercise 8 in Think-tank time.

Here is how the plan for an exercise in our lesson looks:

	Stimulus/Response
Stimulus:	2 bowls, each with 3 pears. A bag
	- -
	with 5 more pears. How many in all?
	- -
Hint/tip:	Use a small "x" for multiplication.
	- -
	Don't forget parentheses.
	- -
Built-in responses:	$(2 \times 3) + 5$
	- - - - - - - - - - - - - -
	$(3 \times 2) + 5$
	- - - - - - - - - - - - - -
	$5 + (2 \times 3)$
	- - - - - - - - - - - - - -
	$5 + (3 \times 2)$
	- - - - - - - - - - - - - -

You must design a minimum of ten exercise sets, although the maximum depends upon your machine and its memory limitations. The memory limitation of your disk is also a consideration. Regardless, you should be able to design as many as 100 exercises and not run into trouble. The DATA STATEMENTS you will use for your lesson begin at line number 2100 in the program. Before entering your statements, you must copy the program onto a blank diskette so that the original MGB program is not damaged. This procedure differs by machine, so you will need to consult the documentation for the MGB Learning Diskette.

Once you have a copy of the program on a new diskette, you are ready to enter your lesson. Line numbers used for your DATA STATEMENTS should be in increments of 10. This allows space to later insert statements that might be overlooked. As you complete the directions, the stimuli, and the responses, you must enter DATA STATEMENTS that read ZZZ DATA "*END*" where the Z's represent the program line number. (It is important that you do not punctuate within the quotation marks for "*END*"; if you do, the program will *not* run.) This statement tells the computer that you are finished with that portion of the lesson. Each time the program runs, ten random exercises are chosen so that there is always variation. Here is how the DATA STATEMENTS for directions and one exercise in our lesson look:

```
                            DATA STATEMENTS
   98 REM ** Directions start here.
   99 REM ******************************************************
 2100 DATA ''MATH TRANSLATION''
 2110 DATA ''The computer is thinking of number''
 2120 DATA ''expressions. It will give you hints''
 2130 DATA ''that will help you guess each one.''
 2140 DATA ''Read the hints carefully, then try''
 2150 DATA ''to guess the expressions. You will''
 2160 DATA ''get 3 tries for each expression.''
 2170 DATA ''Good luck!''
 2180 REM ** Exercises start here.
 2190 REM ******************************************************
 2200 DATA ''2 bowls, each with 3 pears. A bag''
 2210 DATA ''has 5 more pears. How many in all?''
 2220 DATA ''Use a small 'x' for multiplication.''
 2230 DATA ''Don't forget parentheses.''
 2240 REM ** Acceptable answers start here.
 2250 REM ******************************************************
 2260 DATA ''(2 × 3) + 5''
 2270 DATA ''(3 × 2) + 5''
 2280 DATA ''5 + (2 × 3)''
 2290 DATA ''5 + (3 × 2)''
 2300 DATA ''*END*''
```
(Also read the REM STATEMENTS in the program's listing; these statements will also help you design a lesson.)

With a little imagination, a teacher can do a lot with this program. Here are some examples:

- Provide sets of pictures made from keyboard characters, then give the number name for the set. Have students enter the corresponding numeral.
- Provide a row of characters such as UUUUU*UU. Have students identify the ordinal position from the left that the "unlike" character occupies.

- Give students numbers less than 1000. Have them round the numbers to the nearest ten.
- Give number names such as five one-hundredths. Have students write the correct decimal numeral.

Now let's look at the positive qualities of this program. In a nutshell, the most important qualities are as listed below.

- The teacher can design lessons of his or her choice.
- The program monitors student progress. A printed report with traditional scores as well as a list of student errors is provided. Information on errors can help teachers plan lessons based on individual needs.
- The program has the potential to provide more than drill and practice. Creative teachers can plan activities that foster concept development and problem-solving. Generally, however, the program is intended as a practice or evaluation program. It can be used in a variety of ways: homework, make-up quizzes, special practice, and so on.

A commercial program of this type should employ the use of disk files instead of data statements. Had we designed the program in this way, we would not have been able to allow you the experience of modifying the program using DATA STATEMENTS. We feel that the trade-off is a good one. But when you buy a commercial program, you should insist on a more sophisticated method of modifying the program.

Now let's determine characteristics that the program doesn't have:

- The program does not teach anything new. The learner must have some knowledge of the lesson's content to be successful.
- The program is very limited in terms of the exercises that can be delivered.
- Even though more than one right answer can be entered, acceptable answers must be short.

■ The Rule Genie: A program for thinking

The programs we have dealt with so far have not been specifically designed to foster thinking abilities or problem-solving strategies. We have mainly been concerned with concept or skill development, and although both are extremely important, neither is ultimately important if one does not learn to use these abilities in problem-solving modes to manage life events. Many programs that provide instruction should stress the development of higher cognitive skills, providing opportunities for students to find similarities and differences, see relationships, draw conclusions, make generalizations, describe processes, and so on. So, in this section we will present a simple program in game format. This game requires learners to look for patterns and then find and test mathematical rules.

Playing the game is simple. The program's genie announces that it has a mathematical rule that can be applied to whole numbers within certain limits. The player enters numbers to test the rule. After each of the numbers is entered, the program transforms it by using the rule. The computer then gives the player a number and asks the player to try his or her luck at guessing the rule by applying it to the given number and

entering the results. The player is given several opportunities to observe results when the computer's rule is applied and several opportunities to guess and test guesses.

Playing the game goes something like this:

Genie says, "I have a rule. Enter a number between 0 and 10, and I will put it in the Rule-a-hoop machine and change it."
Player enters the number 5.
Genie uses the Rule-a-hoop machine to transform 5 to 15.
Genie says, "Now try to use my rule to transform 7."
Player enters 21.
Genie responds, "Wrong! Wrong! That's not the answer I'm looking for! May your pet camel grow three humps! Try again."
Player enters 7.
Genie uses the Rule-a-hoop Machine to transform 7 to 19, and then gives player the number 8 to transform.
Player enters 21.
Genie says, "Right. I thought you'd never say that!"

This program has two significant qualities. First, the program requires learners to use abilities other than simple recall or translation. The learners must do a little thinking to draw and test conclusions. Second, the program is in a game format. This is complemented by the use of sound and computer conversation, which motivates students to stay on task and excel.

This program does not teach students new content, but the quality and quantity of monitored practice would be difficult to manage with traditional instructional materials.

The file name for the program is GENIE. Run the program, see how it works, and see if you can outwit the genie in the computer.

■ TOYSTORE: A program of simulation

As we have seen, Computer Directed Instruction programs can help achieve a number of instructional goals. With the appropriate characteristics, CDI programs can aid in developing concepts and related skills and can foster the growth of problem-solving abilities. CDI programs can also provide *simulations*, which are very powerful. A computer simulation is a program that allows the user to have artificial experiences that resemble real experiences. Computer simulations are used whenever real experiences are not feasible or desirable. The real experience may be impossible to create, too dangerous, too expensive, technically difficult to arrange, or too hard to manage. For example, it would be hard to let students experience taking off and landing in a plane in the classroom, but a computer program could be used to simulate such experiences. So, in this section we will present a simulation that involves a skill that all citizens should acquire—the ability to make change. It's possible to provide this experience in the classroom without a computer, but it wouldn't be as effective or as efficient because the computer can deliver and monitor a wider variety of experiences and limitless repetitions. Run the program several times. See if you can make accurate change.

Let's review some of the characteristics of the program. The program randomly chooses a toy that a customer is going to buy and a corresponding price for the toy. The learner is told that the customer is going to pay with a certain number of dollars. The learner is to make change, starting with the smallest realistic coin. The program will allow the learner to use a maximum number of each coin—4 pennies, 1 nickel, 2 dimes, and 3 quarters. The program does not deal in half-dollars. The learner is prompted with a "How many" question for each coin, starting with pennies; e.g., "How many pennies will you use?" Often the most efficient way to make change involves *not* using a particular coin. For example, take the problem of making change from a dollar when an item costs $.62. The most efficient way would be to give 3 pennies, 0 nickels, 1 dime, and 1 quarter. In this case, the learner would have to enter a "0" for the number of nickels. At the start of the program, learners are given an example exercise and are warned that 0's may have to be entered. If the learner has difficulty with two consecutive exercises, the computer will make a printed copy of the incorrect exercises and exit the program, telling the student to get help. We leave it as an exercise for the reader to list other important features of this program. The following is a screen showing one of the TOYSTORE activities.

```
┌─────────────────────────────────────────────────────────────┐
│     ┌─────────────────────────────────────────────────┐      │
│     │              THE TOY STORE                      │      │
│     └─────────────────────────────────────────────────┘      │
│                                                              │
│     Robert would like to purchase a...                       │
│                 teddy bear                                   │
│     The price is           $4.55                             │
│     Robert pays you        $5.00                             │
│     Your job is to make the correct change                   │
│     ──────────────────────────────────────────────────       │
│                                                              │
│     How many pennies?   (0-4)    0                           │
│     How many nickels?   (0-1)    0                           │
│     How many dimes?     (0-2)    2                           │
│     How many quarters?  (0-3)    1                           │
│                                                              │
└─────────────────────────────────────────────────────────────┘
```

■ Computer Directed Instruction: A wrap-up

In this section, we have described CDI programs and have illustrated some of the qualities that these programs may have. We have stressed the importance of carefully examining a program's qualities to determine how well particular instructional needs are satisfied. At the risk of being redundant, we close this section by listing a consideration that one should take into account when shopping for instructional software. This list is certainly not exhaustive and the items are not mutually exclusive. We repeat that teachers should not subjectively evaluate software as "good" or "bad." They should decide what the instructional needs are and objectively list the desired qualities of lessons that meet these needs. Then they should look for software and other materials that can be organized to best satisfy those needs. No more ado! Here's the list.

- A CDI program can provide initial instruction. This instruction may involve concept development where understanding is stressed or it may involve skill development where rote procedures are emphasized. In general, a program that develops understanding is more valuable than one that stresses memorization. A program that stresses skill development must have instructional, motivational, or management qualities that justify its purchase and use.

- A CDI program can provide reinforcement or practice. This practice may involve exercises that require learners to apply important concepts or perform memorized routines. Generally, a program is more valuable when it requires the learner to practice applying concepts.

- A program can provide problem-solving opportunities where learners are required to find similarities and differences, discover relationships, draw conclusions, and synthesize information or experiences. Time in the ordinary classroom is limited and the average teacher's time is divided among many students. So, it is obvious that choosing such programs should be a high priority on any teacher's software shopping list.

- A CDI program can provide simulations. Using computer simulations allows teachers to provide educational experiences that might otherwise be impractical or impossible to provide in the classroom. Simulations can provide opportunities for students to apply skills, concepts, and problem-solving abilities in unique ways.

- A CDI program can deliver exercises that are randomly chosen within specified limits. This is convenient because of the large number of different exercises that can be generated and monitored. These programs are often limited and can become boring if they repetitiously drum away at the same type of exercise. On the other hand, if they possess sound motivational or management features, they can provide systematic practice that ordinarily eats up important teacher time.

- A CDI program can be designed so that teachers can input their own lessons. This valuable feature allows teachers to use the program in a wide variety of circumstances with learners having different needs. These programs are often limited because only a small number of exercises can be conveniently entered. To avoid monotony, a teacher must design a variety of lessons, each having a large number of student experiences. Better yet, these programs can be used to let students design and exchange lessons.

- A CDI program can keep track of student performance in a computerized lesson. Scores, records of exercises missed, as well as "wrong answers," can be stored or printed. Using these data, teachers or even computers could plan further instructional experiences for learners.

- A CDI program can check and control learner input. This feature can be used to exit students from programs when it becomes obvious that they cannot perform the necessary skills or concepts. This feature can also be used to provide experiences where only *right answers* are accepted. This becomes very valuable when other features are present that allow learners to intelligently correct errors and arrive at appropriate responses.

Well we could go on, but by now you have the message. The bottom line is that the microcomputer is a miraculous tool that can do things we've only begun to consider. Unfortunately, programs are written by people who may be locked into past experiences. Thus, teachers need to demand that CDI programs lead to the acquisition of substantial instructional goals that can't be efficiently and inexpensively achieved without the use of the computer. We don't need high priced page-turners. To date, most of the software developed expressly for education has been CDI programs. In many instances, the quality has been poor, and more often than not, the programs have emphasized drill at the exclusion of concept development, problem solving, and simulation. Many of the programs that are currently on the market lack many of the features we have discussed. Moreover, educators have not yet realized the potential of using computer programs to produce other types of instructional tools. So, we will end this section and move on to discuss that potential.

Computer Enhanced Instruction

Computers can benefit instruction in many ways other than by directing it. To appreciate this, you can't think of students behind keyboards. Rather, you must think of ways that the computers can improve the instructional environment and the objects of instruction. The term *Computer Enhanced Instruction* (CEI) pertains to applications that indirectly affect instruction and that might be difficult or impractical to access without computer aid. These applications range from utilities for constructing teaching aids and instructional support materials to software tools that support the self-directed studies of students. The best way to get a feel for this category is to examine a few specifics.

To start, let's look at the objects of instruction—that is, those things that are used in the classroom to directly influence learning. Basically, these objects fall into two categories: *teaching aids* and *learning aids*. Teaching aids are objects used to clarify instructional ideas or provide demonstrations. For example, using transparencies provides visual impact for a lesson. But transparencies are not easy to prepare and special skills are often needed. And no matter how expert you are at making them, the job is time consuming.

Learning aids are tools used by students as they carry out a lesson. A learning aid may have visual, auditory, or kinesthetic impact. It may be a model such as a clock with moveable hands, a game that provides educational experiences, or a set of cards that students arrange or stack in certain ways. When you get right down to it, a simple activity sheet reproduced from a black line master is a learning aid.

The distinction between a teaching and learning aid is artificial. What is important is that you become aware of a range of aids for stimulating learning that can be easily constructed using computers.

Classroom aids are either purchased or constructed by the teacher. Most school budgets are limited, so teachers either make their own aids or go without. Many teachers lack confidence in their abilities to construct aids, and often feel that they lack both talent and time. They often resort to last-minute sketches on a transparency or chalkboard. They pick up objects that are easily found in the environment and use these as manipulatives. Many times, the resulting aids are flawed and unattractive. Unfortunately, a lesson supported by poorly constructed aids can lack motivational qualities and can lead to faulty concept development.

With computers, this needn't be the case. The availability of commercial programs makes it easy and inexpensive to create useful instructional tools. If you have programming skills, you can write simple but useful programs to construct teaching materials. If you can't program or if you don't want to, you can enter programs written specifically for the machine available to you. Such programs can be found in a growing number of publications. *Generic software tools* can also be used to make classroom aids. What are these? They are general-purpose programs such as word processors, graphics programs, spreadsheets, and data-base managers.

There are plenty of sources. But to tap them you must take two steps.

1. You must become aware of the types of aids microcomputers can produce.
2. You must learn to use computer tools to produce the aids.

The rest of this section will help you take these two steps! We will describe programs designed for CEI. Some of these generate specific learning aids while others are more general and can be used for a variety of things.

■ Models for fractions: A program that draws pictures

The first program we describe is called FRACTION MODELS. This program can be used to construct aids that illustrate fractional ideas. When you run the program, a menu appears to give you several options:

1. Draw Models—Alike and Unpainted
2. Draw Models—3 Different Denominators
3. Load a Model
4. Get Instructions
5. Exit

Using option "1," you can produce pictures with six regions illustrating fractions for a denominator of your choice. Here is an aid with regions divided into sixths:

Using option "2," you can produce a picture with regions for three different denominators—say 3, 4, and 12. When you use this option, you can elect to have parts of the regions painted.

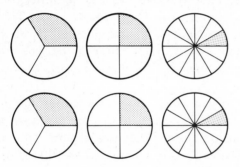

Option "4" provides instructions for operating the program. The program lets you SAVE (S), LOAD (L), or PRINT (P) pictures. Once models have been created, they can be used to make material sheets around which activities can be built. The teacher can also make accurate and attractive artwork for transparencies and bulletin-board displays. By reducing the pictures, artwork for tests or other assignments can easily be constructed. Here are sample activities.

Example 16-1
Demonstrating fractional relationships

Directions: Use Option 1 of FRACTION MODELS and choose a denominator of 12 to make sheets like the one that follows. Also make a set of task cards that indicate fractional relationships that students are to demonstrate.

Teacher directions: Take several material sheets. Color regions of the material sheet to demonstrate the number sentences on the task cards.

Material sheet

Task cards

Here is another activity that the models can be used for. This time, you use option "2" of the program to produce regions for different denominators—say 2, 3, and 12.

Example 16-2
Writing problem situations for fractional models

Materials needed: Cards showing two or more painted regions related to different denominators.

Teacher directions: Examine each card. Make up a story for each one. Be sure the story relates to each of the fraction pictures.

Example:

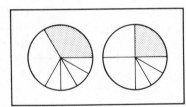

Story

It takes 3 pieces from the pizza cut in twelfths to make 1 piece from the one cut in fourths. It takes 4 pieces from the one cut in twelfths to make one piece from the one cut into thirds.

These are just two of the many activities that could be designed using worksheets constructed from FRACTION MODELS even though the program is not sophisticated. With a good commercial program you could create many more instructional tools.

Using this program, you could also design lessons that demonstrate sums, differences, and products of fractions, as well as probability and statistics concepts. With a little imagination, we could dream up a number of class projects where students would use the program to produce pictures or illustrations. The teacher should not think of the program as simply a "teacher" program.

■ CERTIFICATE: A program for rewarding

The next teachers' aid we will discuss is called CERTIFICATE. Every good boy or girl needs to know that he or she does fine once in a while. So, with the CERTIFICATE program, you can produce on-the-spot certificates for students of your choice. Here is a sample certificate produced with the IBM version of the program.

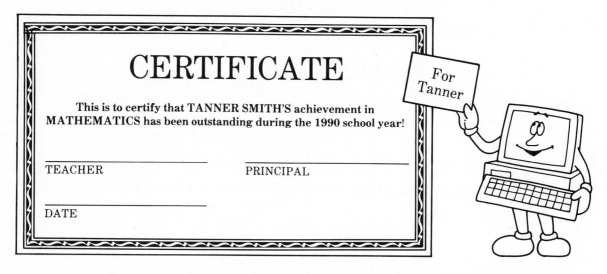

Run the program a couple of times. Notice how easy the certificates are to produce.

■ The Item Database: A program for storing

This program is a small database that can be used to store test items, classroom or homework exercises, brain teasers, and so on. A database is simply a program that lets you store and retrieve information in a systematic way. We have stored several word problems in this database for use with activities presented in Chapter 9. To use the program successfully, more items must be entered (see the Computer corner in Chapter 9). For each item, the following information is stored.

```
Item Number...The computer assigns this number.
Subject...Code, no more than 10 characters.
Learning Objective...Statement, no more than 40 characters.
Objective Number or Code...Code, no more than 10 characters.
Line 1...line 1 of the item, no more than 30 characters.
Line 2...line 2 of the item, no more than 30 characters.
Line 3...line 3 of the item, no more than 30 characters.
Line 4...line 4 of the item, no more than 30 characters.
Line 5...line 5 of the item, no more than 30 characters.
```

Example

```
Item Number:25
Subject:Mathematics
Learning Objective: Solves comparison subtraction probs N<100
Objective: Number or Code Sub-25 level2
Line 1:The fuzzy dog weighs 32 kilograms.
Line 2:The large black cat weighs 15 kilograms.
Line 3:The funny monkey weighs 18 kilograms.
Line 4:How much less does the funny monkey
Line 5:weigh than the fuzzy dog?
```

The database allows you to add new items as well as to change those that have already been entered. When adding an item, you will enter information for all categories (fields) except the first. The computer always assigns the Item Number. You will search for items in terms of subject area and objective number. Since these categories are used as search categories, it is important to design easy-to-remember codes that can be entered with consistency. After adding and changing to suit your fancy, you can SEARCH. To use this option, you enter up to two search codes for each search category. You must be sure to enter search codes exactly as they are entered when items were saved; otherwise your search may come up empty when it shouldn't. When the search is complete, you can select located items to design a student activity sheet. The program allows you to specify your own title and directions. The following is an activity sheet that we designed using the database. We have chosen to prepare a lesson that involves problems that did not have enough information.

Name _____ Date _____

FIXING WORD PROBLEMS:

INSTRUCTIONS: The word problems below need to be fixed. You are to tell what is wrong and rewrite each.

1) 2 cats were sleeping. 1 dog was barking. How many dogs were sleeping?

_____

2) Bill is older than Amy. If Bill is 7 years old then how old is Amy?

_____

3) Josh, who is 5 years old, weighs 52 pounds. His brother Nate weighs 39 pounds. How old is Nate?

_____

4) A cheetah can run up to 74 miles per hour. If it runs 50 miles per hour, how far can it run?

_____

5) A certain ancient Indian tribe of South America ate beetles as part of their diet. If everyone in the tribe ate 37 beetles each day for 10 years, how many beetles were eaten?

_____

6) 5 fish in the pan. Tina put more in the pan. How many fish in the pan?

_____

7) When I went to the zoo I saw 3 monkeys, 2 zebras, and some other animals. How many animals were at the zoo?

_____

8) Mother is cooking 7 hot dogs. They will be finished in 5 minutes. What time will the hot dogs be ready to eat?

_____

There are many different sets of directions that could have been given for this activity.

"Draw a picture for each word problem."
"Read and solve."
"Do not compute, estimate answers."

In fact, you could use the CHANGE in the program to omit all numbers in certain problems. Then, you could prepare an activity sheet with these directions: "Insert numbers in each problem so that the problem makes sense."

By now, it should be pretty clear that such a database could become an important tool for any teacher. We admit that the program is not sophisticated. It was prepared to give you experiences using an instructional database—to store and retrieve items that might be used in planning a mathematics lesson or test. The database limits the number of items you can enter to 200 and does not allow a great variety of search options. Even so, a database with 200 items, search, change, and lesson preparation options can provide a great variety of lessons for a lot of different students. Furthermore, the lessons will be neat and easy to generate.

Presently, the database contains ten word problems. You will be required to enter more word problems if you do the exercises in Chapter 9. When you run the program, you will notice that there is an option that allows you to empty the database and start a new one. You can also copy the database from the MGB Diskette to another diskette; you should do this so that you will have disk space to enter many new items. Run the program, enter a few new items and change some that have been entered. Get the feel of the program, then copy the program to another diskette. Empty the database if desired and start a new one from scratch. Think of ways that you as the teacher might use the program, then think of student projects where students use the database.

To summarize, a teacher who is computer wise is aware that computer databases can be used to store information once and can then make endless use of the information in a variety of instructional ways.

■ MASTER MAKER: Another tool for the teacher

Because microcomputers can repeat the same task without becoming bored or impatient, they are excellent at preparing student lessons for seat work. We prepared a program that would perform this task for Chapter 3. The MASTER MAKER program uses a random generator to construct either place-value models or models for numbers less than 100. Using either model—the teacher's choice—the computer will prepare lessons at the teacher's bidding. The teacher types in the desired directions and title for the lesson, and the computer does the rest. The imaginative teacher can do a lot with these models by simply varying the directions. Here is a part of a lesson with eight exercises generated by the program. (For other examples using this program, see the Computer corner in Chapter 3.)

```
                    REGROUPING TENS AS ONES
Name_____                    Date_____

Directions: Each picture shows tens and ones. Regroup one
            ten as ones. Then write the number of tens and
            ones you have.
```

---

1. SEE HOW

XXXXXXXXXX x   x
XXXXXXXXXX   x

_1_ tens + _13_ ones = _23_

2.

XXXXXXXXXX   x   x
XXXXXXXXXX   x   x
XXXXXXXXXX   x
XXXXXXXXXX   x

__ tens + __ ones = __

3.

XXXXXXXXXX   x   x
XXXXXXXXXX   x
XXXXXXXXXX   x
XXXXXXXXXX   x
XXXXXXXXXX   x

__ tens + __ ones = __

4.

XXXXXXXXXX   x
XXXXXXXXXX   x
XXXXXXXXXX   x
XXXXXXXXXX
XXXXXXXXXX
XXXXXXXXXX

__ tens + __ ones = __

For
Addition

■ DATA: A student utility

The last program we will describe in this section is one we call a *student utility*. Again, the program is not sophisticated. But it will demonstrate a very important concept: the use of computers and "locus of control." A CDI program generally controls the learning environment. Students may have certain options, but the program has specific instructional goals and will lead the student along predetermined learning paths depending upon how options are chosen. In other words, the program drives the student from point A to point B. A student utility in a program, on the other hand, will perform certain tasks only when it is requested to do so. It won't do any leading. This puts the burden of leading on the shoulders of the student, who is thus in control of his or her learning. The student must drive the computer program from point A to point B. Obviously, this may be tougher than being led, but more self-confidence and a feeling of greater control over the machine are the resulting payoffs. The program we will demonstrate is called DATA.

DATA will accept information that has been obtained in an experiment or survey, analyze it, and then provide certain statistics and graphs when requested to do so. For example, suppose we wanted to know how long the average sixth grader could stand on one foot. We could find out by performing an experiment. Twenty students would be tested. Each would be given three attempts, and the best for each student would be accepted. A stopwatch would be used to measure time to the nearest second.

Data for each student would be entered into the program. The program would ask for the name and a short description of the study and the number of participants. The number of seconds (time) for each participant would then be entered. After entering data, "student researchers" would be given the following options:

```
1000 is the maximum number of entries.

      A Enter data from keyboard
      B ..Enter data from a file
      C ...............Sort data
      D .......Save data to file
      E ...........Display data
      F .......Append and delete
      G ............Find Values
      H ..................Graph
      I ....................End


        Enter a selection (A-H)
```

Here is a printed report that was obtained using the program with ten people:

STANDING ON ONE FOOT

This is an experiment involving ten students. The purpose is to determine the average length of time that a student can stand on one foot.

Ordered data: 159, 168, 182, 187, 193, 198, 199, 202, 202, 220

The Mean: 191
The Median: 195.5
The Mode: 202
The Range: 61

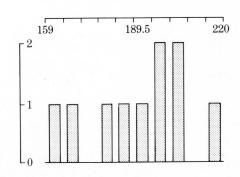

The data collected in this experiment are evenly grouped about the Mean. This tells us that the Mean is a good measure of the average.

Notice that the program examines the data and attempts to select the measure that is best for describing the average. This requires the program to make calculations that are invisible to the student and in most cases beyond the level of instruction. By using the program often in different experiments, students can get a feel for when one of the measures of central tendency—*mean, median,* or *mode*—is better than another for describing the "average."

Before we leave this program, we must decide why this type of program is important. First, the program is a utility—students use the program to accomplish personal goals. This model is analogous to real life experiences where people in professions and businesses, or simply for personal reasons select computer utilities to accomplish desired goals. In this role, students must make decisions and evaluate outcomes. They do the leading. In many situations, this type of experience is far more valuable than sitting behind a computer being told what to do.

Okay, let's summarize the central idea considered in this section. There are many ways a microcomputer can be used to benefit instruction other than being used to tutor. We've only scratched the surface. It's your job now to constantly think, "What computer tools can I find and how can they be used to make instructional activities more interesting, more efficient, more realistic, more powerful, and more substantial?" That's enough to think about!

→ Computer Managed Instruction

Ever since education began in the free world, there has been a dream of providing effective school experiences for individual learners regardless of their social class, race, or sex. Yet until recent times, the expense and unavailability of the necessary technical tools have made this dream unrealistic. The microcomputer revolution, however, brings the potential to realize this dream. But providing so much to so many will require more than advanced tools. Realizing such a dream will require considerable changes in the traditional instructional model and the roles that teachers play. Grading tests, recording and averaging grades, preparing learning prescriptions, submitting student profiles, preparing student rosters, and preparing book inventories would no longer be tasks for teachers to perform. No, the computer would do these and more. The teacher's role would be to orchestrate, to see that important instructional tasks were implemented so that adequate time for direct teacher-student contact was available.

In recent years, technology has improved greatly and costs have continually dropped. Yet in most schools, even those with elaborate computer labs, the potential to revolutionize instructional practices has hardly been tapped. It seems that it is easy to adopt a new tool to teach a specific topic but very difficult to change an institutionalized model of operation. Agreement and support are required from all involved parties. It also requires time for experimentation and the desire to change old habits even when change is painful.

Before the big dream can be realized, school personnel at every level must become aware of and acquire confidence in and reliance on computer management tools. It is beyond the scope of this book to explore CMI in depth. We feel a responsibility, however, to describe types of management tasks, some computer tools for handling these tasks, and instructional goals that could be realized.

■ Instructional management: Bits and pieces

No matter who the teacher is or what group of students he or she is teaching, there are always tasks to be completed that seem very remote to instruction. These usually require collecting, organizing, and reporting student information. Examples include grade reporting; attendance reporting; maintaining classroom inventories of equipment, books, and ancillaries; creating and updating lesson plans; listing learning objectives; establishing banks of test items; and the like. There are other tasks more closely related to instruction. These involve the delivery of instruction where teachers play a managerial role. In this role, teachers are expected to break learning experiences into parts to be delivered by the computer and parts to be delivered by teaching personnel. As managers, teachers organize these parts into meaningful wholes so that learning is not fragmented but effective and enriching. The trick is to require computers to do what they do best and let teachers do what they do best and then make sure that the best of both are properly integrated.

There are many available computer tools that can help teachers perform these tasks. Some are specifically designed for education and some are generic—that is, they are designed for general business or home use. Let's look at some bits and pieces.

■ LIST: A generic tool

Programs that provide lists are common generic microcomputer tools. A good list program should allow you to list, order, and save information you want to keep. The program should put the information in order for you and allow you to add or delete items. Programs that list and store information range from powerful databases that enable users to manipulate data in a great variety of ways to very simple programs that will only generate an alphabetized list. The point is that these tools exist and many are inexpensive. The day is long gone when teachers should be expected to put together long, tedious lists by hand.

Now, what teacher do you know of who doesn't spend a good deal of valuable time creating what are often trivial lists? "The list of students who had more than one tardy last month," or less trivial, "The list of students who forgot lunch money and borrowed from the teacher." So guess what? We have provided a program called LIST for you to experiment with. We want you to use it to make a list of all the lists that a person might make as a consequence of being a teacher.

```
***** LIST *****

1000 IS THE MAXIMUM NUMBER OF ITEMS

    1 ENTER DATA FROM KEYBOARD
    2 ENTER DATA FROM A FILE
    3 SORT DATA
    4 SAVE DATA TO FILE
    5 DISPLAY DATA
    6 APPEND AND DELETE
    7 END PROGRAM

    ENTER A SELECTION (1-7)
```

■ MINI-GRADER: A specific tool

An electronic gradebook is an example of a tool that is specifically designed for education. A good one should have certain essentials. It should be easy to enter data: student names, grades, and so on. It should be easy to correct mistakes that often result when data is being entered. Once data is stored, there should be a variety of ways to sort and list the data—by student number, student name, by class standing, or by assignment or test. The gradebook program should have enough space to hold an adequate number of class records, an adequate number of students for each class, and an adequate number of scores for each student. An efficient program should allow teachers to enter any number of points for an assignment and its comparative weight related to other assignments. The program should compute the percentage right for a given assignment and the teacher should not have to worry about preparing assignments with exactly 100 points. The program should provide a special character to be entered when students miss assignments. This allows the teacher to keep track of all make-up assignments and ensures that inappropriate scores of "0" are not figured into grade averages. Once scores have been entered the program should calculate important statistics—individual student averages, class averages, class median, ranges, and standard deviations. The teacher should be able to specify standards for letter grades, and the program should compare student averages to standard and assign letter grades. After statistics have been computed, the teacher should be able to request reports of various types: Individual Progress Reports, Class Standing Reports, Make-up Assignment Lists, Assignment Scores, and Alphabetized Student Grade Lists. There are other features a gradebook program could have, but the ones we've described would provide an indispensable tool.

To demonstrate some of these features, we provide a program called MINI-GRADER. Run the program several times and enter some data. Then list the features the program has. We warn you in advance that it is a small program that analyzes data for a single student. Here's what a report from MINI-GRADER looks like.

```
     COURSE: MATHEMATICS

 Name: Andria Troutman
 Date: 09-27-1985

 ASSIGNMENTS   WEIGHT   POINTS EARNED   POINTS POSSIBLE
 Homework        5         600              660
 Class Project   2         187              200
 Quizzes         3         570              600
 Examination     3         140              150

 Average:    92.8

 Grade: A
```

Putting the pieces together

Now imagine that you could inventory every learning aid and lesson in a classroom environment regardless of whether it was commercially produced or teacher-made or whether it came from a textbook, an audio or video tape, a film strip, an instructional kit, or a computer program. Suppose, too, that you could classify these materials as skill development, concept development, application, and problem solving. Suppose you could identify the motivational characteristics of these materials. Now, suppose you could tell all this to a computer, and the computer could arrange the materials in terms of instructional objectives so that when the right buttons were pushed, a sequence of lessons were prescribed for a given learning objective. That is, a sequence of lessons having a proper balance between skill development, concept development, and application lessons, a sequence that provided problem-solving experiences, and a sequence providing opportunities for students to guide their own learning as well as being tutored by the teacher or the machine. Even if you overruled some of the computer's choices, wouldn't this be an instructional tool that outpaced any other you've seen?

But let's not stop dreaming. Suppose you could incorporate your testing, grading, and reporting program into this instructional tool. Student performance could be tested, and test data could be electronically scanned and sent to the computer's memory where it could be organized and analyzed. From there, student data could be stored or printed and sequences of lessons could be prepared for the individual learner based on his or her specific difficulties or needs. The computer could maintain files on who did what, when, with what results, and it could amass and organize data on difficulties that students experience. And there's so much more.

Now you might be thinking, "How mechanistic, how lacking in the human touch." Nothing could be further from the truth. There's nothing that says the teacher can't veto a computer decision or choose to perform some tasks the computer performs at any point. The computer is just a well-behaved servant that does what it's told to do. This servant makes life easier for the teacher by doing the time-consuming and routine. The technology for accomplishing this educational goal is now available. Several commercial companies have developed programs that have, at least superficially, many of the characteristics we have described. Yet all of the programs with which we are familiar lack educational substance and are too profit-oriented. In most cases, these programs drain off funds allocated for low achievers in low-income brackets. Characteristically, their management component is awkward and rigid, and the instructional component usually consists of repetitious tutorials or drill-and-practice lessons. It gets worse—the lessons all look the same.

Earlier in this section, we described the dream of providing individual learners with educational experiences tailored to their needs. Now we are telling you that the technical tools exist for getting the job done. Educators can't turn this responsibility over to private industry, and they can't turn away from this important challenge because they have observed commercial programs that lack credibility and integrity even though they claimed to support total learning environments. Teachers, college professors, and school administrators must work together to reorganize the traditional instructional model to take advantage of this technology, constantly guiding the development of commercial products.

Important potpourri

In this chapter, we have discussed three important categories of instructional computing—Computer Directed Instruction, Computer Enhanced Instruction, and Computer Managed Instruction. Though we have alluded to many related issues, there are some important ones that we have not had space to deal with. We do feel, however, that these issues should be called to your attention. So, we will briefly describe some of these issues, and in Think-tank time, we suggest exercises that will help you become more familiar with these issues.

- *Equity:* How can we ensure that instructional opportunities are the same for all children regardless of race, sex, or academic ability? There is some evidence that girls are involved in fewer computer activities than are boys. There is also evidence that minority groups and lower-ability students more often are subjects of tutorials—that is they are led by the computer. On the other hand, white students of higher ability more often participate in computer activities where they gain control over the machine. How will such trends affect the students' self-esteem, and so on?

- *Ethics:* What are the laws concerning the copying of software? What are a teacher's moral obligations as *the model* in the classroom? Where can inexpensive, copyable software be obtained?

- *Computer spaces:* What kind of computer spaces should be provided in the school? Which are more effective for more students? Does it make sense to have a large lab of computers in a room where other activities such as lecturing, testing, seat-work, and the like are taking place?

- *Keeping up with technology:* How can schools get the most from old equipment while steadily acquiring newer equipment? How can program goals be planned that are independent of computer brands?
- *Management systems:* Does it make sense for districts to buy commercial management systems from a variety of vendors that publish textbook series for different subject areas, or should school districts plan total management systems that can be used from school to school and changed from time to time as needed? Which approach in the final analysis would be most cost effective?
- *Programming:* When and to whom should programming skills and concepts be taught?
- *Software review:* With so many different kinds of software appearing in the market place, where can teachers get good reviews? What guidelines should they use when personally reviewing software?
- *Generic software tools:* What are generic software tools? How can teachers use them to make their jobs easier?

☆ ═══ ☆

In closing . . .

Well, there you have it! At least a year's work in one chapter and certainly a lifetime of activities to pursue. Knowing that we could only present a few important ideas related to instructional computing, we tried to organize the ideas into categories that would help you organize new ideas as you encountered them. We also presented some issues that all educators will have to consider as time passes.

We believe that mathematics instruction will change as a direct result of the increased availability of computers in the classroom. Our hope is that not only will the tools for instruction change, but also that the total instructional environment will be remodeled to take advantage of this powerful technology. We hope also that changes extend the capabilities and enrich the lives of both teachers and students.

It's think-tank time

1. What factors might influence a person's views on the improvement of instruction through the use of computers?
2. People view the computer as a tool with different uses. Name some of those uses, then state your own views.
3. Name the four main parts of the computer; give a definition and example of each.
4. Classify the following items as software or hardware (including peripherals). Research the items you are unfamiliar with.
   a. mouse
   b. graphics tablet
   c. cassette tapes
   d. monitor
   e. reel-to-reel tapes
   f. disk drive
   g. printer
   h. floppy disk
   i. keyboard
   j. light pen
   k. hard disk
   l. modem
5. Justify the following sentence: Computers will have a greater effect on education than any other tools before them.
6. There are three categories of Computer Aided Instruction; name, define, and give an example of each.

You will need a computer and a copy of the programs discussed in this chapter to complete activities 7 through 12.

7. Load and run each of the four CDI programs. For each, answer the following questions.
   a. What type of program is it? (drill, tutorial, simulation, etc.)
   b. What are the objectives? Are they related to the CTBS skills?
   c. What are the prerequisite skills?
   d. How would students benefit from this program?
   e. Are there any negative aspects to this program?
8. Using the AUTHOR program, create a series of questions (minimum of five) for the following students:

   a. a second grader
   b. a remedial fourth grader
   c. a fifth grader
   d. a gifted sixth grader

Make several copies of the following form. These will aid you in planning a lesson. Once a lesson is planned, enter it starting at line 2100 in the program code.

Planning Guide for the Program AUTHOR

For each exercise in your lesson, you can enter four lines of text with a maximum of 35 characters in each line. These four lines are for the exercise stimulus and hint. Four more lines of text with a maximum of 20 characters each are available for acceptable responses. Each line of text must begin and end with quotes; however, the quotation marks will not be counted as text characters. When you do not use an available text line, you must enter at least one blank space produced by the computer's space bar in quotes for this line. Use this form to prepare exercises. When complete, enter your lesson into the program code starting at line 2100. When all exercises have been entered, you must enter this statement "*END*". Be sure you do not forget quotation marks in each statement and do not forget the asterisks in the final statement.

Stimulus and Hint

"_____"
"_____"
"_____"
"_____"

Acceptable Responses

"_____"
"_____"
"_____"
"_____"

9. Now, using the CERTIFICATE program, design a reinforcer for each of those students.

10. Load and run the item database. Create your own *rainy day* file for the grade of your choice.

11. Write a small, numerically based survey that you can give to your class. For example, how many gray hairs do you have, or how many hearts have you broken? Use the DATA program and tabulate the results.

12. Load and run MINI-GRADER. Now pretend you are back in sophomore Biology class. Input the names of the students in the class. Type in the assignments and tests, then input hypothetical grades. Does your hypothetical average compare to your remembered Biology grade?

13. Read several articles on equity and computers in education (see Suggested reading list for a few titles). What has the research determined about race, sex, and academic ability in regards to computer use? How would you ensure equity in your classroom? What advice would you give to your building administrator for schoolwide equity?

14. Imagine yourself as a fourth-grade teacher with a class of 20 students. You have two computers in your classroom. Draw a floorplan of your room, positioning the computers, student and teacher desks, centers, and so on for the most effective use of space. Make a mock schedule of how you would use the computers in your class. Which subjects, which students, and how often would you use the computer?

15. Computer programming is being taught at many schools today. Read one article on this practice. Do you believe that each child should be taught programming? At what age should it begin to be taught? Which programming language should be taught? Should it be taught as a separate subject or be integrated with others? State any other views you have on this subject.

16. Check with your university to see if it has an instructional computing department. Find out what criteria are used to determine the educational value of software. Volunteer to look at and write reviews of a few programs. If this is not possible at your university, check with computer stores in the area. Why would it be important for teachers to do software reviews?

➜
Suggested readings

Ascher, Carol. "Microcomputers: Equity and Quality in Education for Urban Disadvantaged Students." *ERIC/CUE Digest*, number 19.

Atkinson, Martha L. "Computer-Assisted Instruction: Current State of the Art." *Computers in the Schools*, Spring 1984, pp. 91–99.

Baillie, Ronald. "Local Area Networks: Industrial Technology Goes to School." *The Computing Teacher*, March 1985, pp. 36–38.

Bardige, Art. "The Problem-Solving Revolution." *Classroom Computer News*, March 1983, pp. 44–46.

Barile, Joe. "One Computer Can Work with Group." *Electronic Education*, May/June 1985, p. 20.

Bitter, Gary. "Computer Labs—Fads?" *Electronic Education*, May/June 1985, pp. 17, 35.

Bork, A. "Computers in Education Today—And Some Possible Futures." *Phi Delta Kappan*, April 1982, pp. 239–243.

Brophy, J. E. "Successful Strategies for the Inner-city Child." *Phi Delta Kappan*, April 1982, pp. 527–529.

Burt, Cynthia. "Software in the Classroom—A Form for Teacher Use." *The Computing Teacher*, May 1985, pp. 16–19.

Caissy, G. A. "Evaluating Educational Software: A Practitioner's Guide." *Phi Delta Kappan*, December 1984, pp. 249–250.

Curran, William S. "Teaching Programming Made Easy." *The Computing Teacher*, May 1985, pp. 32–33.

Del Seni, Donald. "Local School Response to Computer Equity." *The Computing Teacher*, April 1984, pp. 68–69.

di Sessa, A. "Notes on the Future of Programming." *LOGO 84*, June 1984, pp. 149–155.

Dugdale, Sharon. "There's a Green Glob in Your Classroom and Teaching Mathematics Will Never Be the Same." *Classroom Computer News*, March 1983, pp. 40–43.

Feeney, John E. "A Microcomputer Minicurriculum." *The Arithmetic Teacher*, January 1982, pp. 39–42.

Feurzeig, W., and Lukas, G. "LOGO: A Programming Language to Learn Mathematics." *Educational Technology*, March 1972, pp. 49–54.

Finkel, Leroy. "Software Copyright Interpretation." *The Computing Teacher*, March 1985, p. 10.

Halpin, Terry. "Teaching Computer Programming." *The Information Edge*. Third Australian Computer Education Conference. Brisbane: Computer Education Group of Queensland, July 1985.

Harris, Thomas D., III. "You Should Know What the Copyright Law Says." *Classroom Computer Learning*, October 1984, pp. 16, 18–20.

Henderson, Ronald W., Landesman, Edward M., and Kachuck, Iris. "Computer-Video Instruction in Mathematics: Field Test of an Interactive Approach." *Journal for Research in Mathematics Education*, May 1985, pp. 207–224.

Johnson, James P. "Can Computers Close the Educational Equity Gap?" *The Computer and Electronic Graduate*, Spring 1985, pp. 66–69.

Kohl, H. "Who Should Be Evaluating Software?" *Classroom Computer Learning*, September 1984, pp. 30–33.

Laver, Murray. "Computers and Social Change." *Cambridge Computer Science Text 10*. Cambridge, England: Cambridge University Press, 1980, p. 3.

Luehrmann, Art. "The New Trend: Ed-Teching the Computer." *Electronic Learning*, January 1985, p. 22.

Marchionini, Gary. "Teaching Programming: A Developmental Approach." *The Computing Teacher*, May 1985, pp. 12–15.

Papert, Seymour A. *Mindstorms: Children, Computers and Powerful Ideas*. New York: Basic Books, 1980.

Patterson, Marion J. "An Observation of Computer Assisted Instruction on Underachieving, Culturally Deprived Students." *Journal of Secondary Education*, April 1969, pp. 187–188.

Pattison, Linda. "Software Writing Made Easy." *Electronic Learning*, March 1985, pp. 30–36.

Price, Robert. "Care and Feeding of the Micro." *Electronic Education*, March/April 1985, pp. 10–11, 60.

Ring, Geoff, and Selsmark, Jan. "Teacher Participation in the Production of Software." *The Information Edge*. Third Australian Computer Education Conference. Brisbane: Computer Education Group of Queensland, July 1985.

Roland, Leon. "Software Organization." *The Computing Teacher*, March 1985, pp. 39–40.

Salisbury, D. "How to Decide When and Where to Use Microcomputers for Instruction." *Educational Technology*, March 1984, pp. 22–24.

Sawada, Daiyo. "Computer Power in Primary Grades: Mathematics with Big Trak." *The Arithmetic Teacher*, October 1984, pp. 14–17.

Slesnick, Twila. "Should Software Pirates Walk the Plank?" *Classroom Computer Learning*, October 1984, pp. 17–20.

Thompson, Carla J. "Computer Sorting With Kids." *The Arithmetic Teacher*, November 1984, pp. 40–43.

Troutman, Andria P. "Here Goes the Information Age: Save a Seat on the Bus for Me!" *The Information Edge*. Third Australian Computer Education Conference. Brisbane: Computer Education Group of Queensland, July 1985.

Wallis, Barbara J., and Probert, Patricia J. "Course in Computer-Aided Instruction Design and Implementation." *The Information Edge*. Third Australian Computer Education Conference. Brisbane: Computer Education Group of Queensland, July 1985.

Winner, Alice-Ann, and McClung, Margo D. "Computer Game Playing—'Turn-On' to Mathematics." *The Arithmetic Teacher*, October 1981, pp. 38–39.

17

The End . . .
Your Beginning
Toward Efficient Instruction

■ In this chapter, we briefly outline techniques for keeping records, and we give tips for arranging and managing the classroom. We also provide an extensive bibliography that lists resources involving topics that affect the teaching of mathematics but that we could not deal with in depth.

Rarely is a new teacher expected to plan a total instructional program, since most school districts have established programs. The purpose of this chapter is to prepare teachers and provide them with resources to adjust their school district's programs to their own individual styles and to the needs of their students. ■

The ideas presented in this book are based on the belief that children in the elementary grades should develop a variety of concepts from many areas of mathematics. The development of these concepts should be built around seeing-and-doing activities that allow children to go beyond specific information, to discover relationships and generalities, and to solve problems. Activities that encourage skill acquisition are also important. But the acquisition of skills cannot be based on memorizing meaningless routines. Appropriate skill development must follow concept development, and it must be associated with realistic applications and problem-solving activities.

To present this view of mathematics instruction, we've discussed content and methods that we feel you must be familiar with. We've given sample classroom activities, provided material sheets for you to use in making inexpensive teaching aids, and listed objectives for children and the associated goals for teachers.

In many instances, we have used important concepts from areas such as curriculum and instruction, learning theory, special-education motivation theory, and evaluation and testing. These areas greatly affect how mathematics is taught to children in the elementary grades. For example, throughout the book we have relied heavily on the concepts of Jean Piaget, especially in Chapters 1, 12, 14, and 15, where we discuss beginning number concepts, geometry, and measurement. Yet we do not discuss specific studies by Piaget. In almost every chapter, we have described learning activities that are related to specifically stated learning objectives, and we have included lists of learning objectives for both teachers and students. But explaining how to write learning objectives has not been a primary goal. In the chapter on diagnosis, we incorporate ideas involving individualization of instruction, although we do not discuss it as a special topic. In short, as all authors must do when writing a book, we have chosen priorities—topics that we feel must receive explicit attention. Other important topics have not been ignored—far from it. They have been woven into the text, providing substantial support for the priorities.

Our decision to include some topics for specific attention over others was guided by several considerations. First, data from recent national and state tests indicate that students at all levels have difficulty with fractions, decimals, percents, number sentences, and word problems. Since most texts for elementary mathematics education deal primarily with whole-number concepts, we thought it necessary to fill this void by providing substantial materials on these concepts. Of course, this coverage required space that might have been used for other topics.

Second, as authors, we consider it a responsibility to provide a textbook that is substantial and cohesive in terms of the main message: How to teach math to children. We also see a responsibility to analyze, digest, and present relevant supporting material. Our purpose is to keep the main message visible and uncluttered.

Third, we realize that a prospective elementary or junior high school teacher takes many college courses that deal with curriculum design, educational psychology, learning theory, and efficient instruction. Thus, we feel that our space may be used more effectively if it deals with priorities that are not likely to be met in other courses. Nonetheless, we recognize that many readers—professors, teachers, and students—will have specific interests that they would like to pursue. Therefore, the Suggested readings at the close of this chapter list resources on topics that are relevant to mathematics education. We have annotated the list where we thought it would be helpful.

Now, suppose you read this book thoroughly and selected some topics from the Suggested readings for further study. How could you put all this together and teach? It would take another book to deal with this question fully. In closing, however, we will outline briefly some strategies that might help you get started.

→ Decide on a plan

An effective teacher must have a plan, which must include a systematic means of evaluating student progress and diagnosing student difficulties. It must also include procedures for helping children develop concepts, overcome difficulties, and extend special interests. Establishing a plan with these characteristics will involve using a variety of techniques for evaluation and diagnosis, keeping good records, and arranging the classroom for effective instruction. Obviously, the most effective plan will be one that relies heavily on a computer management system.

■ Keeping records

In Chapter 8, we discussed techniques for evaluating student progress and diagnosing difficulties. Once these techniques have been used to collect information, a computer management system should be available that allows the entry of important student data. This management system should manage student records and relate them to appropriate learnings for individual students.

Student folders and profiles
The teacher should organize a folder for each student. The folder should contain computer progress reports, teacher observation reports, and samples of student work. A computerized student profile should also be included in the folder. This profile should reflect the learning objectives the student has acquired and those not acquired. The profile should also include information on specific errors or difficulties the student has experienced.

Class profiles
The management system that is used in the classroom should provide a classroom profile, which should identify small groups of students who have similar instructional needs. From this information, the teacher can plan remedial experiences, enrichment experiences, or activities for developing new skills and concepts.

Arranging and managing the classroom

Evaluation, diagnosis, and record-keeping must be followed by appropriate instructional activities. The key words here are *preparation* and *organization*.

Preparation means that teachers collect and store instructional materials that help students acquire the learning objectives chosen by the teacher. As we have emphasized often in this book, these materials can include games, puzzles, activities requiring manipulation of concrete or pictorial materials, written work from textbooks, and practice exercises. The best materials are those that have high interest for students, are inexpensive to prepare, and can be used over and over again. The materials should be sufficient in supply and should span a range of instructional goals— remediation, development, and extension.

Organization implies that materials are stored in the classroom, so that the teacher and students have ready access to them. Guidelines for using materials also must be established. Students need to know when and how to get materials, how to use materials, and when to put materials away. These guidelines must be flexible but consistent. Although there are frequently reasons to deviate from the established plan, an overall consistency provides students with the security necessary for learning to work independently.

Being prepared and organized is not as difficult as it sounds. Here are some tips that will help to make it happen.

• Make the most of the room that you have. Don't think of the classroom as a static arrangement of tables, desks, cabinets, and file drawers. Decide on the sequence of main topics that you expect to cover during the term, the amount of time that you feel is required to cover the topics, and the materials that you will need. Then arrange the classroom accordingly.

Most of the room should be devoted to activities that involve the greatest number of students. But some special corners should be established. One might contain a couple of tables and the materials needed to provide remedial experiences. It might include a listening center and a slide projector. Another corner might be designed for students who have free time or who are between assignments. This kind of center is usually called an *interest center* and should contain books, puzzles, games, activity cards, and so on. Another corner might be devoted to activities involving the calculator. Here, small groups of students could work independently to solve problems that were unmanageable a few years ago. Whatever the kind of corner you decide to create, it should be attractive. Colorful bulletin boards that read "Try This" or "Show What You Did" will help.

- Don't try to do everything yourself. Teachers often feel overcome by what seems to be an impossible amount of work. That's usually because they try to do everything themselves. Let the students help! Let them develop a sense of responsibility, a sense of worth, and an understanding of group effort. What can students do? Just about anything! They can help you to reteach. They can help you to prepare, inventory, and repair materials. They can even help you collect materials. Students can also help evaluate their own progress. All you need to furnish is firm and consistent guidance.

- Inspire good discussion sessions by asking good questions. Ask questions that require more than simple one-word or yes-or-no answers. Don't answer your own questions. Give children time to think through questions before you move on. Ask questions that require students to look for attributes, find similarities and differences, classify, order, and so on. Control your facial expressions and the tone of your voice so that both remain nonjudgmental. Take answers from several children, and encourage each to defend his or her choice.

- Be a good disciplinarian. No matter how loving you are with children and how friendly you'd like the atmosphere in your classroom to be, don't abandon your post as leader. As leader, you must learn to calmly manage a variety of hectic situations. To do so, you must discipline yourself. This discipline should extend to the level and tone of your voice, your enthusiasm, your consistency, your maintenance of high standards, your honesty, your dependability, and your fairness and firmness.

Develop standards in cooperation with your class, incorporating school policies. Make sure that children are aware of consequences for good or bad behavior. Be sure to have a system for rewarding good behavior, but keep punishment a private matter. Take your time when making decisions, and keep your composure. Children learn to ignore wishy-washy teachers who make snap judgments and frequently lose control. Don't sell your children short. Children will live up to the expectations of a loving but firm teacher who expects good behavior and high standards.

• Be a pack rat. Supplies that you use for making activities, games, and puzzles don't have to be new. Bottle tops make great counting chips, game markers, place-value pieces, and so on. Margarine tubs and coffee cans make excellent containers. Self-sealing plastic bags are fantastic for storage and retrieval. Large plastic bottles are good storage and measuring devices. Gift boxes can be used for making and storing games. (Draw or paste the playing surface on the bottom of the box. The other game materials can be stored in the box.) Pieces of cardboard found in commercially packaged goods can be used to make activity cards, game boards, and pieces. Shoe boxes make good files for activity cards or cassette tapes. Throw-away plastic cups can be dice shakers. Pictures from magazines can be used to illustrate mathematical situations. And don't forget egg cartons. Our society throws away enough material to supply every energetic teacher with the resources for making almost all the instructional materials he or she needs.

• Don't get discouraged. Don't get discouraged if you can't be all things to all students at once, and don't expect every moment of school time to be spent efficiently. If you're working with a small group of students, the others don't have to be involved in prescribed learning activities. Let them play games, wrestle with puzzles, read books, or even make or prepare instructional materials. In other words, take "time-outs." You and your students will be better off for it.

These are just a few tips that will help you to make teaching individual students manageable. After a few years of teaching, you'll no doubt have a long list of your own.

The computer corner

This is the last computer corner in the book. You have come a long way from the beginning. We dedicate this corner to you, giving you a certificate made by the MGB AWARD program. We hope you will use this program to make many an award for the children you teach.

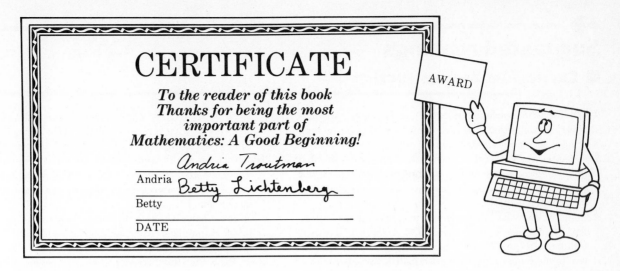

CERTIFICATE

To the reader of this book
Thanks for being the most
important part of
Mathematics: A Good Beginning!

Andria Troutman

Andria

Betty Lichtenberg

Betty

DATE

■ Computer tips

Computers can never take the place of good teachers, but they can make educational experience more exciting, more fulfilling, and more efficient for teachers and students. In the rush of things, the human touch must not get lost. We as teachers must never forget to respect the contributions and growth of students, and we must demonstrate this respect in warm tangible ways.

 ═══

In closing . . .

In this chapter, we've outlined some ideas for keeping records and implementing individual instruction. We have also included a bibliography on topics that support the teaching of mathematics. These ideas, along with other materials provided in the book—objectives for children, sample activities, material sheets, and methods for presenting mathematical ideas—will get you off to a great start. But telling and showing is one thing—it's another thing to do! That's when real learning takes place, so . . . IT'S OUR END AND *YOUR* BEGINNING!

→
Suggested readings
■ Curriculum and instruction

The Arithmetic Teacher, January 1972. This issue provides several articles that deal with many aspects of individualizing mathematics instruction.

The Arithmetic Teacher, February 1979. This issue is devoted to developing a case for the comprehensive mathematics curriculum. It also identifies important topics that are not necessarily basic.

Arnold, William R. "Management by Learning Activities: An Alternative to Objectives." *The Arithmetic Teacher*, October 1977, pp. 50-55. The author provides an alternative to the typical objective-based instructional program.

Brodinsky, Ben. "Back to the Basics: The Movement and Its Meaning." *Phi Delta Kappan*, March 1977, pp. 522-527. The article discusses causes for the back-to-the-basics movement as well as both the merits and weaknesses of the movement.

Bruner, Jerome S. *Toward a Theory of Instruction*. Cambridge, Mass.: Harvard University Press, 1966. The author discusses the development of instruction and variables such as cognitive growth, visual imagery, motivation, individual differences, and evaluation.

Davis, Robert B. "New Math: Success/Failure?" *Instructor*, 1974, *83*(6):53-55. The article discusses historical changes in mathematical content that is taught as well as changes in methods used to teach content. The author questions whether moves in the 1960s to teach "new math" resulted in the neglect of teaching computational skills.

Glennon, Vincent J. "Mathematics: How Firm the Foundation?" *Phi Delta Kappan*, 1976, *57*(5):302-305. The author reviews influences on mathematics curriculum—"the new math," "back-to-the-basics movement," and current psychological theories, including those of Piaget. The author also argues for systematic programs for individualizing mathematical learning.

Houser, Larry L., and Heimer, Ralph T. "A Model for Evaluating Individualized Mathematics Learning Systems." *The Arithmetic Teacher*, December 1978, pp. 54-55. The author describes necessary conditions for adequate individual learning systems in mathematics.

National Council of Supervisors of Mathematics. "Position Paper on Basic Goals." *The Arithmetic Teacher*, October 1977, pp. 19-21. The paper gives a clear and comprehensive view of what is meant by basic skills in mathematics. This view has applications at all levels of education.

Nichols, Eugene D. "Are Behavioral Objectives the Answer?" *The Arithmetic Teacher*, October 1972, pp. 419-476. The author presents objections to objective-based and managed instruction.

O'Daffer, Phares G. "Individualized Instruction—A Search for a Humanized Approach." *The Arithmetic Teacher*, January 1976, pp. 50-53.

Reys, Robert E. "Stop! Look! Think!" *The Arithmetic Teacher*, October 1977, pp. 8-9. The dangers of establishing minimal competencies in mathematics are presented in this article.

Sells, Lucy W. "The Critical Role of Elementary School Mathematics in Equalizing Opportunity." *The Arithmetic Teacher*, September 1981, pp. 44-45.

Smith, William D. "Minimal Competencies—A Position Paper." *The Arithmetic Teacher*, November 1978, pp. 25-26. The author claims that the curriculum must do more than list minimal competencies; it must also describe the conditions related to their attainment.

Trafton, Paul R. "Individualized Instruction: Developing Broadening Perspectives." *The Arithmetic Teacher*, January 1972, pp. 7-12.

Walbesser, Henry H. "Behavioral Objectives, a Cause Celèbré." *The Arithmetic Teacher*, October 1972, pp. 418, 436-440. The author discusses the negative aspects of adopting instructional programs based on behaviorism, and provides a rationale for accepting behavioral objectives as necessary instructional ingredients.

Weaver, J. Fred, and Brawley, Cleo Fisher. "Enriching the Elementary School Mathematics Program for More Capable Children." *Journal of Education*, October 1959, pp. 1-40. The authors provide suggestions on how to provide for students whose performance is better than average.

Yvon, B. R., and Spooner, E. B. "Variations in Kindergarten Mathematics Program and What a Teacher Can Do About It." *The Arithmetic Teacher*, January 1982, pp. 46-52.

■ Evaluation

Anastasi, Anne. *Psychological Testing (4th Ed.)*. New York: Macmillan, 1976.

The Arithmetic Teacher, January 1974. This issue contains several articles on evaluating many variables involved in teaching mathematics.

The Arithmetic Teacher, November 1979. This issue focuses on assessing student learning.

Bloom, Benjamin S., Hastings, Thomas J., and Maddaus, George F. *Handbook on Formative and Summative Evaluation of Student Learning*. New York: McGraw-Hill, 1971.

Buros, O. K. (Ed.). *The Eighth Mental Measurements Yearbook*. Highland Park, N.J.: The Gryphon Press, 1979.

Carpenter, Thomas P., *et al.* "Results and Implications of the Second NAEP Mathematics Assessments: Elementary School." *The Arithmetic Teacher*, April 1980, pp. 10-47.

Carpenter, Thomas P., *et al.* "Decimals: Results and Implications from National Assessment." *The Arithmetic Teacher*, September 1981, pp. 34-37.

Carry, L. Ray. "A Critical Assessment of Published Tests

for Elementary School Mathematics." *The Arithmetic Teacher*, January 1974, pp. 14–18.

Freeman, D. J., Kuhs, T. M., Knappen, L. B., and Porter, A. C. "A Closer Look at Standardized Tests." *The Arithmetic Teacher*, March 1982, pp. 50–54.

Lankford, Francis G., Jr. "What Can a Teacher Learn about Pupils' Thinking through Oral Interviews?" *The Arithmetic Teacher*, January 1974, pp. 26–32.

Long, L. "Writing an Effective Arithmetic Test." *The Arithmetic Teacher*, March 1982, pp. 16–18.

National Council of Teachers of Mathematics. *Evaluation in Mathematics.* Twenty-sixth Yearbook of the National Council of Teachers of Mathematics. Washington, D.C.: The Council, 1961.

Norton, MaryAnn. "Improve Your Evaluation Techniques." *The Arithmetic Teacher*, May 1983, pp. 6–7.

Pearson, Craig. "Teaching Arithmetic Is More Than Marking 'Right' and 'Wrong.'" *Learning*, April 1977, pp. 30–34. The article describes criterion-referenced tests for determining children's ability to apply mathematical operations. Information on nonmathematical aspects of tests that may impede performance is also given.

■ Learning

Ausubel, David P. "Facilitating Meaningful Verbal Learning in the Classroom." *The Arithmetic Teacher*, February 1968, pp. 126–132. Concept development through meaningful verbal learning and didactic teaching modes is discussed in this article.

Biggs, Edith E., and Maclean, James R. *Freedom to Learn.* Don Mills, Ontario: Addison-Wesley, 1969.

Brownell, William A. "Psychological Considerations in the Learning and the Teaching of Arithmetic." In *The Teaching of Arithmetic.* Tenth Yearbook of the National Council of Teachers of Mathematics. New York: Bureau of Publications, Teachers College, Columbia University, 1935. Pp. 1–32. Brownell discusses skill development and incidental learning, and emphasizes concept learning.

Copeland, Richard W. *How Children Learn Mathematics.* New York: Macmillan, 1979. This book discusses the implications of Piaget's research for mathematics teaching in the elementary school.

Dienes, Zolton P. *Building Up Mathematics.* London: Hutchinson Educational, 1960. The author develops a theory of learning based on intuitively developing math concepts by manipulating appropriate concrete models.

Gagne, Robert M. *The Conditions of Learning (3rd Ed.).* New York: Holt, Rinehart and Winston, 1977. Gagne presents his view on how we learn.

Getzels, J. W. "Creative Thinking, Problem-Solving, and Instruction." In Ernest R. Hilgard (Ed.), *Theories of Learning and Instruction.* Chicago: National Society for the Study of Education, The University of Chicago Press, 1964. Pp. 240–267. The author discusses Piaget's studies and describes his four stages of mental development in children.

Juraschek, William. "Piaget and Middle School Mathematics." *School Science and Mathematics*, January 1983, pp. 4–13. This article increases teachers' awareness of Piaget's theory of cognitive development and its implications for mathematics in the middle school.

Labinowicz, Ed. *The Piaget Primer: Thinking, Learning, Teaching.* Menlo Park, Calif.: Addison-Wesley, 1980. The author discusses Piaget's ideas and methods, giving clear presentations of concepts of children's stages of development. Examples from the classroom are also provided.

Lankford, Francis G., Jr. "Implications of the Psychology of Learning for the Teaching of Mathematics." *The Growth of Mathematical Ideas—Grades K-12.* Twenty-fourth Yearbook of the National Council of Teachers of Mathematics. Washington, D.C.: The Council, 1959. Pp. 405–430. Eleven concepts from learning psychology are given as guides for the teaching of mathematics.

Lovell, Kenneth R. "Intellectual Growth and Understanding Mathematics: Implications for Teaching." *The Arithmetic Teacher*, April 1972, pp. 277–282. Implications of Piaget's research for mathematical education are discussed.

Massalski, William J. "Mathematics and the Active Learning Approach." *The Arithmetic Teacher*, September 1978, pp. 10–12. The author describes methods that promote the active learning of mathematics—using learning centers, open-ended instructional tasks, and posing good problems.

Nasca, Donald. "Math Concepts in the Learner-Centered Room." *The Arithmetic Teacher*, December 1978, pp. 448–452. The author discusses levels at which math concepts are acquired and stresses that concepts should be presented on a level comprehensible to the individual learner.

"Practical Applications of Research." *Phi Delta Kappan*, June 1981. This issue deals with teacher enthusiasm and how it affects student learning.

Rappaport, David. "Historical Factors that Have Influenced the Mathematics Program for the Primary Grades." *School Science and Mathematics*, January 1965, pp. 25–33. Theories of learning are discussed, including stimulus response, the meaningful verbal theory, and concepts of Piaget.

Sears, Pauline S., and Hilgard, Ernest. "The Teacher's Role in the Motivation of the Learner." In Ernest R. Hilgard (Ed.), *Theories of Learning and Instruction.* Chicago: National Society for the Study of Education, The University of Chicago Press, 1964. Pp. 182–209. This article discusses and compares different types of rewards and their effect on motivating the learner.

Skemp, Richard R. *The Psychology of Learning Mathematics.* Harmondsworth, England: Penguin Books, 1971. This book describes how mathematics is learned and applies this psychology to the child's development.

Weaver, J. Fred. "Some Concerns about the Application of Piaget's Theory and Research in Mathematics Learning and Instruction." *The Arithmetic Teacher*, April 1972, pp. 263–270. The author discusses situations where Piaget's theories were not relevant but were nonetheless applied.

■ The learning environment

Brush, Lorelei R. "Some Thoughts on Classroom Anxiety." *The Arithmetic Teacher*, December 1981, pp. 37–39. The article makes suggestions for actions by mathematics teachers to improve students' attitudes toward the field.

Bulmahn, Barbara J., and Young, David M. "On the Transmission of Mathematics Anxiety." *The Arithmetic Teacher*, November 1982, pp. 55–56. The role of teacher attitude and elementary students' math anxiety are discussed.

Burns, Marilyn. "The Role of Questioning." *The Arithmetic Teacher*, February 1985, pp. 14–16. How to develop mathematical thinking through questioning is discussed in this article.

Chandler, Theodore A. "What's Wrong with Success and Praise?" *The Arithmetic Teacher*, December 1981, pp. 10–12. The author tells how to avoid the pitfalls of giving success and praise.

Davidson, Patricia. "An Annotated Bibliography of Suggested Manipulative Devices." *The Arithmetic Teacher*, October 1968, pp. 509–524.

Fennema, Elizabeth. "Manipulatives in the Classroom." *The Arithmetic Teacher*, May 1973, pp. 350–352. The author provides a rationale for using manipulatives and identifies when it is and is not appropriate to use manipulatives.

Gersting, Judith L., Kuczkowski, Joseph E., and Alton, Elaine V. "Banish the Boring Bulletin Board." *The Arithmetic Teacher*, January 1978, pp. 46–48. The authors offer suggestions for preparing thought-provoking bulletin boards.

Kennedy, Leonard M., and Michon, Ruth L. *Games for Individualizing Mathematics Learning*. Columbus, Ohio: Merrill, 1973. The article describes skills- and concept-development activities for the mathematics classroom.

Kidd, Kenneth P., Myers, Shirley S., and Cilley, David M. *The Laboratory Approach to Mathematics*. Chicago: Science Research Associates, Inc., 1970. The authors thoroughly describe the mathematics laboratory setting in the classroom.

McCullough, Dorothy, and Findley, Edye. "How to Ask Effective Questions." *The Arithmetic Teacher*, March 1983, pp. 8–9. Developing mathematical thinking skills through effective questions is discussed.

Molina, Norma M. *Mathematics Games for Classroom Use*. Morristown, N.J.: Silver Burdett, 1980.

National Council of Teachers of Mathematics. *Instructional Aids in Mathematics*. Thirty-fourth Yearbook of the National Council of Teachers of Mathematics. Washington, D.C.: The Council, 1973.

National Council of Teachers of Mathematics. *Teacher-Made Aids for Elementary School Mathematics*. Reston, Va.: The Council, 1974.

National Council of Teachers of Mathematics. *Games and Puzzles for Elementary and Middle School Mathematics*. Reston, Va.: The Council, 1975.

National Council of Teachers of Mathematics. *The Mathematics Laboratory: Readings from The Arithmetic Teacher*. Reston, Va.: The Council, 1977.

Post, Thomas, Behr, Merlyn J., and Lesh, Richard. "Interpretations of Rational Number Concepts." *In Mathematics for the Middle Grades*. 1982 Yearbook of the National Council of Teachers of Mathematics. Reston, Va.: The Council, 1982.

Schussheim, Joan Y. "A Mathematics Laboratory, Alive and Well." *The Arithmetic Teacher*, May 1978, pp. 121–155. The author describes aspects of the mathematics lab and gives lists of materials and distributors' addresses.

Stoops, Emery, and King, Joyce. "Discipline Suggestions for Classroom Teachers." *Phi Delta Kappan*, September 1981, p. 58.

Thornton, Carol A. "Math Centers for Young Learners." *Learning*, August/September 1977, pp. 56–57. Aspects and values of learning centers in the classroom are described.

Weber, Larry Jerome, and Todd, Robert M. "On Homework." *The Arithmetic Teacher*, January 1984, pp. 40–41.

■ Research and current issues

Ashlock, Robert B., and Herman, Wayne L., Jr. *Current Research in Elementary School Mathematics*. London: Collier-Macmillan, 1970.

Atweh, Bill. "Developing Mental Arithmetic." In *Mathematics for the Middle Grades*. 1982 Yearbook of the National Council of Teachers of Mathematics. Reston, Va.: The Council, 1982.

Downs, John P., et al. *76 Questions: A Synthesis of the Research on Teaching and Learning Mathematics*. Atlanta, Ga.: Georgia State Department of Education, 1977.

Driscoll, Mark J. *Research within Reach Bulletins*. St. Louis, Mo.: R & D Interpretive Services, Inc., 1981.

National Society for the Study of Education and National Council of Teachers of Mathematics. *Selected Issues in Mathematics Education*. Mary M. Lindquist (Ed.). Chicago: University of Chicago Press, 1981.

Suydam, Marilyn N., and Weaver, J. Fred. *Using Research: A Key to Elementary Mathematics*. Columbus, Ohio: ERIC Information Analysis Center for Science, Mathematics and Environment Education, 1975.

Appendix A
Instructional Objectives Chart

The first number is the chapter number, and the second refers to the number of the objectives for children. (For example, 1–8 refers to Chapter 1 and objective 8.)

Mathematics Activities Strand Chart

	Classification	Number Concepts and Numeration	Addition, Subtraction: Whole Numbers	Multiplication, Division: Whole Numbers	Addition, Subtraction: Rational Numbers
Level One (Grades K–2)	1-1, 1-2, 1-3, 1-4, 1-5, 1-6, 1-7, 1-8, 1-9, 1-10, 1-11	1-12, 1-13, 1-14, 1-15, 1-16, 1-17, 1-18, 1-19, 2-1, 2-2, 2-3, 2-4, 2-5	1-20, 1-21, 1-22, 1-23, 3-1, 3-2, 3-3, 3-4, 3-5, 3-6, 3-7, 3-8, 3-12	Readiness Activities	Readiness Activities
Level Two (Grades 2–4)	Activities That Refine and Reinforce	2-6, 2-7, 2-8, 2-9, 2-10, 2-11, 5-5, 6-1, 6-2, 6-3	3-9, 3-10, 3-11, 3-13, 3-14	4-1, 4-2, 4-3, 4-4, 4-5, 4-6, 4-7, 4-8, 4-16, 5-1	Readiness Activities
Level Three (Grades 4–6)	Activities That Refine and Reinforce	2-12, 2-13, 2-14, 4-7, 4-8, 4-9, 4-10, 4-14, 5-7, 5-8, 5-9, 5-10, 6-4, 6-5, 6-6, 6-7, 6-8, 6-15, 6-16	Activities That Refine and Reinforce	4-9, 4-10, 4-11, 4-12, 4-13, 4-14, 4-15, 4-16, 5-2, 5-3, 5-4, 5-6	6-9, 6-10, 6-11, 6-12, 6-13

Level	Multiplication, Division: Rational Numbers			Geometry			Measurement			Probability and Statistics			Problem Solving		
Level One (Grades K–2)	Readiness Activities			11-1	11-2	11-3	Readiness Activities			Readiness Activities			9-1	9-2	9-3
				11-4									9-11	9-13	9-14
													9-16		
Level Two (Grades 2–4)	Readiness Activities			11-5	11-6	11-7	14-1	14-3	14-4	15-1	15-2	15-3	9-1	9-2	9-3
				11-8	11-9	11-10	14-5	14-6	14-7	15-4	15-9		9-4	9-7	9-9
				11-11			14-8	14-9					9-10	9-11	9-12
													9-13	9-14	9-16
													9-17	9-18	9-19
													9-20		
Level Three (Grades 4–6)	7-1	7-2	7-3	11-12	11-13	11-14	14-1	14-2	14-3	15-5	15-6	15-7	9-1	9-2	9-3
	7-4	7-5	7-6	11-15	11-16	11-17	14-4	14-5	14-6	15-8	15-10	15-11	9-4	9-5	9-6
	7-7	7-8		11-18	11-19		14-7	14-9	14-10				9-7	9-8	9-9
							14-11	14-12	14-13				9-10	9-11	9-12
							14-14	14-15					9-13	9-14	9-15
													9-16	9-17	9-18
													9-19	9-20	

Appendix B
Selected Answers for
Think-Tank Time Exercises

■ Chapter 1

2. Classifying, finding the relations between two sets, conserving relations, finding and using properties of relations, ordering on the basis of the more-than or fewer-than relation, classifying on the basis of the as-many-as relation.
3. a. Selects objects with a specific attribute.   b. sorts on the basis of a general attribute.
   c. duplicates a pattern.   d. selects objects with two attributes.
4. a. Children do not see that the more-than relation is not symmetric.   b. Children use the transitive property of the as-many-as relation.   c. The child does not conserve the as-many-as relation and fails to use the reflexive property of the as-many-as relation.   d. The child does not use the transitive property of the more-than relation.
7. Yes. For every even whole number n, there is exactly one whole number $2n$, and for every even whole number m greater than or equal to 0, there is exactly one whole number $m \div 2$.

■ Chapter 2

2. multiplication and addition.   3. a set of symbols and rules for combining those symbols; used for communicating number ideas.

4. a.
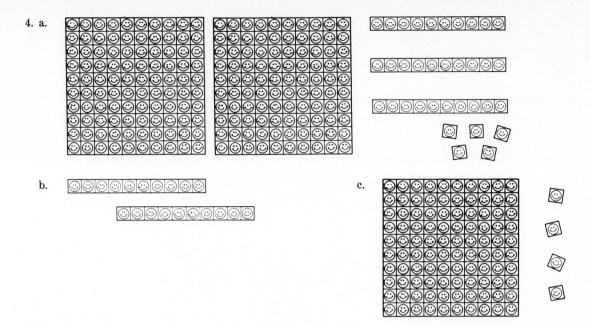

b.

c.

5. 10 000   6. a. 53   7. a. 6   c. 86.2   e. 860
b. 13 000   b. 86   d. 860 and 870

9. a. $9 \times 10^5 + 1 \times 10^4 + 8 \times 10^3 + 7 \times 10^2 + 6 \times 10^1 + 4 \times 10^0$
$9 \times 100\ 000 + 1 \times 10\ 000 + 8 \times 1000 + 7 \times 100 + 6 \times 10 + 4 \times 1$

b. 5×10^2
5×100

c. $3 \times 10^1 + 6 \times 10^0 + 7 \times 10^{-1} + 8 \times 10^{-2} + 2 \times 10^{-3}$
$3 \times 10 + 6 \times 1 + 7 \times \frac{1}{10} + 8 \times \frac{1}{100} + 2 \times \frac{1}{1000}$

d. $1 \times 10^2 + 0 \times 10^1 + 6 \times 10^0 + 0 \times 10^{-1} + 9 \times 10^{-2}$
$1 \times 100 + 0 \times 10 + 6 \times 1 + 0 \times \frac{1}{10} + 9 \times \frac{1}{100}$

e. $0 \times 10^{-1} + 0 \times 10^{-2} + 7 \times 10^{-3}$
$0 \times \frac{1}{10} + 0 \times \frac{1}{100} + 7 \times \frac{1}{1000}$

f. 0×10^0
0×1

10. a. $124_{(five)}$ d. 24
b. $1000_{(five)}$ e. 117
c. $2103_{(five)}$ f. 386

■ Chapter 3

2. conserving the relations, associating sets with numbers, and counting meaningfully.   3. The vertical form is not necessary, since regrouping is not facilitated.   4. It's easier to count to 12 when you start at 9 than it is when you start at 3.

7.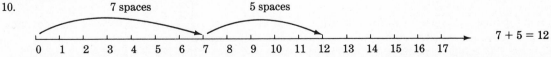
$8 \qquad + \qquad 5 \qquad = \qquad 10 \qquad + \quad 3 \quad = 13$

8. Some of the "near doubles" are: $6 + 7, 3 + 4, 5 + 4, 9 + 8$. They occur in squares adjacent to those on the diagonal in the table.

9. $8 + 6 = \square.\ 8 + 6 = 8 + (2 + 4) = (8 + 2) + 4 = 10 + 4 = 14.$

10.   7 spaces   5 spaces

0 1 2 3 4 5 6 7 8 9 10 11 12 13 14 15 16 17 $7 + 5 = 12$

11. By sliding the top scale, we are able to indicate the appropriate number of spaces for each number involved in the sum.

12.

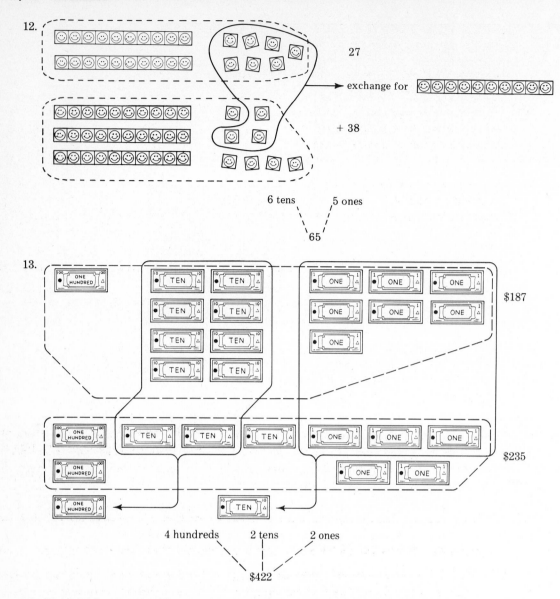

27

exchange for

+ 38

6 tens · · · 5 ones

65

13.

$187

$235

4 hundreds · · · 2 tens · · · 2 ones

$422

14. An algorithm is a specific procedure that can be performed mechanically to produce a certain result. In mathematics, algorithms are used for computational purposes.

15.

Engine	Caboose	Cars
10 + 1	11	2 + 9, 3 + 8, 4 + 7, 5 + 6, 6 + 5, 7 + 4, 8 + 3, 9 + 2
10 + 2	12	3 + 9, 4 + 8, 5 + 7, 6 + 6, 7 + 5, 8 + 4, 9 + 3
10 + 3	13	4 + 9, 5 + 8, 6 + 7, 7 + 6, 8 + 5, 9 + 4
10 + 4	14	5 + 9, 6 + 8, 7 + 7, 8 + 6, 9 + 5
10 + 5	15	6 + 9, 7 + 8, 8 + 7, 9 + 6
10 + 6	16	7 + 9, 8 + 8, 9 + 7
10 + 7	17	8 + 9, 9 + 8
10 + 8	18	9 + 9

16. a. $298 + 167 = 298 + (2 + 165)$
   $= (298 + 2) + 165$
   $= 300 + 165$
   $= 465$

b. $599 + 381 = 599 + (1 + 380)$
   $= (599 + 1) + 380$
   $= 600 + 380$
   $= 980$

c. $989 + 215 = 989 + (11 + 204)$
   $= (989 + 11) + 204$
   $= 1000 + 204$
   $= 1204$

17. knowledge of the interpretation for addition, knowledge of the basic facts for addition.

19. a.

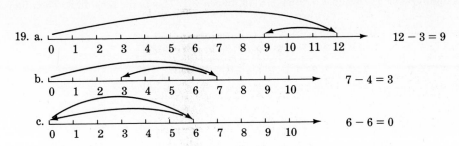

$12 - 3 = 9$

b.

$7 - 4 = 3$

c.

$6 - 6 = 0$

20. Answers will vary. The following is an example: $5 - 3 \neq 3 - 5$; since $2 \neq -2$.

21. a. Show 13. You can't take 6 ones away, so exchange the ten for 10 ones. Now take 6 ones away.

take-away

b. Show 31. You can't take 4 ones away, so you exchange a ten for 10 ones. You have 2 tens and 11 ones. Now take away 4 ones and 1 ten, leaving 1 ten and 7 ones, or 17.

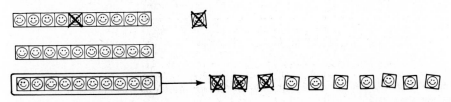

c. Show 106. You can take 4 ones away, but you can't take 9 tens away. Change the hundred-piece for 10 tens. Now take 9 tens away.

22. To find $126 - 78$, regroup as 10 ones, 1 of the tens indicated by the 2 in 126. Subtract the 8 ones in 78. Regroup the 1 hundred as 10 tens and subtract 7 tens. The result is 48.

23. a. Adding 3 to each gives $1503 - 1000$ or 503.   b. Adding 1 to each gives $88 - 30$ or 58.   c. Adding 3 to each gives $1029 - 920$ or 109.   d. Subtracting 1 (or adding $- 1$) from each gives $999 - 185$ or 814.   e. Subtracting 1 from each gives $99 - 53$ or 46.

24. a. 348 is approximately 350, and 796 is close to 800; so the sum is about 1150.

b. 904 is about 900, and 299 is close to 300; so the difference is approximately 600.

25. a. 0   b. $- 2$ c. $- 4$ d. $- 7$

■ Chapter 4

3. a.

$3 \times 4 = 12$

$3 \times 4 = 12$

b.

$5 \times 5 = 25$

$5 \times 5 = 25$

c.

$6 \times 2 = 12$

$6 \times 2 = 12$

d.

$7 \times 1 = 7$

$7 \times 1 = 7$

4. Multiplication facts for 2 always involve a numeral having a ones digit of 0, 2, 4, 6, or 8. Multiplication facts for 5 always involve a numeral having a ones digit of 0 or 5. To find $n \times 9$ for $n < 10$, find $n - 1$. This is the tens digit. Then find $9 - (n - 1)$. This is the ones digit.

5.

$\times$	$3 \times \square = 12$	$2 \times \square = 16$	$3 \times \square = 27$	$a \times b = c$ (for $a, b \neq 0$)
$\div$	$12 \div \square = 3$ $12 \div 3 = \square$	$16 \div \square = 2$ $16 \div 2 = \square$	$27 \div \square = 3$ $27 \div 3 = \square$	$c \div a = b$ $c \div b = a$

8.

$\boxed{\$10}$ $\boxed{\$10}$ $\boxed{\$10}$ $\boxed{\$10}$

$\boxed{\$10}$ $\boxed{\$10}$ $\boxed{\$10}$ $\boxed{\$10}$ $3 \times 40 = 3 \times 4$ tens $= 12$ tens $= 120$.

$\boxed{\$10}$ $\boxed{\$10}$ $\boxed{\$10}$ $\boxed{\$10}$

9.
$$\begin{array}{r} 18 \\ \times\ 13 \\ \hline 24 \\ 30 \\ 80 \\ 100 \\ \hline 234 \end{array}$$

$24 \longrightarrow 3 \times 8$ *(A)*
$30 \longrightarrow 3 \times 10$ *(B)*
$80 \longrightarrow 10 \times 8$ *(C)*
$100 \longrightarrow 10 \times 10$ *(D)*

11. Using the commutative property of multiplication, we can conclude that the two numbers are equal; that is,
$100 \times 1\ 000\ 000\ 000 = 1\ 000\ 000\ 000 \times 100$.

12. *Products:* knowledge of the row-by-column interpretation for multiplication, knowledge of the basic facts for multiplication, skills for computing products of multiples or powers of ten, skills for computing sums, knowledge of place-value ideas. *Quotients:* skills for computing products, skills for subtracting, and skills for estimating.

15. The sign $\div$ can be used after children have developed appropriate interpretations for fractional notation.

16. $4 \times 97 = (4 \times 100) - (4 \times 3) = 400 - 12 = 388$
$6 \times 49 = (6 \times 50) - (6 \times 1) = 300 - 6 = 294$
$7 \times 66 = (7 \times 70) - (7 \times 4) = 490 - 28 = 462$
$3 \times 1196 = (3 \times 1200) - (3 \times 4) = 3600 - 12 = 3588$

17. a. sometimes   b. sometimes   c. never   d. sometimes (when $a \neq 0$)   e. sometimes   f. always   g. always
h. sometimes   i. sometimes   j. sometimes   k. sometimes   l. sometimes

■ Chapter 5

2. 1, 4, 9, 16, 25. Prediction: 49.
3. a. $2 \times 3 \times 3 \times 3$ or 2×3^3
   b. $2 \times 2 \times 3 \times 3 \times 5 \times 5$ or $2^2 \times 3^2 \times 5^2$
   c. 3×19
   d. $2 \times 2 \times 2 \times 2 \times 2 \times 2 \times 2$ or 2^7
   e. $5 \times 5 \times 5 \times 5$ or 5^4
   f. $3 \times 5 \times 5$ or 3×5^2
   g. 3×17
   h. $2 \times 2 \times 2 \times 2 \times 5 \times 5 \times 5 \times 5$ or $2^4 \times 5^4$
   i. $2 \times 3 \times 3 \times 67$ or $2 \times 3^2 \times 67$
   j. $2 \times 5 \times 29$
4. Answers will vary. For example, for any two composite numbers a and b, ab is a greater composite number; or a^2, a^3, a^4, and so on are composite.
5. a. in the second, fourth, and sixth columns.   b. in the third and sixth columns.   c. Starting with 5, these multiples are in a diagonal pattern from right to left. The diagonal pattern starts again at 30, 60, and 90.   d. Starting with 7, these multiples are in a diagonal pattern from left to right. The diagonal pattern starts again at 49 and 91.   e. in the first column or in the fifth column.   f. Primes in the first column can be expressed as $6n + 1$. Primes in the fifth column can be expressed as $6n - 1$.

6. a. 1000 is a multiple of 8.   b. Any number multiplied by 1000 will be a multiple of 8.   c. Any number greater than 1000 can be expressed as the sum of a multiple of 1000 and another number less than 1000. For example, 15 432 = 15 000 + 432.   d. If the second number in this sum is a multiple of 8, then the sum is a multiple of 8. For example: in 15 432, 15 000 is a multiple of 8, since it's 15 × 1000. Since 432 is also a multiple of 8, 15 000 + 432 (or 15 432) is a multiple of 8.

7.

$$108 \\ \diagdown \\ 2 \times 54 \\ 2 \times 2 \times 27 \\ 2 \times 2 \times 3 \times 9 \\ 2 \times 2 \times 3 \times 3 \times 3$$

$$108 \\ 3 \times 36 \\ 3 \times 3 \times 12 \\ 3 \times 3 \times 2 \times 6 \\ 3 \times 3 \times 2 \times 2 \times 3$$

$$108 \\ 4 \times 27 \\ 2 \times 2 \times 3 \times 9 \\ 2 \times 2 \times 3 \times 3 \times 3$$

$$108 \\ 6 \times 18 \\ 2 \times 3 \times 2 \times 9 \\ 2 \times 3 \times 2 \times 3 \times 3$$

$$108 \\ 12 \times 9 \\ 2 \times 6 \times 3 \times 3 \\ 2 \times 2 \times 3 \times 3 \times 3$$

8. a. 3   b. 9   c. 6   d. 8   e. 1   f. 6

9. a. $2 \times 2 \times 2 \times 3 \times 3$ or $2^3 \times 3^2 = 72$.   b. $2 \times 2 \times 3 \times 3 \times 3$ or $2^2 \times 3^3 = 108$.   c. $2 \times 3 \times 11$ or 66.
   d. 17×19 or 323.   e. $2 \times 2 \times 2 \times 2 \times 2 \times 2 \times 3$ or $2^6 \times 3 = 192$.

10. a. When you have an odd number of things and you put them in pairs, you'll have one left over. If you don't have one left over, then your number is even. Now, if you have 0 things and try to put them in pairs, do you have one left over? No, you don't have one left over. So 0 is even.   b. Use the "every other number is even" approach starting with 10 and counting backward to illustrate that 0 fits this pattern for even numbers.

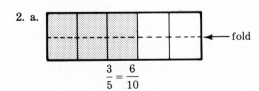

12. a. Not necessarily. The number could be 4 or 12 or lots of other numbers. The number is a multiple of 4, so it is necessarily a multiple of 2. The strongest statement we can make is that it can be expressed as $4n$, a multiple of 4. When n is odd, $4n$ is not a multiple of 8, because it will have only 2 factors of 2. Three factors of 2 are necessary for a multiple of 8. For children, one counter-example is enough . . . 12.
    b. The number is 48, which has 10 factors. 48 is 16×3. The factors of 48 are

$2^0 \times 3^0$ or 1 and $2^0 \times 3^1$ or 3
$2^1 \times 3^0$ or 2 and $2^1 \times 3^1$ or 6
$2^2 \times 3^0$ or 4 and $2^2 \times 3^1$ or 12
$2^3 \times 3^0$ or 8 and $2^3 \times 3^1$ or 24
$2^4 \times 3^0$ or 16 and $2^4 \times 3^1$ or 48

Thus, there are 5×2 or 10 factors. Children, while not using this type of system, can see that 1, 2, 3, 4, and 6 are factors of 48 and that these are paired with 48, 24, 16, 12, and 8. So there are 10 in all.
    c. Yes. Yes. A multiple of 7 can be expressed as $7n$. Since the two numbers are not necessarily the same, they must be expressed as $7n$ and $7m$. Their sum is $7n + 7m$ or $7(n + m)$, and this is a multiple of 7. Their product is $7n \times 7m$ or $7(7nm)$, and this is also a multiple of 7.

13. 1   1   2   3   5   8   13   21   34   55   89   144   233   377   610   987   1597   2584   4181   6765
    a. Every fourth number in the list is a multiple of 3.
    b. Every fifth number in the list is a multiple of 5.
    c. Every sixth number in the list is a multiple of 8.

■ Chapter 6

2. a.

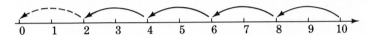

$$\frac{3}{5} = \frac{6}{10}$$

b.

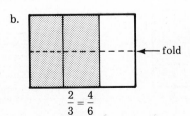

$$\frac{2}{3} = \frac{4}{6}$$

c.

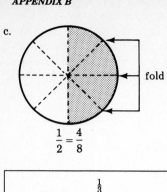

$$\frac{1}{2} = \frac{4}{8}$$

d.

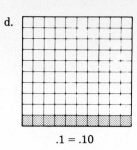

.1 = .10

e.

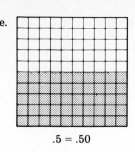

.5 = .50

3.

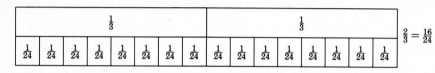

$\frac{1}{3}$
$\frac{1}{24}$ $\frac{1}{24}$ $\frac{1}{24}$ $\frac{1}{24}$ $\frac{1}{24}$ $\frac{1}{24}$ $\frac{1}{24}$ $\frac{1}{24}$

$\frac{1}{3} = \frac{8}{24}$

$\frac{1}{8}$	$\frac{1}{8}$	$\frac{1}{8}$
$\frac{1}{24}$ $\frac{1}{24}$ $\frac{1}{24}$	$\frac{1}{24}$ $\frac{1}{24}$ $\frac{1}{24}$	$\frac{1}{24}$ $\frac{1}{24}$ $\frac{1}{24}$

$\frac{3}{8} = \frac{9}{24}$

$\frac{1}{12}$	$\frac{1}{12}$	$\frac{1}{12}$	$\frac{1}{12}$	$\frac{1}{12}$

$\frac{1}{24}$ $\frac{1}{24}$ $\frac{1}{24}$ $\frac{1}{24}$ $\frac{1}{24}$ $\frac{1}{24}$ $\frac{1}{24}$ $\frac{1}{24}$ $\frac{1}{24}$ $\frac{1}{24}$

$\frac{5}{12} = \frac{10}{24}$

$\frac{1}{2}$

$\frac{1}{24}$ $\frac{1}{24}$ $\frac{1}{24}$ $\frac{1}{24}$ $\frac{1}{24}$ $\frac{1}{24}$ $\frac{1}{24}$ $\frac{1}{24}$ $\frac{1}{24}$ $\frac{1}{24}$ $\frac{1}{24}$ $\frac{1}{24}$

$\frac{1}{2} = \frac{12}{24}$

$\frac{1}{3}$	$\frac{1}{3}$

$\frac{1}{24}$ $\frac{1}{24}$ $\frac{1}{24}$ $\frac{1}{24}$ $\frac{1}{24}$ $\frac{1}{24}$ $\frac{1}{24}$ $\frac{1}{24}$ $\frac{1}{24}$ $\frac{1}{24}$ $\frac{1}{24}$ $\frac{1}{24}$ $\frac{1}{24}$ $\frac{1}{24}$ $\frac{1}{24}$ $\frac{1}{24}$

$\frac{2}{3} = \frac{16}{24}$

$\frac{1}{6}$	$\frac{1}{6}$	$\frac{1}{6}$	$\frac{1}{6}$	$\frac{1}{6}$

$\frac{1}{24}$ $\frac{1}{24}$ $\frac{1}{24}$ $\frac{1}{24}$ $\frac{1}{24}$ $\frac{1}{24}$ $\frac{1}{24}$ $\frac{1}{24}$ $\frac{1}{24}$ $\frac{1}{24}$ $\frac{1}{24}$ $\frac{1}{24}$ $\frac{1}{24}$ $\frac{1}{24}$ $\frac{1}{24}$ $\frac{1}{24}$ $\frac{1}{24}$ $\frac{1}{24}$ $\frac{1}{24}$ $\frac{1}{24}$

$\frac{5}{6} = \frac{20}{24}$

4.

.01
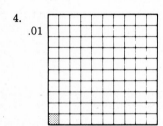

.1
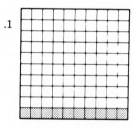

.17
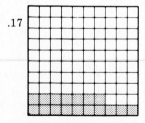

.7
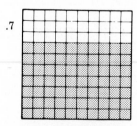

.71
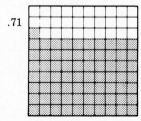

1
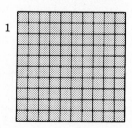

5. a. $\frac{7}{10} > \frac{1}{2}$ and $\frac{5}{12} < \frac{1}{2}$, so $\frac{7}{10} > \frac{5}{12}$.   b. $\frac{13}{22} > \frac{1}{2}$ and $\frac{11}{24} < \frac{1}{2}$, so $\frac{13}{22} > \frac{11}{24}$.

6. $\frac{15}{31} > \frac{14}{29}$. 15×29 and 14×31. We are comparing $\frac{15 \times 29}{31 \times 29}$ to $\frac{14 \times 31}{29 \times 31}$. The fractions have common denominators, so no computation is necessary for the denominators.

7. $3\frac{3}{8}$ is the same as $3 + \frac{3}{8}$. 3 can be renamed as $\frac{24}{8}$. So $3 + \frac{3}{8} = \frac{24}{8} + \frac{3}{8} = \frac{27}{8}$.

10.
$$\left(\frac{3}{4} + \frac{5}{6}\right) + \frac{5}{8} = \frac{3}{4} + \left(\frac{5}{6} + \frac{5}{8}\right)$$
$$\left(\frac{9}{12} + \frac{10}{12}\right) + \frac{5}{8} = \frac{3}{4} + \left(\frac{20}{24} + \frac{15}{24}\right)$$
$$\frac{19}{12} + \frac{5}{8} = \frac{3}{4} + \frac{35}{24}$$
$$\frac{38}{24} + \frac{15}{24} = \frac{18}{24} + \frac{35}{24}$$
$$\frac{53}{24} = \frac{53}{24}$$

11. $\frac{2}{4} + \frac{1}{4} = \frac{1}{4} + \frac{2}{4}$
$$\frac{3}{4} = \frac{3}{4}$$

12. $\frac{5}{8} - \frac{1}{4} \neq \frac{1}{4} - \frac{5}{8}$
$$\frac{5}{8} - \frac{2}{8} \neq \frac{2}{8} - \frac{5}{8}$$
$$\frac{3}{8} \neq -\frac{3}{8}$$

13. A. a. $5\frac{11}{15}$ b. $7\frac{5}{12}$ c. 6.96   d. $1\frac{3}{4}$ e. .328   f. 1.02   g. $2\frac{7}{12}$ h. .017

B. Example b may be answered as follows: The computation for $5\frac{5}{6} + 1\frac{7}{12}$ looks like this:
$$5\frac{5}{6} = 5\frac{10}{12}$$
$$1\frac{7}{12} = 1\frac{7}{12}$$
$$6\frac{17}{12} = 7\frac{5}{12}$$

These concepts are involved: $5\frac{5}{6}$ means $5 + \frac{5}{6}$ and $1\frac{7}{12}$ means $1 + \frac{7}{12}$. Now, to add $5 + \frac{5}{6}$ and $1 + \frac{7}{12}$, we can add the whole numbers and add the fractional numbers (using the associative and commutative properties of addition). To find the sum of $\frac{5}{6}$ and $\frac{7}{12}$, we must have a common denominator. The least common multiple of 6 and 12 is 12, so the least common denominator of $\frac{5}{6}$ and $\frac{7}{12}$ is 12. $\frac{5}{6}$ is equal to $\frac{10}{12}$, since multiplying 5 by 2 and multiplying 6 by 2 gives an equivalent fraction. Now, $\frac{10}{12} + \frac{7}{12} = \frac{17}{12}$, and this can be written as $\frac{12}{12} + \frac{5}{12}$, or $1 + \frac{5}{12}$, or simply $1\frac{5}{12}$. The sum of our whole numbers is 6; the sum of the fractional numbers is $1\frac{5}{12}$, and the sum of these is $7\frac{5}{12}$. Example h may be answered as follows: The computation for .037 − .02 looks like this:
$$.037$$
$$-.020$$
$$.017$$

These concepts are involved: .02 is the same as .020. $\frac{2 \times 10}{100 \times 10} = \frac{20}{1000}$. Multiplying both 2 and 100 by 10 gives an equivalent fraction. Subtracting 0 thousandths from 7 thousandths gives 7 thousandths. Subtracting 2 hundredths from 3 hundredths gives 1 hundredth, and subtracting 0 tenths from 0 tenths gives 0 tenths. So, we have 0 tenths, 1 hundredth, and 7 thousandths, and this is .017.

14. a. False: $\frac{a}{a} = 1$; $\frac{b}{b} = 1$; and $\frac{a+b}{a+b} = 1$. $1 + 1 \neq 1$.   b. False $2\frac{3}{4} = 2 + \frac{3}{4} = \frac{8}{4} + \frac{3}{4} = \frac{8+3}{4}$.   c. True.   d. False: $\frac{a}{b} \neq \frac{a-1}{b-1}$ if $a \neq b$.   e. True.   f. True.   g. True.   h. False: $.472 = (4 \times \frac{1}{10}) + (7 \times \frac{1}{100}) + (2 \times \frac{1}{1000})$.   i. True!   j. False! Reread this chapter to see why.

15. a. 1 dime can be exchanged for 10 pennies.   b. 2 dimes and 7 pennies combined with 3 dimes and 2 pennies will be 5 dimes and 9 pennies.   c. 1 dollar, 5 dimes, and 8 pennies combined with 2 dollars, 4 dimes, and 3 pennies will be 3 dollars, 9 dimes, and 11 pennies. This can be exchanged for 3 dollars, 10 dimes, and 1 penny and then 4 dollars and 1 penny.   d. If you have 5 dimes and have to give someone 3 pennies, you can exchange one of your dimes for 10 pennies, and you'll end up with 4 dimes and 7 pennies.   e. If you have 5 dollars and have to give someone 4 dollars and 2 dimes and 1 penny, you can exchange one of your dollars for 10 dimes and then one of these dimes for 10 pennies. You'll be left with 7 dimes and 9 pennies.   f. 27 cents can be exchanged for 2 dimes and 7 pennies.

18. a. 5.346×10^3 c. 1.00034×10^5
    b. 6.78902×10^5 d. 2.3456789×10^7

■ Chapter 7

2. $\left(\frac{1}{2} \times \frac{3}{5}\right) \times \frac{7}{8} = \frac{3}{10} \times \frac{7}{8} = \frac{21}{80}$
$$\frac{1}{2} \times \left(\frac{3}{5} \times \frac{7}{8}\right) = \frac{1}{2} \times \frac{21}{40} = \frac{21}{80}$$

3. Answers will vary. Example: $\frac{2}{3} \div \left(\frac{1}{3} \div \frac{1}{6}\right) = \frac{2}{3} \div 2 = \frac{1}{3}$
$$\left(\frac{2}{3} \div \frac{1}{3}\right) \div \frac{1}{6} = 2 \div \frac{1}{6} = 12$$

4. Answers will vary. Example: $\dfrac{2}{3} \times \dfrac{1}{5} = \dfrac{2}{15}$

$\dfrac{1}{5} \times \dfrac{2}{3} = \dfrac{2}{15}$

6. $1\dfrac{1}{2} \times 2\dfrac{2}{5} = 1\dfrac{1}{2} \times \left(2 + \dfrac{2}{5}\right) = \left(1\dfrac{1}{2} \times 2\right) + \left(1\dfrac{1}{2} \times \dfrac{2}{5}\right)$

$\qquad = \left[\left(1 + \dfrac{1}{2}\right) \times 2\right] + \left[\left(1 + \dfrac{1}{2}\right) \times \dfrac{2}{5}\right]$

$\qquad = \left[(1 \times 2) + \left(\dfrac{1}{2} \times 2\right)\right] + \left[\left(1 \times \dfrac{2}{5}\right) + \left(\dfrac{1}{2} \times \dfrac{2}{5}\right)\right]$

$\qquad = (2 + 1) + \left(\dfrac{2}{5} + \dfrac{2}{10}\right)$

$\qquad = 3 + \dfrac{3}{5}$

$\qquad = 3\dfrac{3}{5}$

7.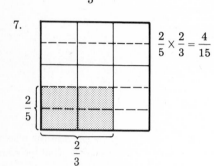

$\dfrac{2}{5} \times \dfrac{2}{3} = \dfrac{4}{15}$

8. $\dfrac{9}{10} = \dfrac{3}{5} \times \boxed{\dfrac{3}{2}}$ ⟵ since $3 \times 3 = 9$.
⟵ since $5 \times 2 = 10$.

9.

$\dfrac{2}{3} \qquad \dfrac{2}{3} \qquad \dfrac{2}{3}$

0   1   2

How many $\dfrac{2}{3}$s are in 2? 3!

10. a.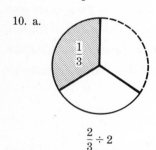

$\dfrac{2}{3} \div 2$

b.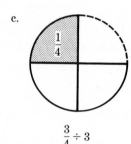

$\dfrac{1}{4} \div 2$

c.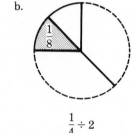

$\dfrac{3}{4} \div 3$

d.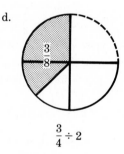

$\dfrac{3}{4} \div 2$

11. a. This is about $36 \div 2$, so the quotient is 15.73.   b. This is about 140×2, so the product is 281.802.

13. $.3 \times .6 = \dfrac{3}{10} \times \dfrac{6}{10} = \dfrac{18}{100}$ and $.2 \times .004 = \dfrac{2}{10} \times \dfrac{4}{1000} = \dfrac{8}{10\,000}$.

15. a. repeating decimal   c. terminating decimal
    b. repeating decimal   d. terminating decimal

■ Chapter 8

1. a. does not regroup when adding.   b. does not regroup when dividing.   c. has difficulty computing sums and differences when numerals don't contain the same number of digits.   d. does not regroup in subtraction when zeros are involved.   e. does not use the distributive property when computing products.   f. omits zeros in the quotient. g. ignores zeros in the divisor.   h. does not use place-value concepts to align the partial products; probably thinks that the product involves "4×7" rather than "40×7."   i. does not complete a division computation when a remainder of 0 is encountered.   j. ignores 0 in the factor 603.   k. does not use place-value concepts (multiples of ten, hundred, and so on) to find the first place-value position of the quotient to be represented and subtracts the smaller digit from the larger.   l. adds $n + 0$ to obtain 0 ("brings down" the 0).   m. ignores the place-value position in a numeral, does not add tenths to tenths, ones to ones, and so on.   n. ignores whole number components when computing with mixed numerals.   o. does not think of each factor as having a denominator of 10 so that a product of the denominators would yield a denominator of 100, thus a product represented by a "two-place" decimal numeral.   p. does not use the equivalent fraction rule correctly.   q. does not associate appropriate meaning with addition.   r. does not use the equivalent

fraction rule correctly to find the equivalent fractions. s. relates a fraction to division, but uses the numerator as the divisor—does not use the equivalent fraction rule correctly. t. "cancels" incorrectly-again, the equivalent fraction rule is used incorrectly. u. does not use the equivalent fraction rule to find the correct placement of the decimal in division. v. multiplies instead of dividing. w. "inverts" the wrong one—applies definition of division incorrectly.

■ Chapter 9

2. Possible answers:

a.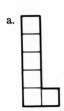

b.

c.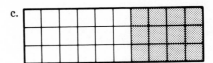

3. a. $(8 \times 3) + 1 = 25$ b. $(4 \times 5) + 4 = 24$ or $(5 \times 5) - 1 = 24$
   c. $(2 \times 6) + (3 \times 3) = 21$ or $(5 \times 6) - (3 \times 3) = 21$ d. $(2 \times 7) + (4 \times 4) + 6 = 36$
4. $3 \times [5 - (2 + 1)] = 6$ 5. $(3 \times 5) - (2 + 1) = 12$
6. a. $3 \times [(2 \times 3) + 8]$ b. $[(2 \times 15) + (3 \times 5) + 60] \div 2$
   c. $\frac{1}{2}(n - \frac{1}{3}n)$
8. a. $n = 5$ b. $n = 30$ c. $n = 6\frac{2}{3}$

■ Chapter 10

1. Answers will vary. Here are some possibilities. (Don't fudge now; ends of toothpicks must meet, and you can't break them!)

a.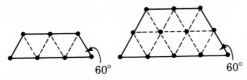

b.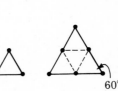

c.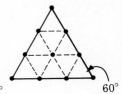

d. impossible

e.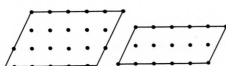

f.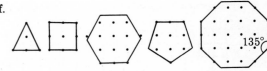

g.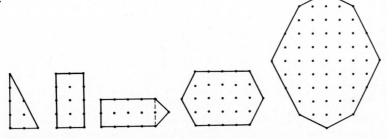

2. a.

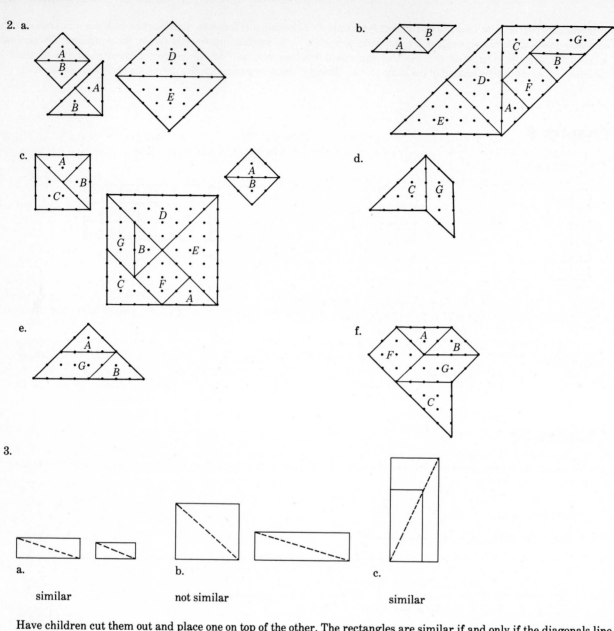

3.

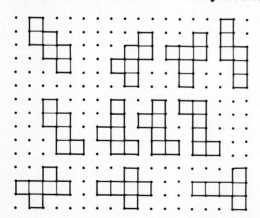

a.   b.   c.

similar   not similar   similar

Have children cut them out and place one on top of the other. The rectangles are similar if and only if the diagonals line up as in picture c.

4. There are 35 different ways to arrange a set of 6 connected squares so that they are joined along their edges. These arrangements are called *hexominoes*. Of these arrangements, 11 are covers for cubes. These are as follows:

8. a. FGJLNPQRSYZ   b. ABCDEKMTUVW   c. HIX   d. none   e. 0
11. a. cone   b. bell   c. doughnut

■ Chapter 12

3. Measurement requires the isolation and description of an attribute. Measurement requires the use of a number and an appropriate standard unit. Measurement involves a comparison of the unit to the attribute. Measurement is approximate.
4. meter, kilogram.
5. A derived unit is a unit that is defined in terms of another unit. For example, a unit used in the measurement of area, the square centimeter, is derived from the centimeter; and a unit used in the measurement of volume, the cubic centimeter, is derived from the square centimeter.
6. one thousandth; one hundredth; one tenth; one thousand.
8. 160 kilometers; 800 kilometers; 40 kilometers.
11. Temperatures that are below 32° Fahrenheit will be below 0° Celsius and will be designated by negative numbers.
14. See the section "Getting ready to measure."
15. The top of my desk has the same area as the top of my desk. If the top of my desk has the same area as the top of your desk, then the top of your desk has the same area as the top of my desk. If the top of my desk has the same area as the top of John's desk and the top of John's desk has the same area as the top of Bob's desk, then the top of my desk has the same area as the top of Bob's desk.
16. a. Iteration of a unit involves the separation of an attribute to be measured into a number of subunits. The separation occurs through the selection of a smaller unit, and the smaller unit must be compared to the attribute to be measured.
    b. To determine the volume of a box with a height of 2 cm, a length of 3 cm, and a width of 4 cm, we can imagine the volume of the box as being separated into centimeter cubes. Then we decide how many centimeter cubes will "fit" inside the box.

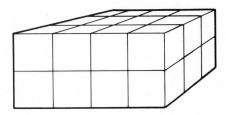

17. We don't want to give this one away!
18. a. 168 cm   b. 380 mm   c. .48 kg   d. 5000 m   e. 17 000 mL   f. .258 L   g. .001 g   h. 40 000 cm²
    i. 7 000 000 cm³   j. 178 mm²   k. 1 000 000 000 mm³   l. 60 000 000 mg   m. .003 m   n. 25 000 000 m²
    o. 86 000 mm³   p. 179 cm   q. 95° F   r. 10 000 mm²   s. 5° C   t. 1000 cm³

■ Chapter 14

2. a. $P = 3s$, where P is the perimeter and s is the length of a side. All three sides of an equilateral triangle are the same length, so to find the distance around an equilateral triangle, we can multiply the length of one side by 3.
   b. $A = s^2$, where A is the area and s is the length of a side. The length and width of a square are the same, so we can multiply this number by itself to find the number of square units for the area.   c. $P = ns$, where P is the perimeter, n is the number of sides, and s is the length of a side. All of the sides of a regular polygon are the same length, so we can multiply the number of sides by the length of a side to find the distance around a regular polygon.
3. Answer will vary for the procedures.   a. A stack of notebook paper that is 1 centimeter thick contains approximately 120 sheets. So 1 sheet is $\frac{1}{120}$ cm, or .0083 cm.   b. A stack of dimes that is 1 centimeter high contains 8 dimes. One kilometer is equal to 100 000 centimeters. So you would have 100 000 × 8 dimes, or $80,000.

4. a.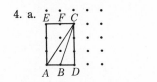

The area of rectangle $BDCF$ is 3 square units, so the area of triangle BDC is $1\frac{1}{2}$ square units. The area of rectangle $AECD$ is 6 square units, so the area of triangle AEC is 3 square units. Then, the area of triangle ABC is $6 - 1\frac{1}{2} - 3$ square units, or $1\frac{1}{2}$ square units.

b. $A = \frac{1}{2} bh$. The base is 1 unit. The height is 3 units. So, $A = \frac{1}{2}(1)(3)$, and the area is $1\frac{1}{2}$ square units.

c.

6. a. $(1000)^3$, or 1 000 000 000 cubic millimeters.   b. 1000 kilometers (from here to ?).

8. a. 1) The longest side is twice the length of the shortest side. 2) The two shorter sides are the same length. 3) All three sides are the same length.   4) The two shorter sides are the same length.

■ Chapter 15

3. There are three ways to select 2 blue marbles, six ways to select 1 blue and 1 green marble, and one way to select 2 green marbles. So the probability of drawing 2 green marbles is $\frac{1}{10}$. A good prediction for 100 draws would be $\frac{1}{10}$ of 100, or 10.

6. a. HHH   b. $\frac{1}{8}$ c. $\frac{3}{8}$
   HHT
   HTH
   HTT
   THH
   THT
   TTH
   TTT

7. a. $\frac{1}{6}$ b. $\frac{1}{12}$ c. $\frac{5}{12}$ d. $\frac{1}{2}$

8. a. $\frac{1}{2}$ b. $\frac{1}{2}$ c. $\frac{1}{4}$ d. $\frac{7}{64}$

Appendix C
Material Sheets

■ Material Sheet 1 Geo-Pieces

Use four colors of construction paper (red, blue, orange, and green). Use as a pattern to cut out 32 Geo-Pieces.

■ Material Sheet 2 Classification Pieces

Copy on four colors of construction paper (red, blue, orange, and green). Cut out your 32 classification pieces.

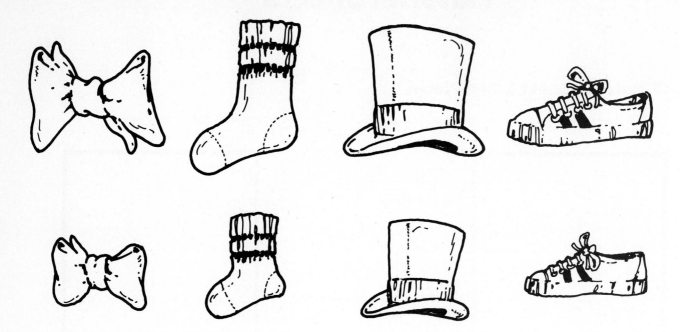

■ One-to-One Matching Pieces

Duplicate on heavy paper and cut out. Select 3 more appropriate patterns to use for matching.

■ Material Sheet 3 Happy Face Pieces

Reproduce as needed on heavy paper. Cut out hundreds, ten-strips, and singles.

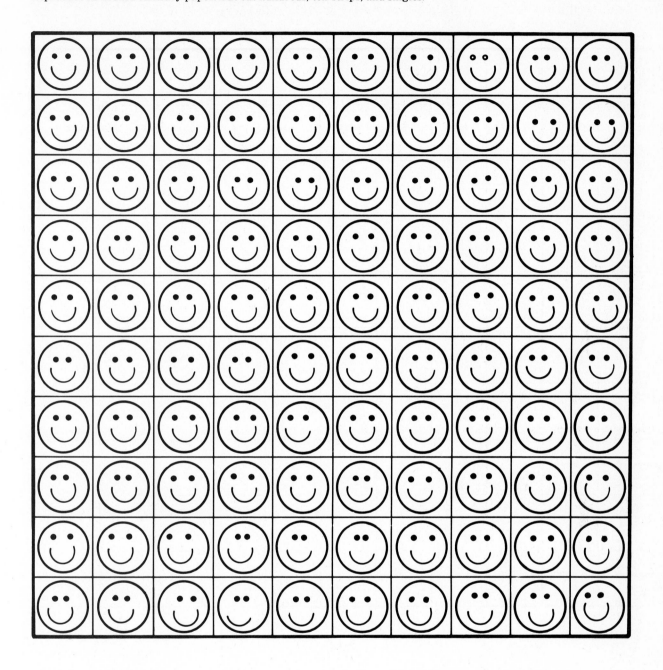

■ Material Sheet 4 Table for Addition or Multiplication

Use with Chapter 3 or 4.

	0	1	2	3	4	5	6	7	8	9
0										
1										
2										
3										
4										
5										
6										
7										
8										
9										

■ Material Sheet 5 Play Money

Reproduce as needed on green paper. Cut out money.

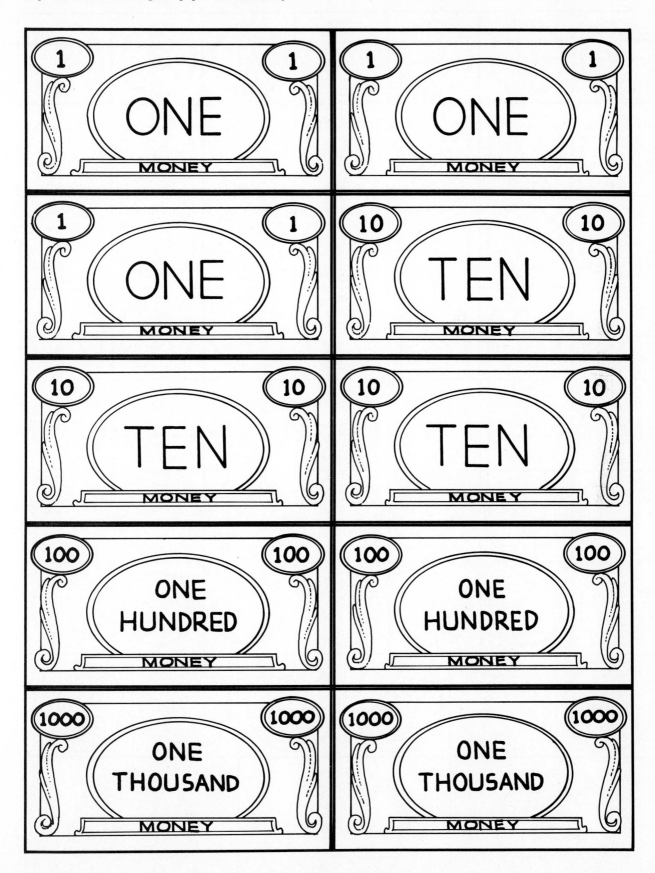

■ Material Sheet 6 Coin Cards

Reproduce on colored card stock. Pennies should be tan.

■ Material Sheet 7 Lost in Space (see Example 2-10)

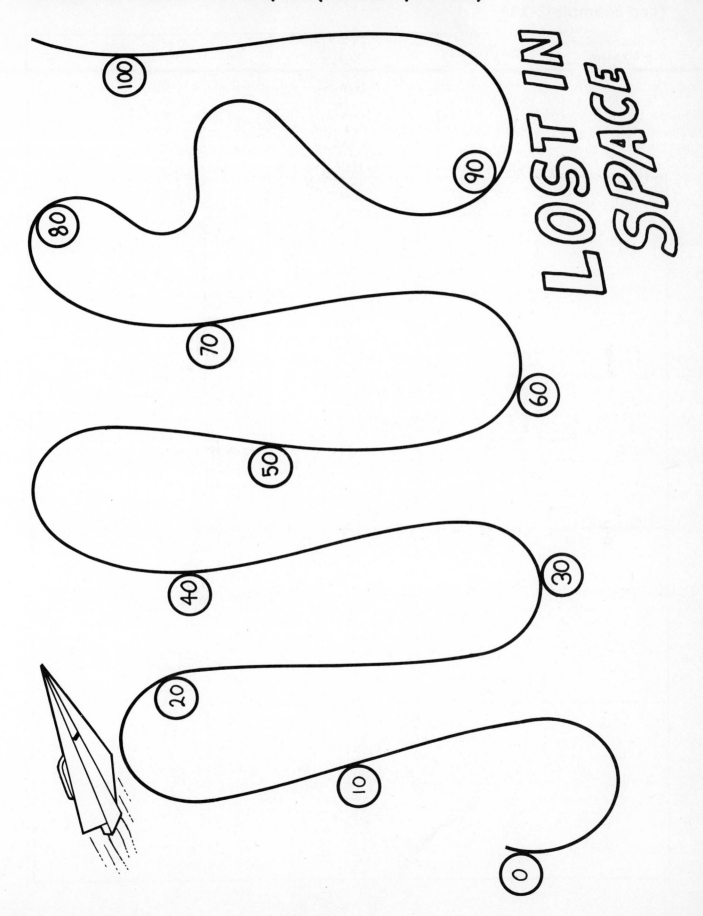

■ Material Sheet 8 Place-Value Displayer
(see Example 2-11)

For best results, reproduce
on heavy paper.

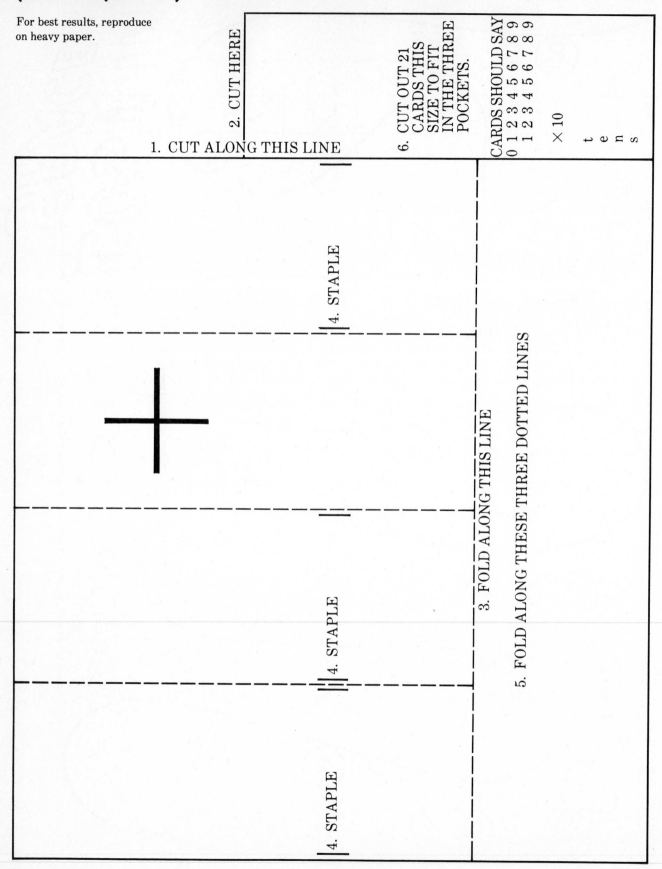

■ Material Sheet 9A Tap-Down Cards
(see Example 2-12)

Reproduce on heavy paper. Cut in horizontal strips and cut vertically to the right of the points.

1.	1	0.	1	0	0	0	0.
2.	2	0.	2	0	0	0	0.
3.	3	0.	3	0	0	0	0.
4.	4	0.	4	0	0	0	0.
5.	5	0.	5	0	0	0	0.
6.	6	0.	6	0	0	0	0.
7.	7	0.	7	0	0	0	0.
8.	8	0.	8	0	0	0	0.
9.	9	0.	9	0	0	0	0.

■ Material Sheet 9B Tap-Down Cards
(see Example 2-12)

1	0	0.	1	0	0	0.
2	0	0.	2	0	0	0.
3	0	0.	3	0	0	0.
4	0	0.	4	0	0	0.
5	0	0.	5	0	0	0.
6	0	0.	6	0	0	0.
7	0	0.	7	0	0	0.
8	0	0.	8	0	0	0.
9	0	0.	9	0	0	0.

■ Material Sheet 10 Groovy Board (see Example 3-2)

Reproduce on heavy paper. Use stick-on circles. Cut out these Groovy Boards for 5 and 6. On the back write 5 and five and 6 and six. Make Groovy Boards for other numbers through 9. Use a rubber band in the grooves to show combinations. Use a transparent file folder to slide over circles and cover them, to show subtraction ideas.

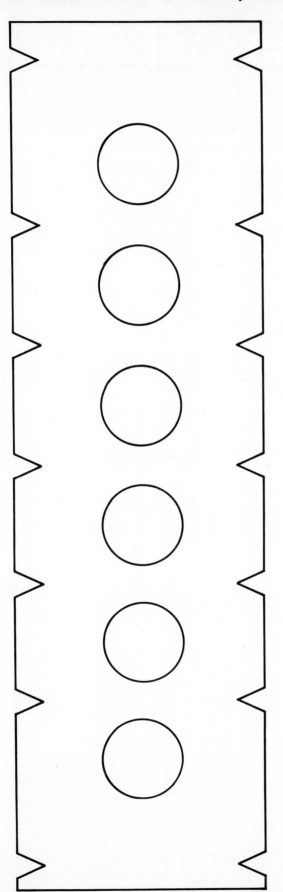
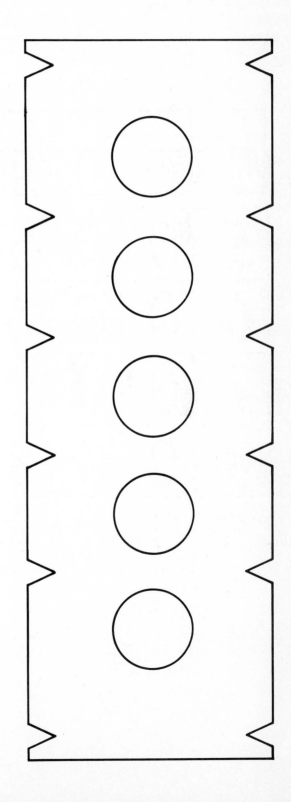

■ Material Sheet 11 Addition Slide Rule
(see Examples 3-5 and 3-15)

You need two colors of construction paper and scissors.
Cut the scale into two parts and glue the parts on the paper.

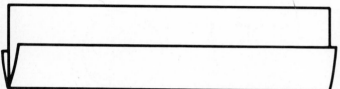

Fold so that the two pieces of paper fit together like this.

A 18 17 16 15 14 13 12 11 10 9 8 7 6 5 4 3 2 1 0

18 17 16 15 14 13 12 11 10 9 8 7 6 5 4 3 2 1 0 B

■ Material Sheet 12 Karrie Kangaroo
(see Example 3-7)

Karrie can be glued on a file folder as is. Cut lots of little baby kangaroos. Label them with expressions such as $2 + 4$ $3 + 5$ $6 + 1$ $6 + 0$ $2 + 3$. (These should have sums less than ten.) Make Karrie a flag to hold for the numbers 5, 6, 7, 8, and 9.

■ Material Sheet 13 Train Sets (see Example 3-8)

You need an engine, some cars, and a caboose for each number between 10 and 20. (17 has Engine 10 + 7, cars 9 + 8 and 8 + 9, and Caboose 17.)

Engine is powerful with place-value representation. Caboose is the end, with the standard name.

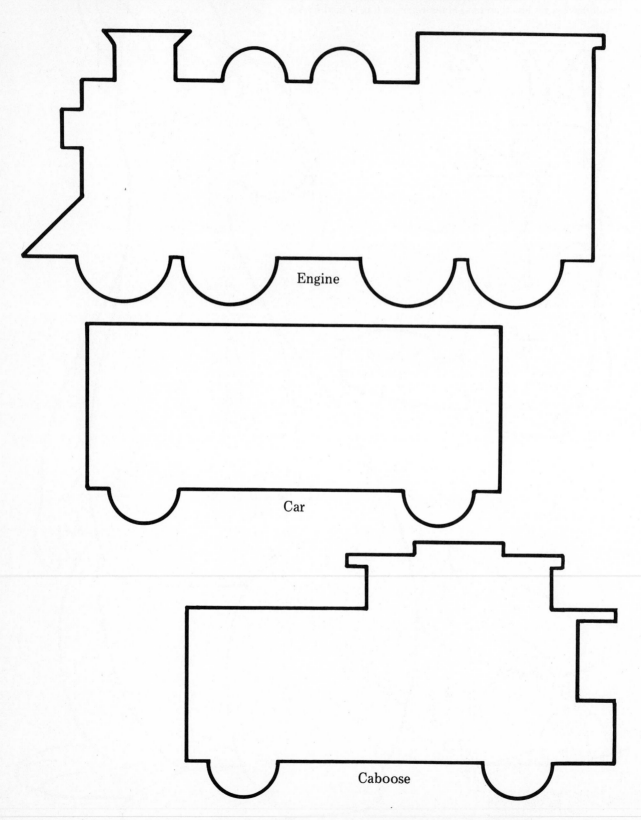

Engine

Car

Caboose

■ Material Sheet 14 Subtract As You Go
(see Example 3-16)

Reproduce on heavy paper or glue on construction-paper backing or glue on file folder. Put numbers from 0 through 9 in envelope and attach to game.

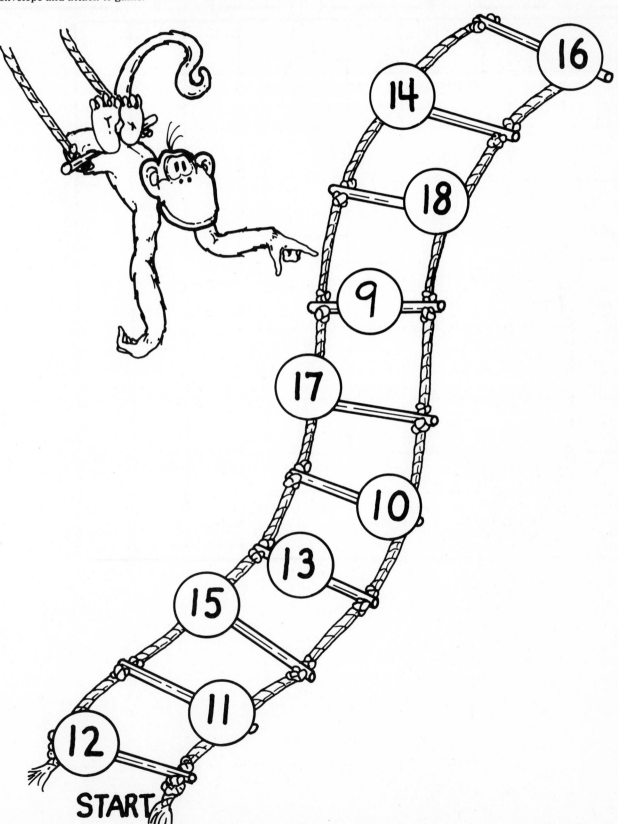

■ Material Sheet 15 Fact Finder (see Example 4-5)

Use a large L-shaped piece of construction paper to cover part of the grid. A 3×4 rectangle will display 12 squares.

	1	2	3	4	5	6	7	8	9	10
1										
2										
3										
4										
5										
6										
7										
8										
9										
10										

■ **Material Sheet 16 Dot Paper**

■ **Material Sheet 16 Dot Paper**

■ Material Sheet 17A Grid with Centimeter Squares

■ Material Sheet 17B Grid with Inch Squares

■ Material Sheet 17C Squared Paper

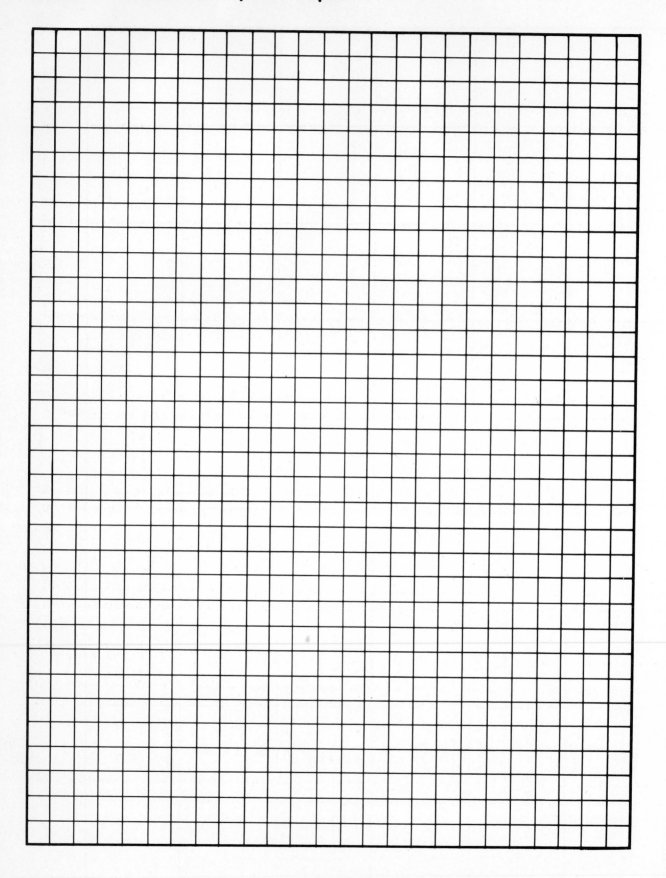

■ Material Sheet 18A Fraction Chart

1

$\frac{1}{2}$	$\frac{1}{2}$

$\frac{1}{3}$	$\frac{1}{3}$	$\frac{1}{3}$

$\frac{1}{4}$	$\frac{1}{4}$	$\frac{1}{4}$	$\frac{1}{4}$

$\frac{1}{6}$	$\frac{1}{6}$	$\frac{1}{6}$	$\frac{1}{6}$	$\frac{1}{6}$	$\frac{1}{6}$

$\frac{1}{8}$	$\frac{1}{8}$	$\frac{1}{8}$	$\frac{1}{8}$	$\frac{1}{8}$	$\frac{1}{8}$	$\frac{1}{8}$	$\frac{1}{8}$

$\frac{1}{12}$	$\frac{1}{12}$	$\frac{1}{12}$	$\frac{1}{12}$	$\frac{1}{12}$	$\frac{1}{12}$	$\frac{1}{12}$	$\frac{1}{12}$	$\frac{1}{12}$	$\frac{1}{12}$	$\frac{1}{12}$	$\frac{1}{12}$

$\frac{1}{24}$	$\frac{1}{24}$	$\frac{1}{24}$	$\frac{1}{24}$	$\frac{1}{24}$	$\frac{1}{24}$	$\frac{1}{24}$	$\frac{1}{24}$	$\frac{1}{24}$	$\frac{1}{24}$	$\frac{1}{24}$	$\frac{1}{24}$	$\frac{1}{24}$	$\frac{1}{24}$	$\frac{1}{24}$	$\frac{1}{24}$	$\frac{1}{24}$	$\frac{1}{24}$	$\frac{1}{24}$	$\frac{1}{24}$	$\frac{1}{24}$	$\frac{1}{24}$	$\frac{1}{24}$	$\frac{1}{24}$

■ Material Sheet 18B Fraction Chart

1

$\frac{1}{2}$	$\frac{1}{2}$

$\frac{1}{4}$	$\frac{1}{4}$	$\frac{1}{4}$	$\frac{1}{4}$

$\frac{1}{5}$	$\frac{1}{5}$	$\frac{1}{5}$	$\frac{1}{5}$	$\frac{1}{5}$

$\frac{1}{10}$	$\frac{1}{10}$	$\frac{1}{10}$	$\frac{1}{10}$	$\frac{1}{10}$	$\frac{1}{10}$	$\frac{1}{10}$	$\frac{1}{10}$	$\frac{1}{10}$	$\frac{1}{10}$

$\frac{1}{10} = .1$

$\frac{1}{20}$	$\frac{1}{20}$	$\frac{1}{20}$	$\frac{1}{20}$	$\frac{1}{20}$	$\frac{1}{20}$	$\frac{1}{20}$	$\frac{1}{20}$	$\frac{1}{20}$	$\frac{1}{20}$	$\frac{1}{20}$	$\frac{1}{20}$	$\frac{1}{20}$	$\frac{1}{20}$	$\frac{1}{20}$	$\frac{1}{20}$	$\frac{1}{20}$	$\frac{1}{20}$	$\frac{1}{20}$	$\frac{1}{20}$

$\frac{1}{100} = .01$

■ Material Sheet 19 Fraction Slide Rule and Decimal Slide Rule (see Examples 6-11 and 6-20)

The fraction slide rule looks like "Freddie—the Fraction Fish."

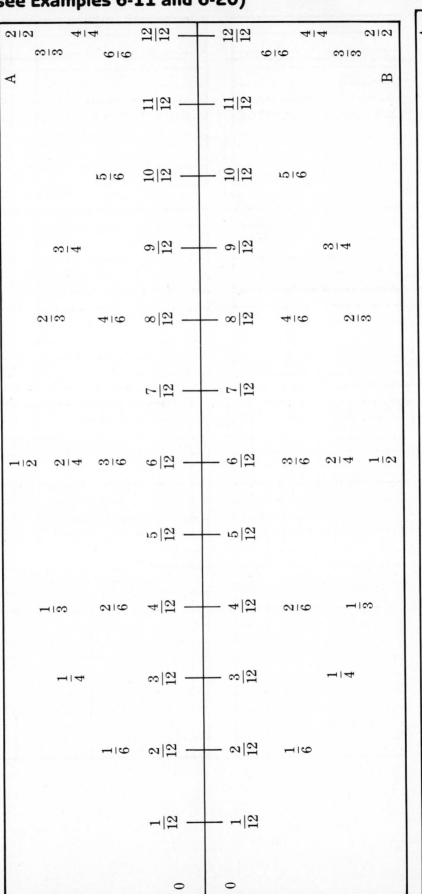

■ Material Sheet 20 Decimal Paper

■ Material Sheet 21 Centimeter Ruler

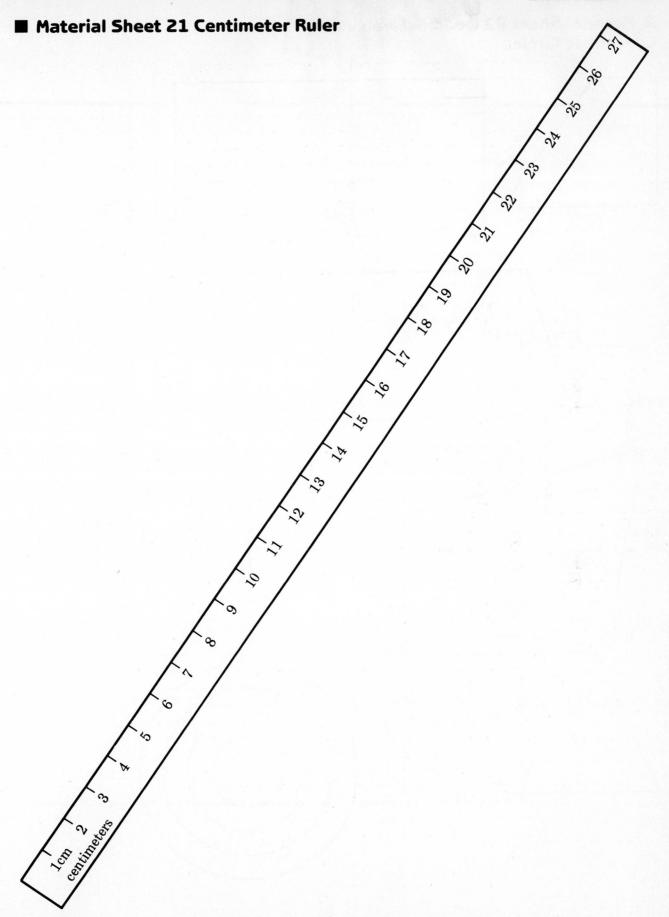

■ Material Sheet 22 Quadrilaterals and Concentric Circles

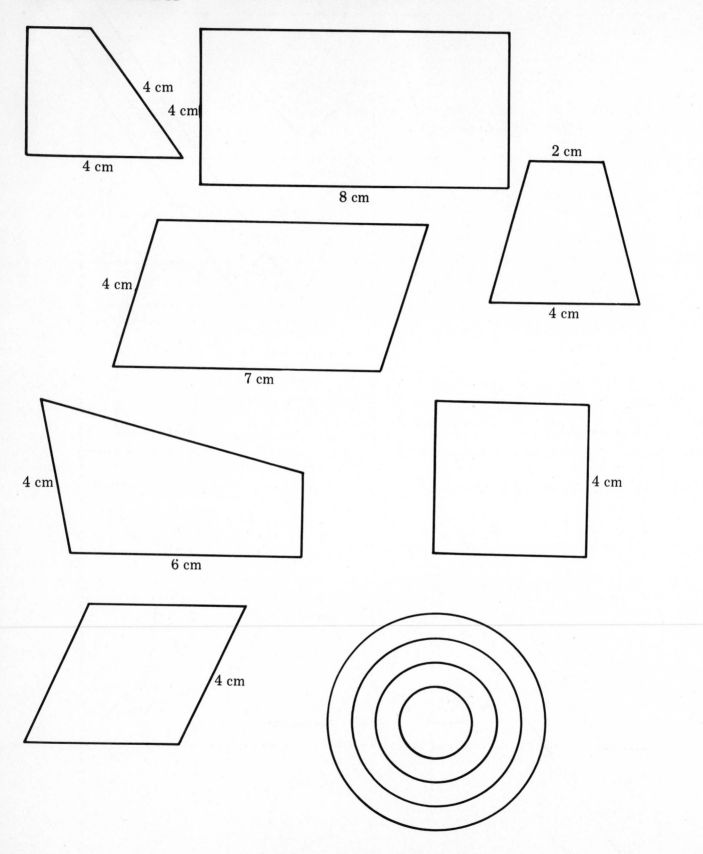

■ Material Sheet 23 Plane Figures

Duplicate on heavy paper. Cut out along the lines.

■ Material Sheet 24 Geo-Cards

Prepare Geo-Cards on poster board. Make them nice and big.

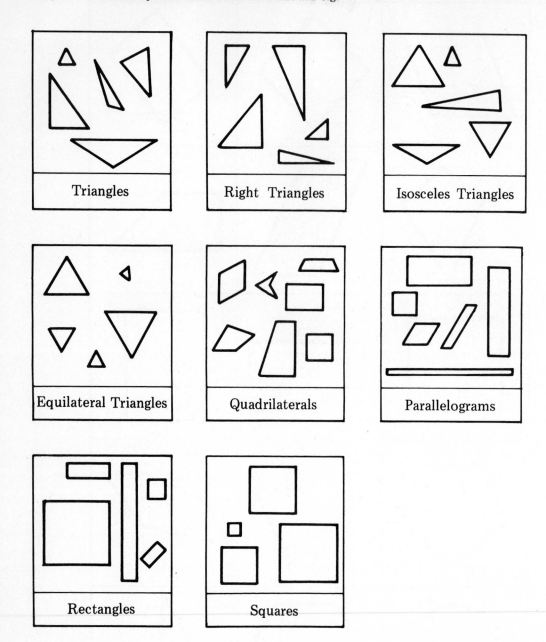

■ Material Sheet 25 United States Map

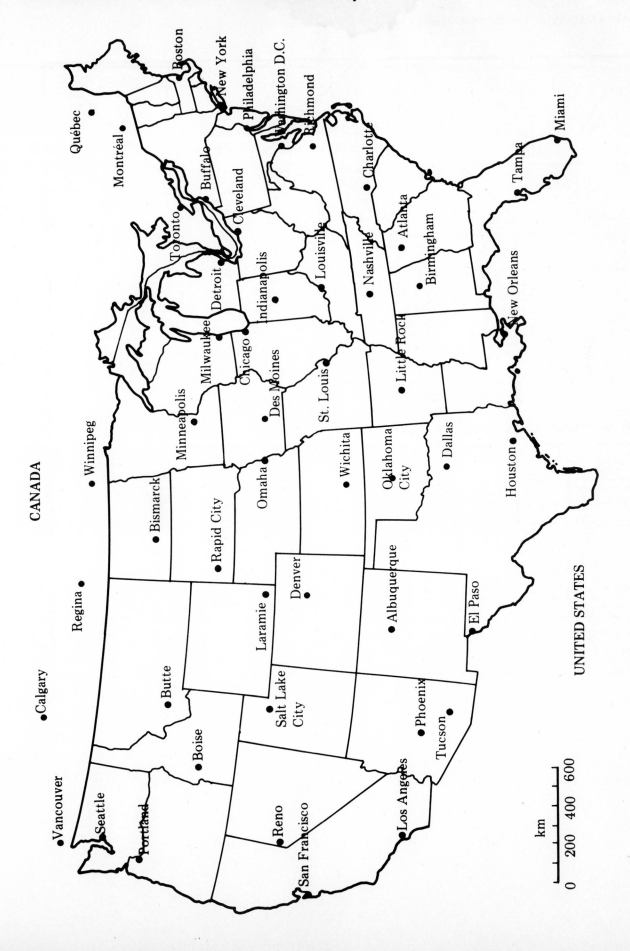

■ Material Sheet 26 Tangram Puzzle

Reproduce on heavy paper. Cut out the seven pieces.

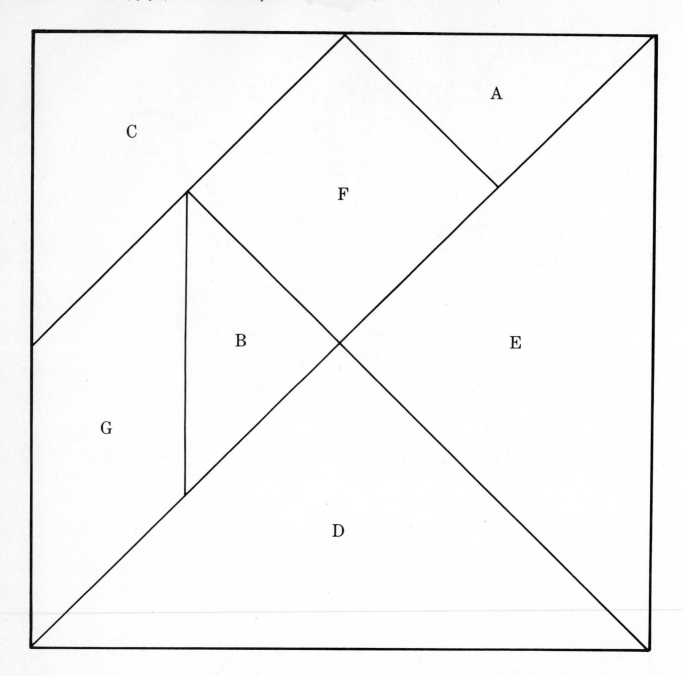

INDEX

49
84
297
300

2   4   8   16 | 32   64   128   256